# Outlook and Impacts of Global Warming

# Outlook and Impacts of Global Warming

Edited by **Vivian Moritz**

R CALLISTO REFERENCE

New York

Published by Callisto Reference,
106 Park Avenue, Suite 200,
New York, NY 10016, USA
www.callistoreference.com

**Outlook and Impacts of Global Warming**
Edited by Vivian Moritz

International Standard Book Number: 978-1-63239-499-6 (Hardback)

# Contents

# Preface

The world is advancing at a fast pace like never before. Therefore, the need is to keep up with the latest developments. This book was an idea that came to fruition when the specialists in the area realized the need to coordinate together and document essential themes in the subject. That's when I was requested to be the editor. Editing this book has been an honour as it brings together diverse authors researching on different streams of the field. The book collates essential materials contributed by veterans in the area which can be utilized by students and researchers alike.

Global warming is described as a gradual increase in the overall temperature of the earth's atmosphere generally attributed to the greenhouse effect. It has acquired the position of one of the most complex issues being encountered by world leaders. Therefore, it calls for significant concern from numerous modern societies, power and energy engineers, scholars, researchers and stakeholders. The consensus in the past century regarding anthropogenically induced global warming, has recently been questioned by a rising number of climate change panelists. Irrespective of the uncertainty of climate models, humankind needs to continuously struggle towards decreasing the amount of greenhouse gases released into the atmosphere for the conservation of natural resources and living organisms by presenting new advanced methods on alternative fuels and other related technologies. This book thoroughly addresses the fundamentals on the origin of global warming and associated technologies which can be used to decrease human impact on climate change as well as presents innovative strategies which activists and practitioners should adopt. This book is a collective effort of various authors compiled in a chronological and simple manner in order to strike the correct balance between breadth and depth of coverage of several topics.

Each chapter is a sole-standing publication that reflects each author's interpretation. Thus, the book displays a multi-facetted picture of our current understanding of application, resources and aspects of the field. I would like to thank the contributors of this book and my family for their endless support.

**Editor**

# Global Warming and Its Impact

# The Impact on Global Warming of the Substitution of Refrigerant Fluids in Vapour Compression Plants: An Experimental Study

C. Aprea, A. Greco and A. Maiorino

Additional information is available at the end of the chapter

## 1. Introduction

The development of vapour compression refrigerating units was strictly related to the characteristics of the working fluid from the beginning of their commercial diffusion.

Initially, natural substances were employed, such as ethyl ether, methyl ether, dimethyl ether, carbon dioxide, ammonia, sulphuric anhydride and methyl chloride. Potential users of the refrigerating equipment were somewhat diffident because of their toxicity and/or flammability. Most first generation refrigerant fluids were retired for safety reasons. Some, such as ammonia and hydrocarbons, survived or were later revived for limited applications in which their risks were manageable, such as industrial or small-charge systems.

Apparently, all safety problems were overcome with the appearance of the first, non toxic, non-flammable chloro-fluoro-hydrocarbon (CFC). Further chloro-fluoronated compounds followed, originating either from methane or from ethane by partial HCFC or total CFC substitution of the hydrogen atoms.

The first global environment problem with the second generation refrigerants was the depletion of stratospheric ozone. The problem arises from destruction of ozone molecules in the upper atmosphere, primarily by bromine and chlorine from anthropogenic chemicals. The chlorine and bromine react catalytically to destroy ozone molecules, thereby reducing the natural shield from incoming ultraviolet-B radiation. Molina and Rowland [1] identified CFCs and HCFCs as a source for chlorine in the stratosphere and the potential for more serious ozone depletion, with projected growth in use of these chemicals. The index used to indicate the relative ability of a refrigerant or other chemical to destroy stratospheric ozone is the Ozone Depletion Potential (ODP).

ODP is defined for any given substance as the ratio between the ozone consumption per unit mass released in the atmosphere and that consumed by the CFC R11 [2].

Chlorinated and brominated refrigerants, along with similar solvents, foam blowing agents, aerosol propellants, fire suppressants, and other chemicals are being phased out under the Montreal Protocol, a landmark international treaty to protect the ozone layer [3].

Therefore the Montreal Protocol forced abandonment of ozone-depleting substances (ODSs) as refrigerants in current vapour-compression refrigeration systems. CFCs have been banned since 1996, with HCFCs for interim use. Indeed, their ODP, though lower than that of the CFC, is different from zero [2]. The transition from HCFCs also is underway. The Montreal Protocol sets limits for the HCFC consumption (or cap), defined as production plus imports less exports and specified destruction: in 1996 (freeze at calculated cap), 2004 (65% of cap), 2010 (25%), 2015 (10%), and 2020 (0.5%) with full consumption phase-out by 2030 in non-Article 5 countries [3,4]. Individual countries adopted different response approaches. Most western and central-European countries accelerated HCFC phase outs, while the majority of other developed countries set limits by phasing out propellant and blowing agent uses early, requiring phase-out of R-22 by 2010, and then banning all HCFC use in new equipment by 2020. The schedule for Article 5 countries begins with a freeze in 2013 (based on 2009–2010 production and consumption levels) with declining limits starting in 2015 (90%), 2020 (65%), 2025 (32.5%), and 2030 (2.5%) followed by phase-out in 2040. Again, continued future use and service, even after 2040, are allowed for existing equipment employing HCFC refrigerants until otherwise retired except as restricted by national regulations. Exports from Article 5 countries into non-Article 5 countries are effectively restricted to meet the more stringent non-Article 5 schedules. To avoid separate domestic and export products and to exploit newer technologies derived from joint ventures and licensing agreements, some products in Article 5 countries incorporate replacements earlier than required.

The HFCs are a new family of substances that are candidates for substitution of both CFCs and HCFCs. HFCs, in fact, are entirely harmless towards the ozone-layer, since they do not contain chlorine.

Now a further problem must be considered, the so-called greenhouse effect stemming from the capture of infrared radiation by some components of the atmosphere [5,6]. The average temperature at the surface of our planet results from an equilibrium between incoming solar energy and heat radiated back into space. Most of the latter is in the infrared range of emissions. Gases that absorb this infrared energy enhance the greenhouse effect of our atmosphere, leading to warming of the Earth. Human activities have increased substantially the concentration of greenhouse gases. As a result, a substantial warming of the earth surface and atmosphere occurred that might adversely affect the natural ecosystem. Over the last hundred years, the mean temperatures have increased by 0.3-0.6 °C. Doubling the amount of carbon dioxide in the atmosphere is likely to yield a further temperature increase by 1.5-4.5 °C [7,8]. Refrigerants have been identified as greenhouse gases. The impact of a given greenhouse gas on global warming is quantified by its GWP (Global Warming

Potential). GWP is defined as the mass of $CO_2$ that would yield the same net impact on global warming as the release of a single unit (kg) of the given atmospheric component. The GWP values used in this paper are relative to a 100 years Integration Time Horizon. Shorter integration periods emphasize near-term effects, while longer intervals better reflect the total impact of a release. Carbone dioxide is used as the reference chemical for GWPs because it is the one between natural gases with the greatest net impact. Other chemicals, including most refrigerants, are more potent as greenhouse gases.

In December 1997 more than 160 nations met in Kyoto, Japan, to negotiate binding limitations on greenhouse gases for the developed nations, pursuant to the objectives of the United Nations Framework Convention on Climate Change (UNFCCC). The outcome of the meeting was the Kyoto Protocol [9], in which the developed nations agree to limit their greenhouse gases emissions, relative to the levels emitted in 1990. The Protocol is subjected to ratification, acceptance, approval or accession by Parties to the Convention. Due to the voluntary characteristics of the Protocol some countries have no targets under the Protocol, but the protocol reaffirms the commitments of the Framework Convention by all Parties to formulate and implement climate change mitigation and adaptation programs. Even countries such as the United States of America that have not formally signed onto the Kyoto Accord have introduced their own plans to aggressively reduce greenhouse gas emissions.

National laws and regulations implementing the Kyoto Protocol differ from one another, but they typically prohibit avoidable releases of HFC refrigerants. In some countries, their use undergoes control and/or taxation. More recent measures (either already adopted or proposed) at local level (regional, national, municipal) are even more stringent. These restrictions are forcing the shift to a fourth generation of refrigerants with both ODP and GWP regulations [10].

In the field of the mobile refrigeration systems, the European Parliament already set a regulation of F-Gases phase out [11] that bans the use of refrigerants having GWPs exceeding 150 (based on 100 years integration time horizon). Such regulation begins in 2011, and will be effective for all air conditioners of new automobiles in 2017 .

However, the EU Parliament [12] rejected recommended measures that would have banned HFCs as aerosol propellants by 2006, as foam blowing agents by 2009, and as refrigerants in stationery air conditioners and refrigeration by 2010.

## 2. The TEWI concept

The US Energy Information Authority projects that world carbon dioxide emissions will increase from 25.028 million metric tons in 2003 to 33.663 million metric tons in 2015 and 43.676 million metric tons in 2030. Electricity actually causes more carbon dioxide emissions than all other anthropogenic sources. According to the Energy Industry Administration in the United States, electricity generates 39% of the total anthropogenic carbon dioxide emissions. These emissions are expected to grow by almost 45% over the next 25 years and grow to 42% of the total carbon dioxide emissions.

In order to reduce the production of greenhouse gases, it is reasonable to assume that we should focus on the reduction of emissions in both electricity and transportation.

According to US Department of Energy [13-15], the largest single use for electricity is lighting (27%) followed by cooling (refrigeration and air conditioning) at 15%, as shown in Figure 1.

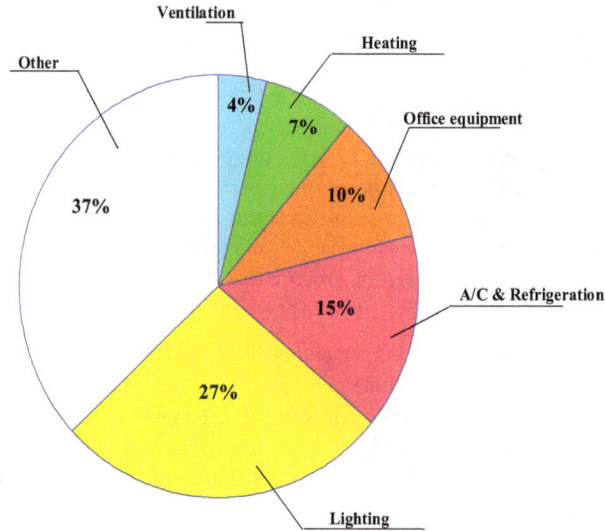

**Figure 1.** USA Electricity Consumption

Therefore in order to reduce the $CO_2$ emissions caused by electricity a good option is to reduce the refrigeration and air conditioning emissions.

Vapour compression plants produce both a direct and an indirect contribution to global warming. The former depends on the GWP of refrigerant fluids and on the fraction of refrigerant charge which is released into the atmosphere during operation and maintenance, or is not recovered when the system is scrapped [16,17]. The indirect contribution is energy-related. In fact, a vapour compression refrigerator requires electrical energy produced by a power plant that typically burns a fossil fuel releasing $CO_2$ into the atmosphere.

The concept of total equivalent warming impact (TEWI) was developed to combine the effect of direct refrigerant emission with those due to energy consumption and the related combustion of fossil fuels for the electric energy production. TEWI provides a measure of the environmental impact of greenhouse gases originating from operation, service and end-of-life disposal of the equipment. TEWI is the sum of the direct contribution of the greenhouse gases used to make or to operate the systems and the indirect contribution of carbon dioxide emissions resulting from the energy required to run the systems along their normal lifetime [18].

The TEWI is calculated as [19-21] :

$$TEWI = CO_{2,dir} + CO_{2,indir} \qquad \left[ kg\ CO_2 \right]$$

$$CO_{2,dir} = RC \left[ P_L + \left( \frac{1-P_R}{V} \right) \right] V \cdot GWP \qquad \left[ kg\ CO_2 \right] \qquad (1)$$

$$CO_{2,indir} = \alpha \cdot \frac{\dot{Q}_{ref}}{COP} \cdot H \cdot V \qquad \left[ kg\ CO_2 \right]$$

The direct global warming effect of refrigerant fluids, stemming from the absorption they produce of long-wave radiations, depends on their GWP and on the fraction of refrigerant charge released into the atmosphere. The last is mainly due to leakage during the plant operational life time ($P_L$) and to the residual amounts which, according to the current state of technology, are not recyclable and thus are released into the atmosphere when taking the plant out of operation (1-$P_R$).

As already stated, the indirect contribution to TEWI consists in the so-called energy-related contribution. Indeed, an electrical refrigerator requires electrical energy from a $CO_2$ releasing power plant that typically burns a fossil fuel. The amount of $CO_2$ emitted is a function of the refrigerator COP, of the power plant efficiency and of the fuel used in the conversion plant that affect the emissions per unit energy converted [22]. When a fuel is burnt, energy is produced and carbon dioxide and other chemicals, mostly water, are produced. The ratio of $CO_2$ emitted to the electricity generated differs according to the type of fuel used. Electricity is generated from a range of fuels including nuclear, gas, oil, coal and in some cases waste. Besides burning a fuel there are several other alternative to produce electricity like hydroelectric plants, wind power, geothermal energy sources, tidal power, photovoltaic panels etc.

The relationship between $CO_2$ production and electricity generation may vary significantly, depending on the approach to be followed. In Figure 2 are reported the best values of $CO_2$ emissions by primary energy sources taken from the reference providing the most recent data.

The typical power-plant technology adopted varies, therefore each country and each region inside each country has its own mix of primary sources for electricity generation. This mix can change significantly from country to country and even from one region to another in the same country.

The literature provides some indicative, average levels of $CO_2$ release per KWh of electrical energy ($\alpha$) for various countries [23-26]. Table 1 reports a values for different continents and, in each continent for different countries. In Table 1 are reported a range of $\alpha$ values and the "best value" for each country.

Figure 3 illustrates the range of average country emission rates for several western European nations and compares those with relative percentages of electricity produced by each countries.

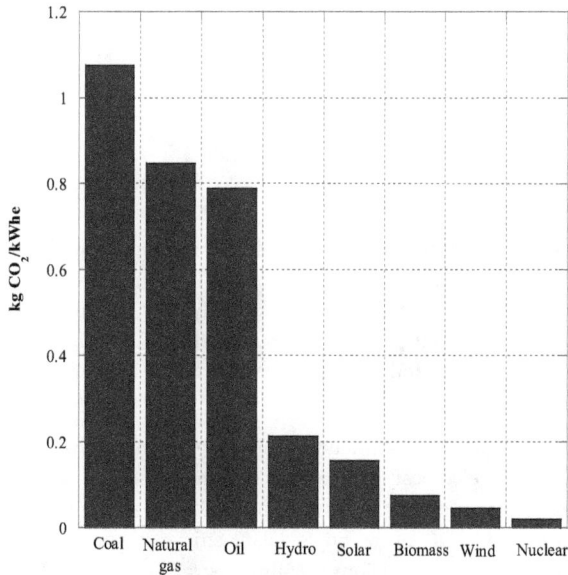

**Figure 2.** Average $CO_2$ emissions by primary energy source.

In order to understand direct and indirect contributions impact on the greenhouse gases emissions typical household and light commercial refrigeration systems (cooling capacity below 1 kW at LBP ASHRAE conditions) were chosen for three representative market around the world (North America, Europe and Asia). Each system was tested following the energy consumption standards of each region and the results are presented in Table 2.

The concept of TEWI in this study is used to identify the global warming impact for each equipment and market region.

For calculation purposes an annual leakage rate of 5% was adopted. This value was an intermediary value for household and light commercial applications. A mean value of life expectancy for refrigeration equipment in this study was 10 years. A recycling of 75% was considered for both household and light commercial applications. This means that 75% of the refrigerant charge is recovered at the end of the equipment useful life.

The results are shown in Figure 4.

In figure is also reported the influence of indirect and direct contribution to global warming in terms of $CO_2$ emissions. The figure allows a comparison between the TEWI values for household and light commercial refrigeration systems.

The Impact on Global Warming of the Substitution of Refrigerant Fluids in Vapour Compression Plants: An Experimental Study

9

With reference to the previous study, in Figure 5 is reported the percentage of $CO_2$ emissions of the direct and of the indirect contribution to global warming for each application.

| Continent | Country | Energy Mix by Primary Source % | | | | Total Energy Production | $\alpha$ (kgCO$_2$/kWhe) | | |
|---|---|---|---|---|---|---|---|---|---|
| | | Thermal | Hydro | Nuclear | Renewables | (billion kWhe) | Max | Min | Best |
| Africa | South Africa | 93 | 1 | 7 | 0 | 196 | 1.06 | 0.620 | 0.840 |
| | Egipt | 81 | 19 | 0 | 0 | 72 | 0.998 | 0.543 | 0.770 |
| | Average | 90 | 5 | 5 | 0 | 287 | 1.049 | 0.604 | 0.827 |
| South America | Argentina | 59 | 34 | 7 | 0 | 85 | 0.814 | 0.402 | 0.608 |
| | Brazil | 6 | 89 | 1 | 4 | 339 | 0.440 | 0.056 | 0.248 |
| | Average | 17 | 78 | 3 | 3 | 431 | 0.514 | 0.125 | 0.320 |
| North America | Canada | 28 | 58 | 13 | 1 | 566 | 0.562 | 0.198 | 0.380 |
| | Mexico | 79 | 14 | 4 | 3 | 199 | 0.960 | 0.530 | 0.745 |
| | USA | 71 | 6 | 21 | 2 | 3719 | 0.846 | 0.481 | 0.663 |
| | Average | 66 | 13 | 19 | 2 | 4484 | 0.815 | 0.447 | 0.631 |
| Europe | Germany | 62 | 4 | 30 | 4 | 545 | 0.700 | 0.540 | 0.610 |
| | Spain | 50 | 18 | 27 | 4 | 223 | 0.400 | 0.530 | 0.480 |
| | France | 8 | 14 | 77 | 1 | 520 | 0.270 | 0.080 | 0.090 |
| | Italy | 79 | 18 | 0 | 3 | 259 | 0.520 | 0.630 | 0.590 |
| | United Kingdom | 74 | 1 | 24 | 2 | 361 | 0.890 | 0.640 | 0.640 |
| | Sweden | 4 | 51 | 43 | 2 | 153 | 0.230 | 0.040 | 0.040 |
| | Average | 45 | 18 | 34 | 3 | 2382 | 0.510 | 0.130 | 0.470 |
| Asia | China | 82 | 17 | 1 | 0 | 1288 | 0.999 | 0.548 | 0.773 |
| | India | 83 | 14 | 3 | 0 | 512 | 1.000 | 0.555 | 0.777 |
| | Japan | 60 | 8 | 30 | 2 | 1037 | 0.729 | 0.405 | 0.567 |
| | Russia | 66 | 19 | 15 | 0 | 835 | 0.835 | 0.447 | 0.641 |
| | Average | 72 | 15 | 12 | 1 | 3788 | 0.888 | 0.487 | 0.688 |
| Pacific | Australia | 91 | 8 | 0 | 1 | 198 | 1.069 | 0.609 | 0.839 |
| | New Zeland | 32 | 58 | 0 | 11 | 38 | 0.613 | 0.223 | 0.418 |
| | Average | 81 | 16 | 0 | 2 | 236 | 0.996 | 0.547 | 0.772 |

**Table 1.** $\alpha$ values: $CO_2$ emissions from power plants.

| Region | Application | System model | Internal Volume Capacity | Refrig. | Amount (g) | Energy Consumption (kWh/month) | Ambient Temperature (°C) |
|---|---|---|---|---|---|---|---|
| North America | Household | Top Mounted | 600 lt | HFC134a | 110 | 40 | 32 |
| | | Side by Side | 800 lt | HFC134a | 150 | 50 | 32 |
| | | Chest Freezer | 500 lt | HFC134a | 150 | 45 | 32 |
| | | Vertical Freezer | 500 lt | HFC134a | 120 | 55 | 32 |
| | Light Commercial | | 600 cans | HFC134a | 330 | 210 | 32 |
| | | Glass Door Merchandiser Vending Machine | 600 cans | HFC134a | 400 | 300 | 32 |
| Europe | Household | Small Refrigerator | 250 lt | HC600a | 30 | 15 | 25 |
| | | Combined | 430 lt | HC600a | 60 | 30 | 25 |
| | | Refrigerator | 200 lt | HC600a | 40 | 25 | 25 |
| | Light Commercial | Vertical Freezer | 200 lt | HFC134a | 120 | 100 | 30 |
| | | | 550 lt | HFC134a | 500 | 480 | 30 |
| | | Chest Freezer Display Case | | | | | |
| Asia | Household | Compact Refrigerator | 80 lt | HC600a | 25 | 20 | 25 |
| | | Combined Refrigerator | 350 lt | HC600a | 50 | 25 | 25 |
| | Light Commercial | | 600 cans | HFC134a | 400 | 300 | 32 |
| | | Vending Machine | | | | | |

**Table 2.** Typical refrigeration equipment per market region.

Figure 5 clearly shows the strong influence of indirect effect due to energy consumption on the total $CO_2$ emissions. In the North American market the direct contribution in the household applications ranges between 16 and 23%, in light commercial between 11 and 12%. In Europe, due to the use of hydrocarbons as refrigerants in household applications, direct contribution has practically no contribution to the TEWI. In the light commercial applications the direct contribution ranges between 2 and 25%, with the higher values associated to the refrigerant fluid with higher GWP. The situation in Asia for household applications is very similar to Europe due to the use of hydrocarbons. Light commercial applications in Asia show a direct contribution of 10 %.

Therefore regardless of the market region and type of refrigerant system, indirect effect to global warming always represents the prevalent contribution to global warming.

From the previous data a sensitivity analysis can be carried out employing two different scenarios. In the first one the refrigerants were replaced by a new refrigerant with a low GWP, for example 1 and the energy consumptions was held constant.

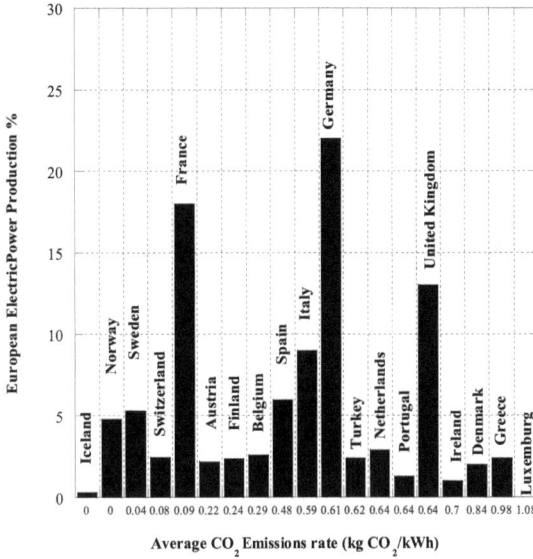

**Figure 3.** Average power plant emission rates and electricity production for European nations.

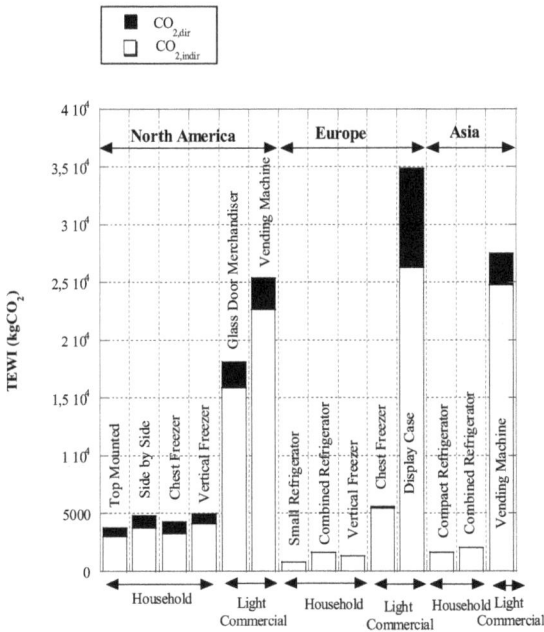

**Figure 4.** TEWI values for different application and different market region.

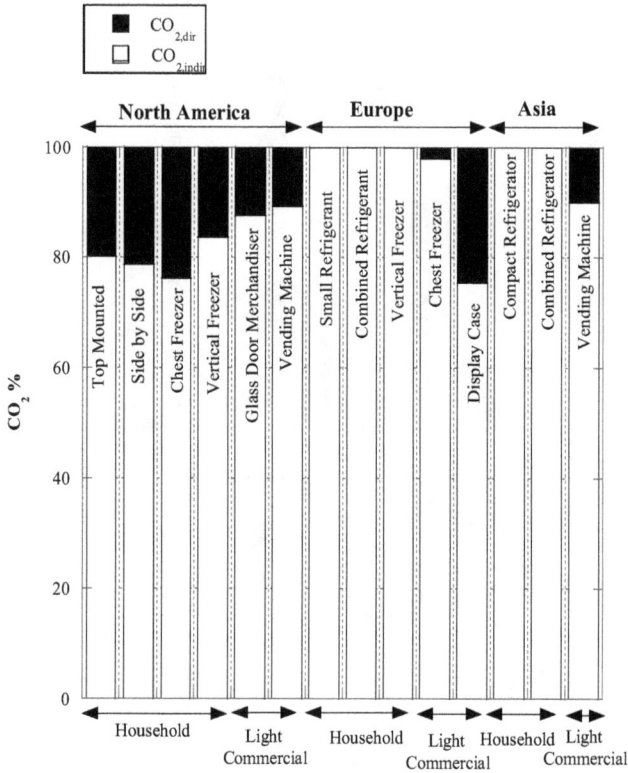

**Figure 5.** Direct and indirect CO₂ emissions percentage for different application.

In Figure 6 are reports the results of the first scenario in terms of ΔTEWI/TEWI.

The Figure 6 clearly shows that in the North America market the replacement of the refrigerant fluid with a unitary GWP refrigerant decreases the TEWI from -24 to -16 % in household application, from-12 to -11 % in light commercial applications. In the European and Asian market, due to current use of hydrocarbons in household applications, the replacement of the refrigerant fluid does not provide any additional benefit. Whereas in the light commercial application in the European market the TEWI decreases between -24 and – 2%, in the Asian market decreases of about -10%.

A second scenario the energy consumption was reduced by 30% and the refrigerant fluid was not changed. In Figure 7 are reports the results of the first scenario in terms of ΔTEWI/TEWI.

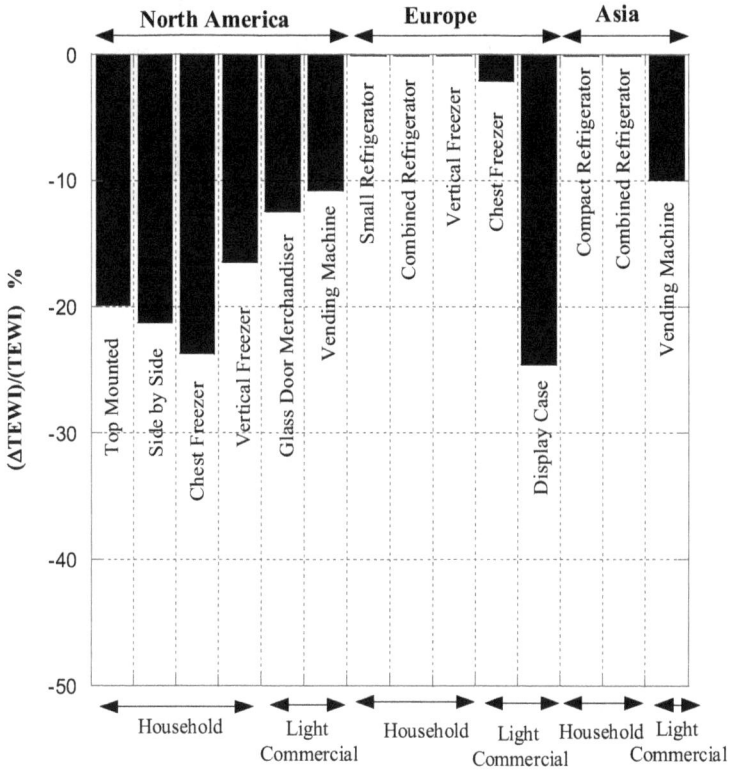

**Figure 6.** ΔTEWI/TEWI replacing the refrigerant fluid with a low GWP fluid.

The figure clearly shows that improving the energy efficiency by 30%, in the North America Market the TEWI decreases between -47 and -38 %, in the European market between -47 and -30%, in the Asian market between -37 and -30%.

The previous analysis clearly shows that the indirect effect on TEWI is stronger than the direct one in the household and light commercial applications regardless of the market region. Therefore it seems more effective, in order to decrease the global warming impact of a refrigeration system, to focus the attention on the improvement of the energy efficiency. To this aim the replacement of a refrigerant fluid with one with low GWP but that introduces penalties to the equipment efficiency, must be avoided.

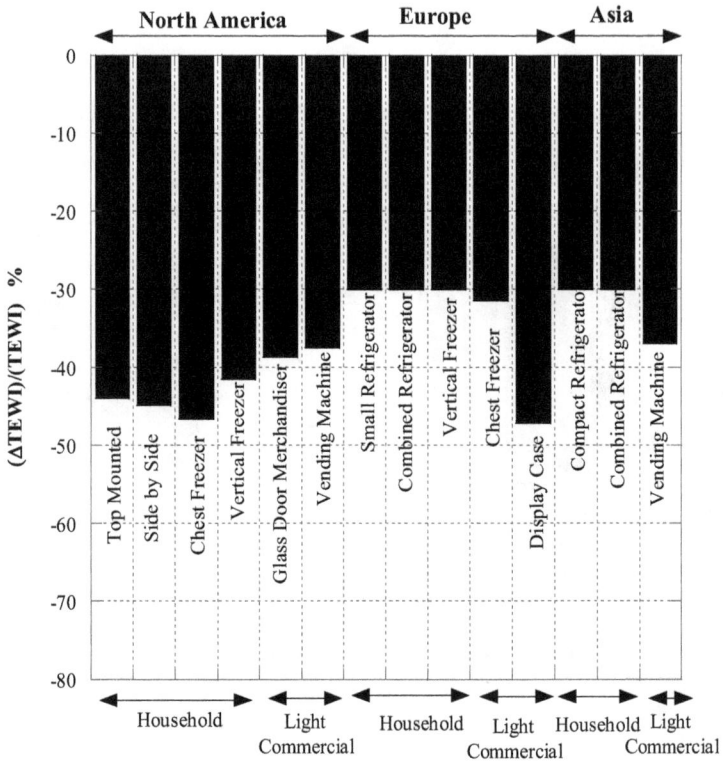

**Figure 7.** ΔTEWI/TEWI decreasing energy consumption by 30%.

## 3. An experimental evaluation of the greenhouse effect in R22 substitution with R407C

### 3.1. R22 substitution with R407C

R22 is an HCFC with an ODP of 0.05 and a GWP of 1700 and is the most widely used refrigerant today in commercial, domestic and industrial applications, and its phase-out will impact a large number of plants in the world. Therefore retrofitting these systems would alleviate the R22 phase-out problem. This opportunity could be cheaper than the installation of new plants, especially for supermarkets, data centers, factories and hospitals. Refrigerant replacement candidates have been checked for environmental and safety requirements, compatibility with lubricant oil, filters, and sealing. In order to establish the best substitute in a specified system among the candidates, it is necessary to estimate energetic performances after refrigerant replacement. In the last years, many companies have expended much effort to develop and characterize refrigerants able to increase the energetic efficiency of a refrigeration system, depending on its application. Similarly many researchers have investigated the energetic performances of the newest substitutes of R22 [27,28].

During this transition period, many industrial and commercial applications used R407C to retrofit R22. The latter is the only drop-in substitute, and is an HFC with zero ODP and a GWP (1700) lower than that of R22. R407C is a zeotropic mixture of R32/R125/R134A (23/25/52 % in weight).

Even if R407C has thermo-physical properties similar to those of R22 and it is non-flammable and non-toxic, it is incompatible with mineral or alkyl-benzene oil. Consequently, the R22 retrofit with R407C implies the adoption of polyester oil.

In the preset paper an experimental comparison between TEWI of R22 and R407C has been carried with a vapour compression pilot plant [29].

## 3.2. The experimental apparatus

The experimental apparatus is reported in Figure 8 and consists of three loops: the refrigerant loop, the condensing water loop and the evaporating water-glycol loop [30].

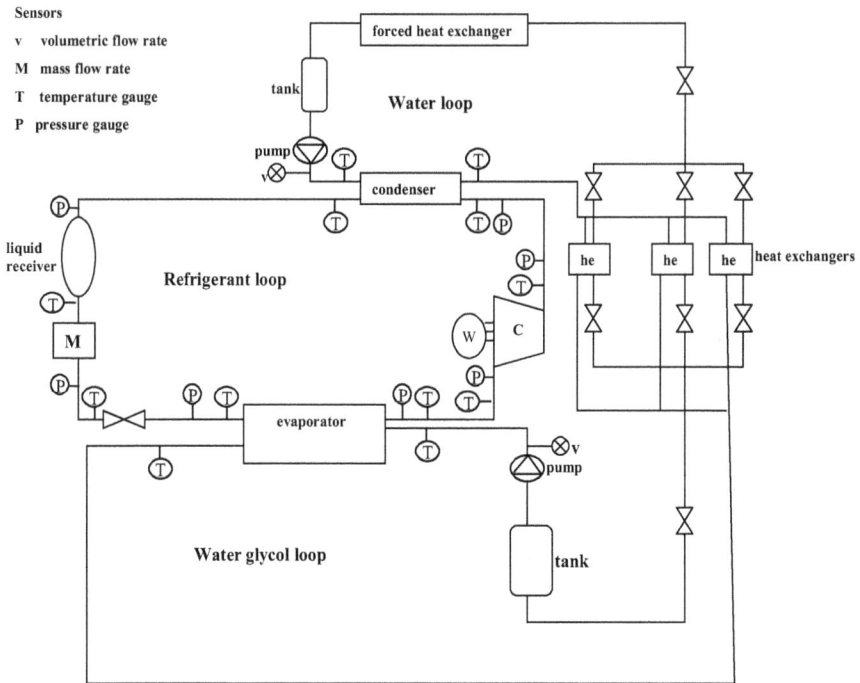

**Figure 8.** Flow-sheet of the experimental apparatus.

The refrigerant loop is a vapour compression plant consisting in a semihermetic compressor, a plate condenser connected to a liquid receiver, a thermostatic valve and a plate evaporator.

The water loop is used for cooling the refrigerant flowing in the plant. It consists in the condenser, a circulation pump, an inertial tank, an air operated cooling exchanger and three plate heat exchangers. The refrigerant duty consists in a water-glycol mixture (70/30 % in weight) continuously heated in the three intermediate plate heat exchangers by means of the water condensing loop. Whenever required, additional heating is provided by three electrical resistances inserted into an inertial tank (maximum power 12 kW). The water-glycol loop consists in the evaporator, a circulation pump and an inertial tank.

In the main loop measurements are performed of: the pressure and f the temperature at the inlet and at the outlet of each device, the mass-flow rate at the outlet of the liquid receiver and the electrical power supplied to the compressor. As regards the water-glycol loop, the temperature at the inlet and outlet of the evaporator, at the intermediate heat exchangers, and at the inertial tank are determined. Measurements are carried out of the volumetric flowrate at the inlet of the tank and also of the electrical power supplied to the resistances. The secondary loop at the condensed has been instrumented in a similarly way. A detailed indication of sensor position is provided in Figure 8.

Pressures are measured by piezoelectric transducers (P) in the pressure ranges are 0 – 7 bar and 0 – 30 bar with an accuracy of $\pm$ 0.5 % F.S. Temperature are measured by means of four-wire 100 $\Omega$ platinum resistance thermometers (T) with an accuracy of $\pm$ 0.15 K. The electric power input to the compressor and that required by the electrical resistances are measured by a Watt transducer with an accuracy of $\pm$ 0.2 %. Refrigerant mass flow rate is measured by a Coriolis effect mass flow-meter (M) with an accuracy of $\pm$ 0.2 %.

Two turbine flow-meters are employed for measuring the condensing water volumetric flow rate (v) and that of the water-glycol mixture circulating in the evaporator (v) with an accuracy of $\pm$ 0.2 5%.

According to Tables 3 and 4, different experimental situations are determined by varying the operating conditions.

Table 3 reports $T_{w,in,co}/T_{w,out,co}$ (the inlet and the outlet temperature at the condenser water side) and $TMT_{wg}$ (the mean thermodynamic temperature of the water glycol mixture). Table 3 reports three different set of runs. In each set of runs, the values of the inlet and outlet temperature at the condenser water side have been kept constant, whereas $TMT_{wg}$ has been varied. The tests conducted at 30 /35 and 35/40 °C refer to a refrigeration plant, whereas the tests at 45/50 °C refer to a heat pump.

Table 4 reports $TMT_{wg}$ and $T_{co}$, i.e. the condensing temperature of the refrigerant fluid (as regards R407C, that is an azeotropic mixture, a mean temperature between dew and bubble point was considered). Table 4 reports three different sets of run. In each set the values of $TMT_{wg}$ and the refrigerant duty has been kept constant, whereas $T_{co}$ has been varied.

| Evaporator | Condenser | |
|---|---|---|
| TMTwg (°C) | Tout,w | Tin,w (°C) |
| -7.9 | | |
| -4.9 | | |
| 6.2 | 35 | 30 |
| 9.8 | | |
| 13.2 | | |
| -5.7 | | |
| -5.2 | | |
| 7.5 | 40 | 35 |
| 10.2 | | |
| 12.1 | | |
| 7.2 | | |
| 10.3 | | |
| 12.1 | 0 | 45 |
| 14.3 | | |

**Table 3.** The different operating conditions.

| Condenser | Evaporator |
|---|---|
| Tco (°C) | TMTwg (°C) |
| 41.8-42.1-44.7-45.9-47.3-49.4-50.4-52.9 | 8 |
| 42.7-44.2-44.8-45.9-46.5-48.3-51.7-53.3 | 10 |
| 43.7-44.0-44.6-46.0-47.2-48.0-50.2-53.0 | 13 |

**Table 4.** The different operating conditions

Table 5 reports the parameters adopted for the TEWI evaluations.

| Parameter | Value |
|---|---|
| H | 950 h/year |
| $P_L$ | 5%/year |
| $P_R$ | 25% |
| V | 10 years |
| $\alpha$ | 0.6 kg $CO_2$/kWhe |

**Table 5.** Parameters in TEWI evaluation.

## 3.3. Results and discussion

### 3.3.1. The direct contribution

This contribution has been evaluated, referring to the experimental plant, on the basis of the measured charge of the plant pertaining to R22 and to R407C and by assuming refrigerant leaks in terms of fraction of refrigerant charge.

The direct contributions to the greenhouse effect to R22 and to R407C during the plant useful life corresponds to the same net impact on global warming as the release of 6375 and 5400 kg of carbon dioxide, respectively.

The R22 direct contribution is about 15 % greater than the R407C one, a direct consequence of the greater GWP and charge pertaining to R22.

### 3.3.2. The indirect contribution

The thermodynamic parameter that affects this contribution is the coefficient of performance of the plant. In the following Figs 9-11, is shown the behaviour of the COP for different experimental situations. Those figures are referred to the test conditions pertaining to different inlet and outlet water temperatures in the condenser (30/35, 35/40, 45/50 °C), and then to different refrigerant condensing temperatures. The first two diagrams refer to a refrigerant plant, whereas the third diagram accounts for a heat pump. In all instances, COP increases with water glycol mean thermodynamic temperature. Indeed, with increasing the latter, both the refrigerating duty and the mechanical power consumption at the compressor increase. The relative increase of the former, however, exceeds that of the latter. As a consequence, COP increases. For all the experimental runs conducted, the performance coefficient of R407C is lower than that pertaining to R22. Therefore, the indirect contribution to the greenhouse effect of R407C is always higher than of R22.

The broken line reported in the previous diagrams estimates a theoretical COP value for R407C (COP407C*) that might compensate for the higher direct contribution of R22 to the greenhouse effect for this plant in the same operating conditions.

Indeed, if the R407C COP* were about 5% lower than that or R22, the greater direct contribution of the latter would be balanced by employing the former in the summer cycle (refrigeration). Unfortunately, as shown in figure 9 and 10, the actual value of the COP pertaining to R407C is lower than that corresponding to this theoretical value (COP407C*). Therefore, under these experimental conditions, R407C has a greater greenhouse effect than R22. At this stage, however, the difference can not be determined quantitatively. In runs performed with water inlet and outlet temperature of 30 and 35 °C, respectively, the COP of R407C is lower than that pertaining to R22 by 8-11 %. In runs with water inlet and outlet temperature of 35 and 340 °C, respectively, the COP of R407C is lower by 16-19%. Therefore, the difference between the actual COP value and COP* of R407C is remarkable, especially in the 35/40 °C runs.

On the contrary, the best performance is achieved with inlet and outlet water temperature of 45 and 50 °C, respectively, corresponding to a winter cycle (heat pump). In this case, the COP* pertaining to R407C that might compensate the higher direct contribution of R22 were about 3% lower than that of R22. Indeed, the actual COP value of R407C is always about 3.2% lower than that pertaining to R22, except for the run at TMTwg=7.4 °C. In this case, it is 6.6% lower. The difference between the actual COP value and COP* of R407C is in this case lower.

The less satisfactory situation corresponds to the run with water inlet and outlet temperature of 35 and 40°C, respectively.

A set of runs have been conducted by keeping the mean thermodynamic water glycol temperature and the refrigerant duty constant and by varying the condensation temperature of the refrigerant fluids in order to better clarify the reason of this behaviour.

The results of the runs obtained with TMTwg =10 °C are summarized in figure 12. Similar results are obtained in all the experimental runs performed with other water glycol mean temperatures reported in Table 5.

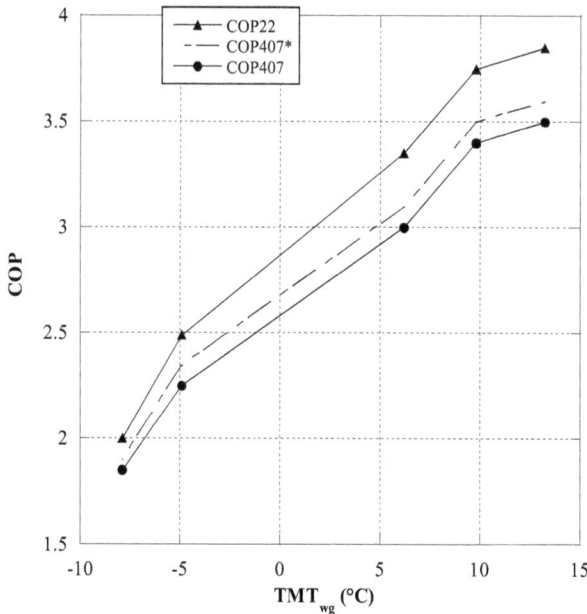

**Figure 9.** COP vs mean thermodynamic temperature of water-glycol mixture, Tin,w,co = 30°C, Tout,w,co = 35 °C.

For both fluids, the COP decreases with the condensation temperature. The shape of the two curves, however, is such that the maximum COP difference is achieved at intermediate value of the condensation temperatures (46-48°C). The value of the condensation temperature pertaining to the corresponding run performed with water inlet and outlet temperatures of 35/40 °C fall in this range.

The general behaviour of the experimental results can be explained in the light of the dependence of the compressor duty on the compression ratio. The compressor duty increases nonlinearly with the compression ratio. The compression ratio increases with the condensing temperature at equal mean thermodynamic temperature and, therefore, at equal evaporating pressure. Therefore, COP decreases with increasing compression duty at equal refrigerant duty.

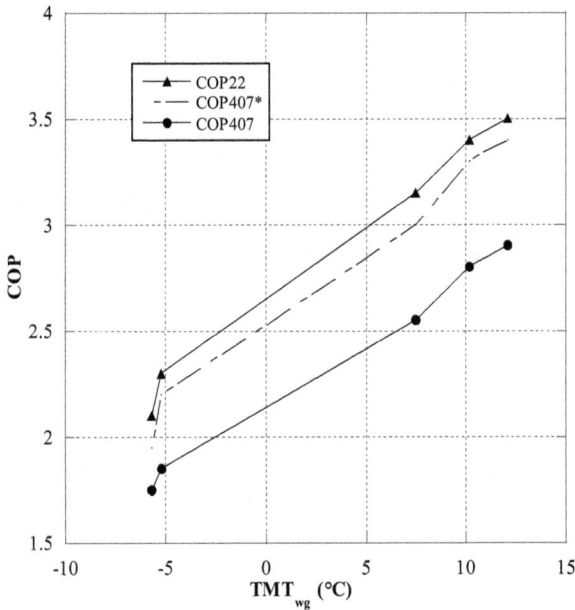

**Figure 10.** COP vs mean thermodynamic temperature of water-glycol mixture, Tin,w,co = 35°C, Tout,w,co =40 °C.

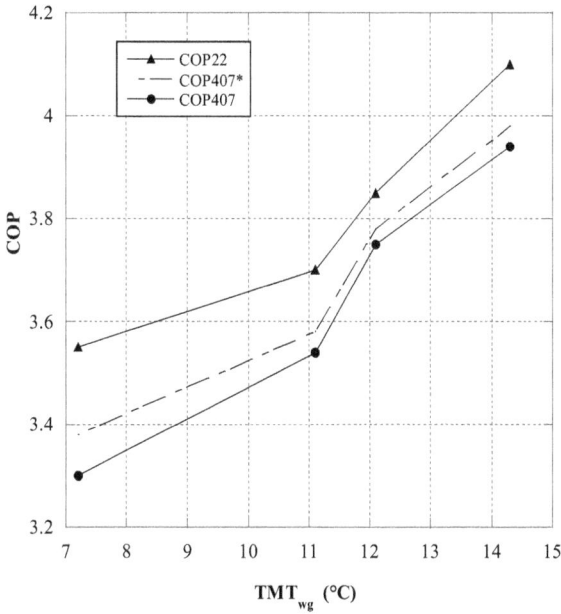

**Figure 11.** COP vs mean thermodynamic temperature of water-glycol mixture, Tin,w,co =40°C, Tout,w,co =45 °C.

### 3.3.3. Overall evaluation

The TEWI has been evaluated as a function of the mean thermodynamic temperature of the water glycol mixture at the evaporator.

The results are summarised in Figs 13-15.

According to the previous discussion, in Figs. 13 and 14, the TEWI pertaining to R407C is greater than that corresponding to R22. The difference ranges from a minimum of 2.3 to a maximum of 4.3 % in the runs performed at 30/35 °C. The effect is even more marked in the runs performed at 35/40 °C (about 11%). This yields an increased CO2 emission of about 320 kg/year for the plant under consideration when working as a refrigerator with R407C as compared to R22.

On the contrary, the two fluids are virtually equivalent in terms of greenhouse effect, when the plant acts as a heat pump in the winter cycle (40/45 °C runs, Fig. 15).

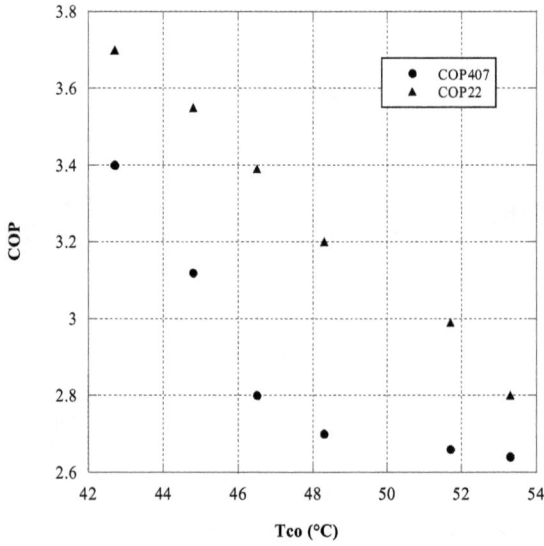

**Figure 12.** COP vs condensing temperature of the refrigerant fluid.

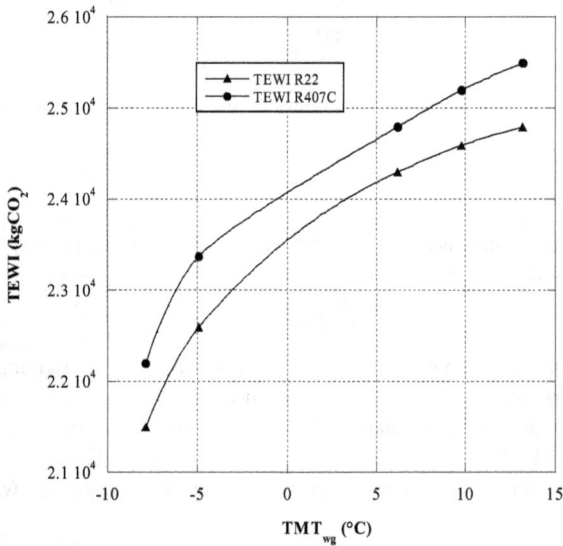

**Figure 13.** TEWI vs mean thermodynamic temperature of water-glycol mixture, Tin,w,co =30°C, Tout,w,co =35 °C.

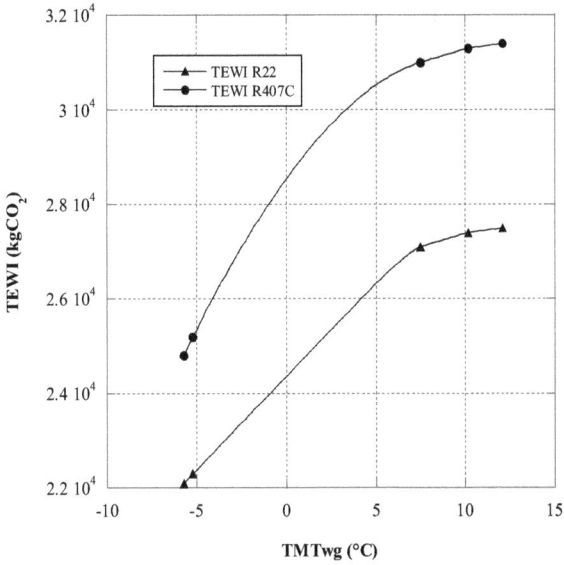

**Figure 14.** TEWI vs mean thermodynamic temperature of water-glycol mixture, Tin,w,co =35°C, Tout,w,co =40 °C.

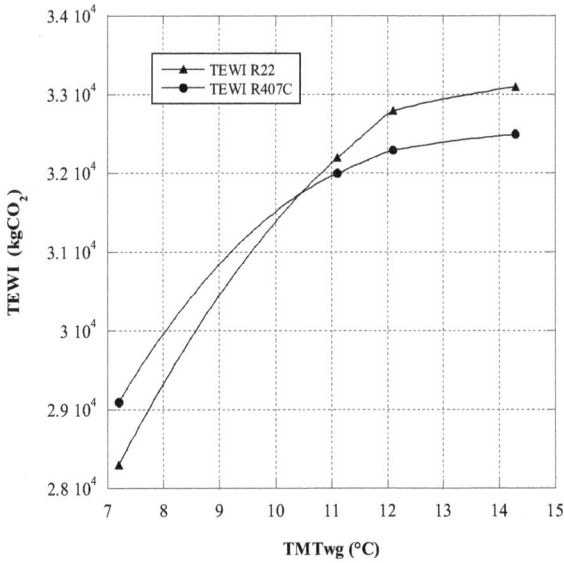

**Figure 15.** TEWI vs mean thermodynamic temperature of water-glycol mixture, Tin,w,co =40°C, Tout,w,co =45 °C.

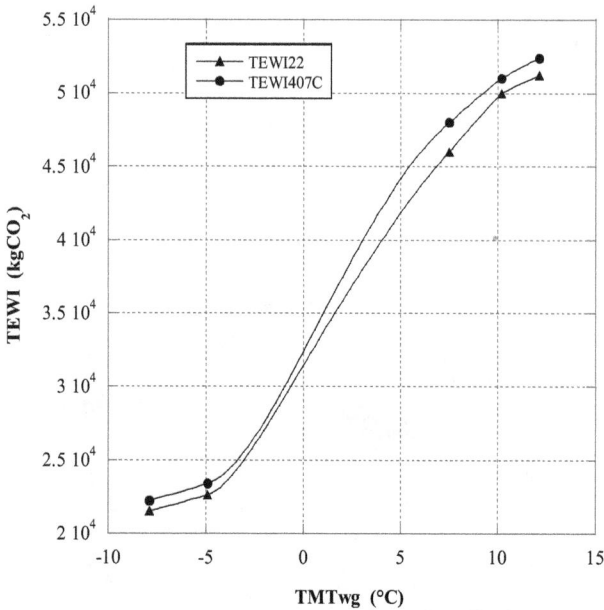

**Figure 16.** TEWI vs mean thermodynamic temperature of water-glycol mixture for the reversible pump working for the whole year.

Figure 16 reports the overall results for the reversible heat pump working for the whole year (both in winter and summer cycle). It can be seen that the TEWI pertaining to R407C is slightly larger than that of R22 (2.7-4.7%). This corresponds to an increased $CO_2$ production of about 87 kg/year.

In Figure 17, the TEWI difference ($\Delta$TEWI) between the two fluids is reported as a function of the condensation temperature for three different values of mean thermodynamics temperature of the refrigerant duty (8 10, 13 °C respectively). The $\Delta$TEWI attains a maximum in the intermediate range of condensation temperatures. On the contrary, for high condensation temperature, $\Delta$TEWI becomes negative. In the condensation temperature range explored, the $\Delta$TEWI decreases with increasing mean thermodynamic temperature of the refrigerant duty.

### 3.4. Concluding remarks

The problem of R22 substitution with R407C in terms of global warming effect has been examined. R407C is harmless as far as ozone is concerned. It yields, however, a negative contribution to the greenhouse effect, both directly with a GWP of 1600 and indirectly, because of the lower energetic efficiency that results in higher $CO_2$ emissions.

Experimental readings have been carried out on a vapour compression pilot plant. The contribution to the greenhouse effect has been taken account by evaluating the TEWI values.

The analysis of the experimental runs leads to the conclusion that the R22 direct contribution to the greenhouse effect is greater than that pertaining to R407C (+15%). The COP corresponding to R407C is 3.3-19% lower than that pertaining to R22. The indirect contribution of the former is, therefore, always higher than that of the latter. Therefore, the final result depends on the operating conditions adopted in the experimental runs.

The most unfavourable conditions are reached for the water inlet and outlet temperature at the condenser of 35-40 °C, respectively. This situation corresponds to a condensation temperature in the range 43-47 °C and to an evaporation temperature in the range -12/-7 °C. Under these conditions, the TEWI pertaining to R407C exceeds that of R22 by about 11%. Therefore, the substitution of R22 with R407C should be unacceptable in this specific situation if specific reference is made to the greenhouse effect.

On the contrary, the most favourable conditions are met for the water inlet and outlet temperature at the condenser of 45-50 °C, respectively. This situation corresponds to a condensation temperature in the range 53-58 °C and to an evaporation temperature inlet range 2 – 10 °C. In this case, the two fluids behave in a similar way as regards the greenhouse effect.

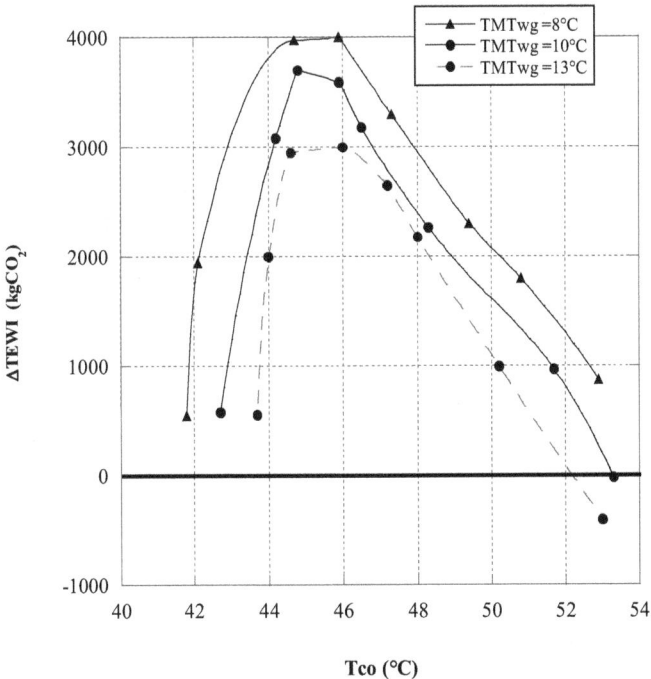

**Figure 17.** Difference between R407C and R22 TEWI vs condensing temperature of the refrigerant fluid for three different thermodynamic temperature of water-glycol mixture (8, 10 and 13 °C).

For high evaporation temperatures, the TEWI of R407C is even slightly lower than that of R22. As a consequence, the substitution of R22 with R407C is favourable, since no harm is produced to the ozone layer and no increase in $CO_2$ emission is made. As a final remark the substitution of R22 with R407C is convenient from the point of view of the greenhouse effect for high condensation (over 50°C) and evaporation temperatures of the refrigerant fluid.

## 4. An experimental evaluation of the greenhouse effect in R22 substitution with R422D

### 4.1. R22 substitution with R422D

As aforementioned, R22 is the most widely used refrigerant today in commercial, domestic and industrial applications, and its phase-out will impact a large number of plants in the world. Therefore retrofitting these systems would alleviate the R22 phase-out problem. This opportunity could be cheaper than the installation of new plants, especially for supermarkets, data centers, factories and hospitals. Refrigerant replacement candidates have been checked for environmental and safety requirements, compatibility with lubricant oil, filters, and sealing. In order to establish the best substitute in a specified system among the candidates, it is necessary to estimate energetic performances after refrigerant replacement. In the last years, many companies have expended much effort to develop and characterize refrigerants able to increase the energetic efficiency of a refrigeration system, depending on its application.. Even if R407C has thermo-physical properties similar to those of R22 and it is non-flammable and non-toxic, it is incompatible with mineral or alkyl-benzene oil. Consequently, the R22 retrofit with R407C implies the adoption of polyester oil, which requires a difficult and expensive conversion. Furthermore, in comparison with R22, experimental tests carried out with R407C have pointed out a reduction in the energetic performances. Recent additions to the aforementioned alternative refrigerants for R22 are R422A, R422B, R422C and R422D. The U.S. environment protection agency reported these alternative refrigerants could be used for household and light commercial air conditioning applications. In particular, R422D is a zeotropic mixture of R134a/ R125/ R600a (31.5/65.1/3.4 % in weight) with no ODP and a GWP of 2230. This refrigerant fluid, originally was designed to replace R-22 in existing direct expansion water chiller systems. It can also be used in residential and commercial air conditioning and medium-temperature (and low) refrigeration systems. Minor equipment modifications (replacement of the filter drier and elastomeric seals/gaskets that are exposed to refrigerant, refill of oil if required) or components tuning may be required in some applications. It is also compatible with mineral oil and there is no need to replace it with synthetic oil. Field experience has shown that R422D provides performance that meets customer requirements in most retrofitted systems. It provides similar cooling capacity to R22 and it is capable to operate at significantly lower compressor discharge temperature.

In this scenario, it is seems sensitive to inquire what is the actual impact of R422D on the environment, when it is employed in retrofitted R22 devices. It is well known that the GWP of R422D is higher than that of R22, but not much is known about the energy efficiency of systems retrofitted by R422D. For this purpose, a comparison of the energy consumptions of R22 and R422D for a direct expansion refrigerator applied to a commercial cold store is proposed. The experimental investigation has been carried out for different application conditions: medium temperature refrigeration for meat, fish, and dairy cases and high temperature refrigeration for air conditioning and cooling of preparation room. Subsequently, we investigated into the possibility of reducing the environmental impact of retrofitting R22 systems with R422D by means of a sensitivity analysis of some of the functional parameters.

## 4.2. Experimental facility

The experimental vapor compression refrigeration plant, applied to a commercially available cold store as shown in Fig. 18, consists of a semi-hermetic reciprocating compressor, an air condenser followed by a liquid receiver, a R22 mechanic thermostatic expansion valve to feed an air-cooled evaporator inside the cold store.

The compressor, as declared by the manufacturer, can operate with the fluid R22 and it is lubricated with mineral oil. With an evaporation temperature range between -20 to 10°C, a 35°C condensing temperature, and utilizing R22 at the nominal frequency of 50 Hz, the compressor refrigerating capacity is in the range of 1.4-4.4 kW. A blower drives the airflow through a thermally insulated channel where some electrical resistances are located with the objective of controlling the temperature of the airflow crossing the condenser. To fix the temperature of the airflow in accordance with the values sought, we changed the voltage supply of the electrical resistances by means of a PID controller. The cooling load in the cold store is emulated by means of additional electrical heaters wired to a voltage regulator. To keep the air temperature reasonably constant in the cold store, an on/off refrigeration control system has been implemented. This is done by turning on/off the compressor and the fan of the heat exchangers.

Table 6 reports the transducers specifications used (Coriolis effect mass flow rate meter, RTD 100 4 wires thermo-resistances, piezoelectric absolute pressure gauge, wattmeter). The thermo-resistances are located outside the pipe, with a layer of heat transfer compound (aluminum oxide plus silicon) placed between the sensor and the pipe in order to provide good thermal contact. The whole pipe is covered with 25 mm thick flexible insulation. The system of temperature measurement was checked against a sensor positioned in pocket in a similarly insulated pipe work. For various test conditions, the difference between the two measurements has been always less than 0.3 °C. The wattmeter is able to measure the electrical power absorbed by the compressor, the blowers and any kind of accessory installed for operation of the device. The energy consumption of the refrigeration system is measured by means of an energy meter. The test apparatus is equipped with 32 bit A/D converter acquisition cards linked to a personal computer that allows a high sampling rate (10 kHz)

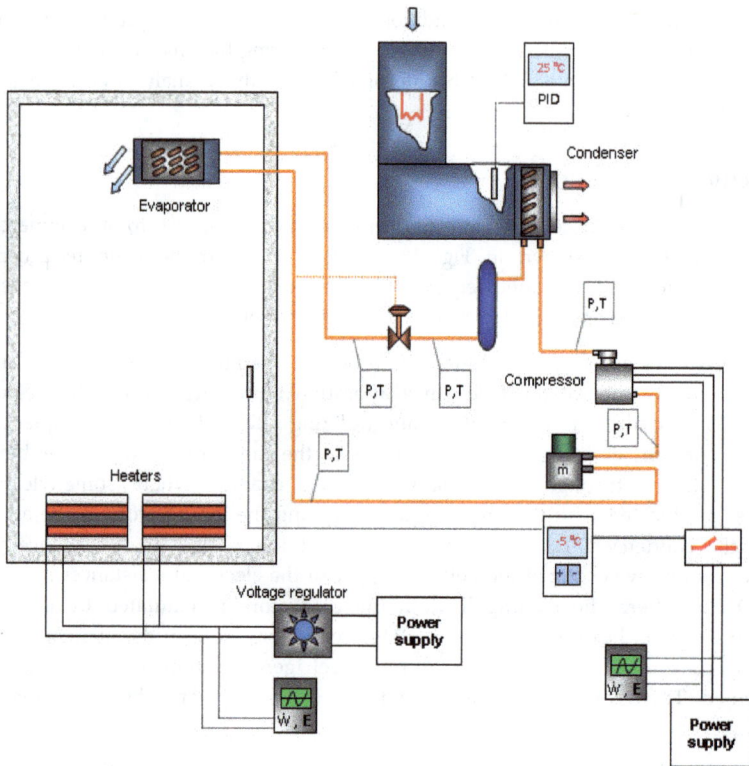

**Figure 18.** Sketch of the experimental plant

| Transducers | Range | Uncertainty |
|---|---|---|
| Coriolis effect flowmeter | 0 ÷ 2 kg/min | + 0.2 % |
| RTD 100 4 wires | -100 ÷ 500 °C | + 0.15 °C |
| Piezoelectric absolute pressure aguge | 1 ÷ 10 bar | + 0.2 % |
| | 1 ÷ 30 bar | + 0.5 % F.S. |
| Wattmeter | 0 ÷ 3 kW | + 0.2 % |
| Energy meter | 0 ÷ 1 MWh | + 1% |

**Table 6.** Transducers specifications

## 4.3. Experimental procedure

We started the experimental investigation by analyzing the operation of the plant with R22. Subsequently, we retrofitted the refrigeration system with R422D in accordance with [31]. During the retrofitting operations, we changed the factory setting of the R22 thermostatic expansion valve in order to keep the operating superheat value for R422D in the same range used for R22, which was performed by turning the adjusting screw of the valve.

For both refrigerants, we used the same experimental procedure. Firstly, we proceeded with identifying the refrigerant charge necessary to guarantee that the fluid adequately wets the evaporator. For this purpose, we set the temperature of the air blown through the condenser to 24°C and the air inner to cold store to -5°C, while the cooling load was kept at 1000 W. By means of a vacuum pump, we evacuated the circuit, and then we proceeded with introducing 0.40 kg of gas in the refrigerant circuit while the system was shut off to preserve the electrical motor of the compressor from overheating. Subsequently, we turned on the plant and the electrical heaters inside the cold store. During the operation of the plant, we monitored the value of the operating superheat, defined as the difference between the temperature at end of the evaporating process (considering the pressure drop into the evaporator) and temperature at the compressor inlet. Then, we systematically continued: additional 0.10 kg of refrigerant was introduced until, under steady state conditions, the operating superheat was not included in the range 7.0 – 10 °C.

| Parameter | Value |
|-----------|-------|
| H | 950 h/year |
| $P_L$ | 10% |
| $P_R$ | 1 |
| V | 1 year |
| $\alpha$ | 0.6 kg $CO_2$/kWhe |
| $RC_{R22}$ | 2.50 kg |
| $RC_{R422D}$ | 2.30 kg |

**Table 7.** Parameters in TEWI evaluation.

Once the system was charged to the specified value we proceeded with the evaluation of the energy consumption due to a year of operation (storage investigation). Usually, the service life (V) of refrigeration system refers to the operation time of the equipment. If one is investigating the change of the environmental impact due to the retrofitting operations, the service life to use for the comparing analysis should not be equal to the actual life of the equipment, but one has to consider a reference service life. In this paper the reference service life is one year. In this scenario, the leakage rate per year has not to include the disposal percentage but only the accidental percentage due to operating conditions. Table 7 reports the parameters for TEWI evaluation according to eq.(1).

In particular, we investigated four different storage applications: -5, 0, 5, 10°C. We considered as external air temperature reference the values reported in Table 8, which represent typical conditions in Milan, Italy, used for our scenario.

Since this table provides the daily change of external air temperature for each month, we planned 12 experiments each 24 hours long. The data reported in Table 8 were loaded as database for the PID controller, which modulates the voltage supply to the electrical resistances used to warm the air intake. For each experiment, recording of the energy consumption values started when the refrigeration control began to operate.

To evaluate the energetic performance, we needed further experiments aimed to analyze the behavior of the plant under steady state conditions (*performance investigation*). For this purpose, we planned 4 experiments at different storage temperatures: -5, 0, 5, 10°C. During the experiments, we shunted the refrigeration control and we set the temperature of the air blown through the condenser to a reference value. We choose as reference the mean of the data reported in Table 8: 21 °C. Usually, the start-up time required was about one hour. Steady state conditions were assumed to hold when the deviations of all controlled variables from their corresponding mean values were lower than 0.5°C for temperatures and 15 kPa for pressures. At this stage, the test started and the logging of data with 0.5 Hz acquisition frequency was performed on all channels for 60 s. For each channel, the 120 samples recorded were averaged. After 180 s, each sample during 60 s was checked against the corresponding mean values of the two previous samples; when the mean values of the temperatures and of the pressures were within the range reported above, the steady state was reached.

| | January | February | March | April | May | June | July | August | September | October | November | December |
|---|---|---|---|---|---|---|---|---|---|---|---|---|
| Hours | [°C] | [°C] | [°C] | [°C] | [°C] | [°C] | [°C] | [°C] | [°C] | [°C] | [°C] | [°C] |
| 0 | 11.4 | 12.7 | 15.7 | 17.5 | 19.9 | 21.6 | 22.2 | 22.35 | 21 | 18.6 | 15.1 | 12.2 |
| 1 | 10.8 | 12.1 | 15.9 | 17.8 | 20.2 | 21.9 | 21.6 | 21.75 | 20.4 | 18 | 14.5 | 11.6 |
| 2 | 10.2 | 11.5 | 14.5 | 16.3 | 18.7 | 20.4 | 21 | 21.15 | 19.8 | 17.4 | 13.9 | 11.0 |
| 3 | 9.8 | 11.1 | 14 | 15.9 | 18.3 | 19.9 | 20.5 | 20.65 | 19.4 | 17 | 13.5 | 10.5 |
| 4 | 9.4 | 10.7 | 13.7 | 15.5 | 17.9 | 19.6 | 20.1 | 20.25 | 19 | 16.6 | 13.1 | 10.1 |
| 5 | 9.3 | 10.6 | 13.5 | 15.4 | 17.8 | 19.4 | 20 | 20.15 | 18.9 | 16.5 | 13 | 10.0 |
| 6 | 9.5 | 10.8 | 13.8 | 15.6 | 18 | 19.7 | 20.2 | 20.35 | 19.1 | 16.7 | 13.2 | 10.3 |
| 7 | 10.1 | 11.4 | 14.4 | 16.2 | 18.6 | 20.3 | 20.8 | 20.95 | 19.7 | 17.3 | 13.8 | 10.9 |
| 8 | 11.2 | 12.5 | 15.5 | 17.3 | 19.7 | 21.4 | 21.9 | 22.05 | 20.8 | 18.4 | 14.9 | 11.9 |
| 9 | 12.8 | 14.1 | 17 | 18.9 | 21.3 | 22.9 | 23.5 | 23.65 | 22.4 | 20 | 16.5 | 13.5 |
| 10 | 14.6 | 15.9 | 18.8 | 20.7 | 23.1 | 24.7 | 25.3 | 25.45 | 24.2 | 21.8 | 18.3 | 15.3 |
| 11 | 16.6 | 17.9 | 20.9 | 22.7 | 25.1 | 26.8 | 27.3 | 27.45 | 26.2 | 23.8 | 20.3 | 17.3 |
| 12 | 18.5 | 19.8 | 22.8 | 24.6 | 27 | 28.7 | 29.2 | 29.35 | 28.1 | 25.7 | 22.2 | 19.3 |
| 13 | 20.1 | 21.3 | 24.2 | 26.1 | 28.5 | 30.1 | 30.7 | 30.85 | 29.6 | 27.2 | 23.7 | 20.7 |
| 14 | 20.9 | 22.2 | 25.2 | 27 | 29.4 | 31.1 | 31.6 | 31.75 | 30.5 | 28.1 | 24.6 | 21.7 |
| 15 | 21.3 | 22.6 | 25.5 | 27.4 | 29.8 | 31.4 | 32 | 32.15 | 30.9 | 28.5 | 25 | 22.0 |
| 16 | 20.9 | 22.2 | 25.2 | 27 | 29.4 | 31.1 | 31.6 | 31.75 | 30.5 | 28.1 | 24.6 | 21.7 |
| 17 | 20.1 | 21.4 | 24.3 | 26.2 | 28.6 | 30.2 | 30.8 | 30.95 | 29.7 | 27.3 | 23.8 | 20.8 |
| 18 | 18.8 | 20.1 | 23 | 24.9 | 27.3 | 28.9 | 29.5 | 29.65 | 28.4 | 26 | 22.5 | 19.5 |
| 19 | 17.2 | 18.5 | 21.5 | 23.3 | 25.7 | 27.4 | 27.9 | 28.05 | 26.4 | 24.4 | 20.9 | 17.9 |
| 20 | 15.6 | 16.9 | 19.9 | 21.7 | 24.1 | 25.8 | 26.4 | 26.55 | 25.2 | 22.8 | 19.3 | 16.4 |
| 21 | 14.3 | 15.6 | 18.6 | 20.4 | 22.8 | 24.6 | 25 | 25.15 | 23.9 | 21.5 | 18 | 15.1 |
| 22 | 13.1 | 14.4 | 17.4 | 19.2 | 21.6 | 23.3 | 23.8 | 23.95 | 22.7 | 20.3 | 16.8 | 13.9 |
| 23 | 12.2 | 13.5 | 16.4 | 18.3 | 20.7 | 22.3 | 22.9 | 23.05 | 21.8 | 19.4 | 15.9 | 12.9 |

**Table 8.** Change of the external air temperature for Milan (Italy).

## 4.4. Data elaboration

To manage the large number of data, we developed a software application called FrigoCheck v.1.0. It is able to show in real time the coefficient of performance, the entropy and the enthalpy values of all points of the thermodynamic cycle. In addition, it shows the whole cycle on p-h diagram and it establishes the achievement of the steady state condition.

Since we measured the daily energy consumption for each month (Ed,i), we calculated the monthly energy consumption (Em,i) by means of the following equation

$$E_{m,i} = E_{d,i} \times ND_i \qquad (2)$$

where the subscript $i$ refers to the generic month and ND is equal to the number of days for the i-th month. We obtained yearlong energy consumption (E) by summing the monthly energy consumptions (Em,i):

$$\frac{\dot{Q}_{ref}}{COP} = E = \sum_i E_{m,i} \qquad (3)$$

The uncertainty of the yearlong energy consumption is equal to ±1% .

Under steady state conditions, the overall efficiency performance of the plant is defined by means of the evaluation of COP, calculated as the ratio between the refrigeration capacity and the electrical power supplied to the plant (compressor, blowers and accessories):

$$COP = \frac{\dot{m}\left(h_{out,EV} - h_{in,EV}\right)}{\dot{W}_{el}} . \qquad (4)$$

The COP accuracy has been equal to ± 2.5%. In order to identify the efficiency of the plant, we considered the following ratio:

$$\varepsilon = \frac{COP}{\dfrac{1}{\dfrac{T_{ex}}{T_{cold}} - 1}} . \qquad (5)$$

## 4.5. Scenario and sensitivity analysis for the TEWI difference

Once completed the experimental investigation and the data elaboration, we focused our attention on the following question: after the retrofitting operations, how can one reduce the R422D TEWI? For this purpose, we considered a scenario and sensitivity analysis for the TEWI difference aimed to individuate the operating conditions (scenario) leading to the reduction of the TEWI. In equation (1), we individuated two parameters:

- Leakage rate per year, which acts on the direct effect;
- Energy saving, which acts on the indirect effect.

For our first analysis, we kept the yearlong energy consumptions equal to those measured and varied the leakage rate per year, in accordance with the range 5-10%. For each leakage rate value, we calculated the new R422D TEWI and compared it with the R22 TEWI by means of the following equation:

$$\Delta TEWI = \frac{TEWI_{R422D} - TEWI_{R22}}{TEWI_{R22}} .. \tag{6}$$

We repeated this procedure for each test condition; therefore we obtained the change of the TEWI difference as a function of the leakage. As a second analysis, we kept the leakage rate equal to that reported in Table 7 and we varied the yearlong energy consumptions by considering an energy saving included in the range 0-100%. Following a similar procedure to that above mentioned, we obtained the change of the TEWI difference at each test conditions as a function of the energy saving.

Since the one parameter analysis could lead to scenarios technically not feasible or very expensive, we deemed appropriate to consider the simultaneous change of both parameters.

### 4.6. Results and discussions

The first step of the experimental investigation led to identifying the correct charge for both refrigerants. As reported in Table 7, from experimental evaluation, the mass of R22 resulted 0.20 kg larger than that of R422D, which means an 8% reduction of refrigerant mass.

By means of the storage investigation we carried out the daily energy consumptions ($E_{d,i}$) for each test conditions (Table 9).

| | Daily energy consumption ($E_{d,i}$) | | | | | | | |
| | -5 °C | | 0 °C | | +5 °C | | 10 °C | |
| | R22 | R422D | R22 | R422D | R22 | R422D | R22 | R422D |
| | (Wh) | (Wh) | (Wh) | (Wh) | (Wh) | (Wh) | (Wh) | (Wh) |
|---|---|---|---|---|---|---|---|---|
| January | 4'848 | 5'080 | 3'051 | 3'392 | 1'800 | 2'097 | 1'039 | 1'334 |
| February | 5'017 | 5'280 | 3'166 | 3'529 | 1'873 | 2'188 | 1'086 | 1'395 |
| March | 5'430 | 5'772 | 3'448 | 3'868 | 2'053 | 2'416 | 1'202 | 1'547 |
| April | 5'701 | 6'098 | 3'635 | 4'093 | 2'173 | 2'567 | 1'279 | 1'648 |
| May | 6'075 | 6'551 | 3'893 | 4'405 | 2'339 | 2'779 | 1'387 | 1'791 |
| June | 6'348 | 6'884 | 4'082 | 4'636 | 2'461 | 2'937 | 1'468 | 1'897 |
| July | 6'436 | 6'991 | 4'143 | 4'710 | 2'500 | 2'987 | 1'494 | 1'931 |
| August | 6'461 | 7'022 | 4'161 | 4'731 | 2'512 | 3'002 | 1'501 | 1'941 |
| September | 6'246 | 6'759 | 4'011 | 4'549 | 2'415 | 2'878 | 1'437 | 1'857 |
| October | 5'864 | 6'295 | 3'747 | 4'229 | 2'245 | 2'660 | 1'326 | 1'710 |
| November | 5'346 | 5'671 | 3'390 | 3'798 | 2'016 | 2'369 | 1'178 | 1'516 |
| December | 4'944 | 5'193 | 3'116 | 3'470 | 1'842 | 2'149 | 1'066 | 1'369 |

**Table 9.** Daily energy consumption measured.

Considering equation (2) we converted the results shown in Table 9 in monthly energy consumptions ($E_{m,i}$) and then in yearlong energy consumptions (Fig. 19).

It can be seen that the yearlong energy consumption pertaining to R422D is larger than that of R22 (7.10-28.9%) for each test conditions. Furthermore, for both refrigerants the energy consumption diminishes with the increase of the air temperature inner cold store. This is easily understandable if one considers the reduction of the operating temperature span, which is defined as the mean difference between the hot thermal sink (external ambient) and the cold thermal sink (cold store).

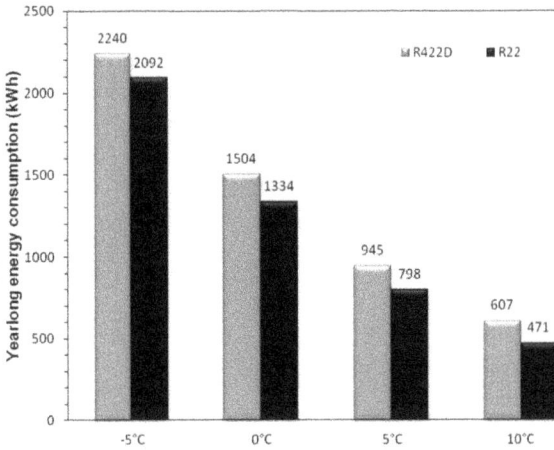

**Figure 19.** Yearlong energy consumption vs. air temperature inner to cold store (± 1%).

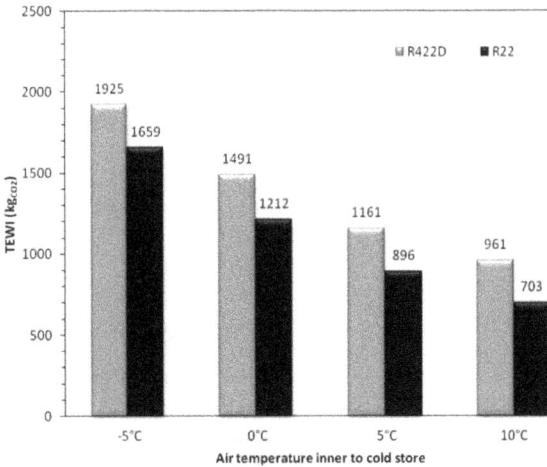

**Figure 20.** TEWI vs. air temperature inner to cold store (±10%).

| | \-5 °C | | 0 °C | | +5 °C | | 10 °C | |
| --- | --- | --- | --- | --- | --- | --- | --- | --- |
| | R22 (Wh) | R422D (Wh) | R22 (Wh) | R422D (Wh) | R22 (Wh) | R422D (Wh) | R22 (Wh) | R422D (Wh) |
| January | 150'295 | 157'473 | 94'568 | 105'144 | 55'809 | 64'994 | 32'220 | 41'354 |
| February | 140'481 | 147'833 | 88'637 | 98'814 | 52'451 | 61'272 | 30'406 | 39'060 |
| March | 168'338 | 178'940 | 106'893 | 119'904 | 63'650 | 74'888 | 37'250 | 47'951 |
| April | 171'040 | 182'945 | 109'037 | 122'777 | 65'177 | 77'024 | 38'368 | 49'452 |
| May | 188'323 | 203'066 | 120'670 | 136'553 | 72'494 | 86'164 | 43'002 | 55'517 |
| June | 190'455 | 206'521 | 122'471 | 139'068 | 73'833 | 88'105 | 44'029 | 56'907 |
| July | 199'505 | 216'713 | 128'433 | 145'995 | 77'511 | 92'610 | 46'299 | 59'862 |
| Agust | 200'298 | 217'685 | 128'984 | 146'668 | 77'868 | 93'070 | 46'534 | 60'173 |
| September | 187'370 | 202'755 | 120'329 | 136'463 | 72'449 | 86'327 | 43'119 | 55'708 |
| October | 181'794 | 195'150 | 116'157 | 131'084 | 69'589 | 82'448 | 41'105 | 53'019 |
| November | 160'377 | 170'141 | 101'711 | 113'953 | 60'490 | 71'070 | 35'335 | 45'467 |
| December | 153'268 | 160'987 | 96'591 | 107'557 | 57'091 | 66'604 | 33'037 | 42'425 |

**Monthly energy consumption ($E_{m,i}$)**

**Table 10.** Monthly energy consumption calculated by means of the Eq. 2

In Fig. 20, for both refrigerants we have drawn the TEWI relative to one year of operation as function of the air temperature inner to cold store. Since the first term of the equation (1) is constant for both refrigerants and for each test condition, it is noticeable that the change of TEWI is directly influenced by that of the energy consumption.

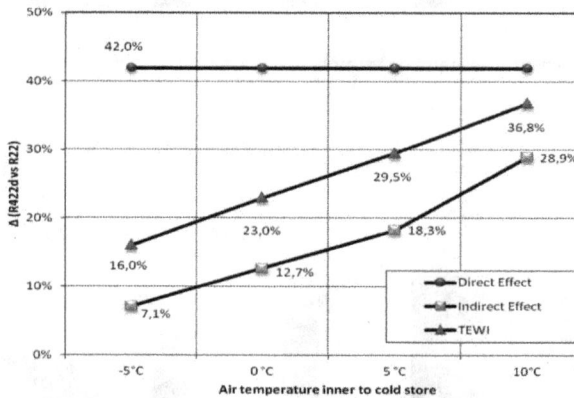

**Figure 21.** Difference percentage of indirect, direct and TEWI vs. air temperature inner to cold store.

As remarked in Fig.21, the adoption of R422D leads to a worsening of the environmental impact. Because the GWP of R422D is 50% higher than that of R22, the 8% reduction of refrigerant mass does not impact significantly the decrease of the direct effect of R422D, which results equal to 42% higher than that of R22. The indirect effect (Fig. 21) of R422D is higher than that of R22; in particular, this difference grows with the increase of the storage temperature. Consequently, the release of $CO_2$, due to the adoption of R422D, grows from a minimum of 7.1% to a maximum of 28.9%.

The augmentation of the energy consumption has been also confirmed by performance investigation. In Fig. 22, we report both the change of the COP and that of the efficiency as a function of the air temperature inner to cold store. The results there illustrated highlights a lower efficiency of the plant due to the adoption of R422D. It is possible to observe that COP increases with the increase of the air temperature inner to cold store and that the COP for R22 is higher than for R422D.

In particular, the difference between the COP for R22 and that for R422D is, on average, 20% and it increases with the increase of the air temperature inner cold store. The worsening of the energy performance, due to the use of R422D as substitute of R22, is remarked by the efficiency values showed in Fig. 22.

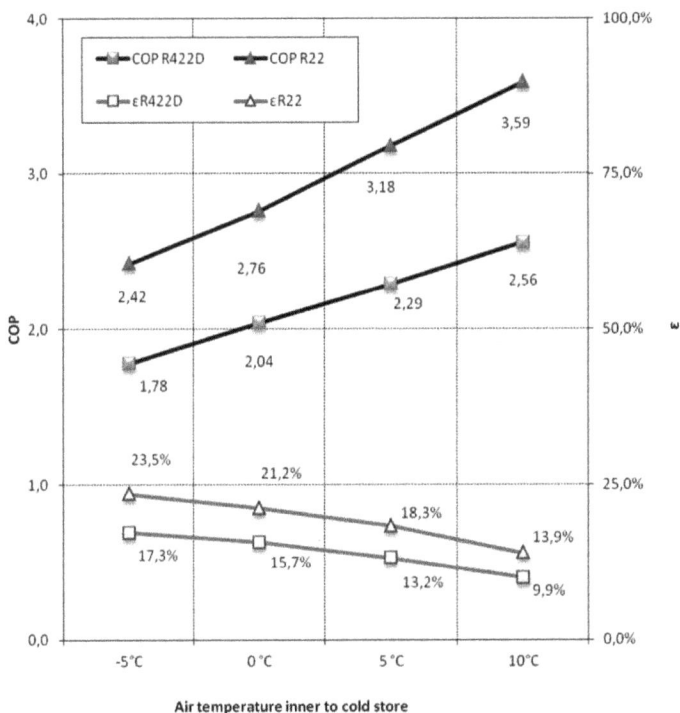

**Figure 22.** COP and efficiency of the plant vs. air temperature inner to cold store (±2.5%).

The lower R422D efficiency values represent a warning on the inefficiency of the plant subsequent to the retrofitting operations. During the performance investigation, we observed a substantially different behavior of the condenser. As reported in Fig. 23 the condensing pressure of R422D was higher than that of R22, while the evaporating pressure for both refrigerants was similar. This gain in terms of pressure at condenser shows that, when R422D is used as refrigerant, the heat exchange surface of the condenser is insufficient to reject the thermal power. Furthermore, a more high condensing pressure leads to an increase of electrical power absorbed by the compressor further reducing the COP. However, it is our interest to understand how to improve the energy performance of the plant operating with R422D, with the intent of reducing the indirect impact on the environment. Particular attention should be given to the condenser, since, as highlighted above, the heat exchange surface required by R22 is less than that required by R422. Referring to the operation with R422D, an improvement of the heat exchange at condenser could lead to reduce the condensing pressure, and then the specific work of compressor. Furthermore, the fluid leaving the condenser would be sub-cooled further, allowing a gain of the specific heat of evaporation. Consequently, a lower mass flow rate could be required, and then the power absorbed by the compressor could be reduced, guarantying an improvement of the COP.

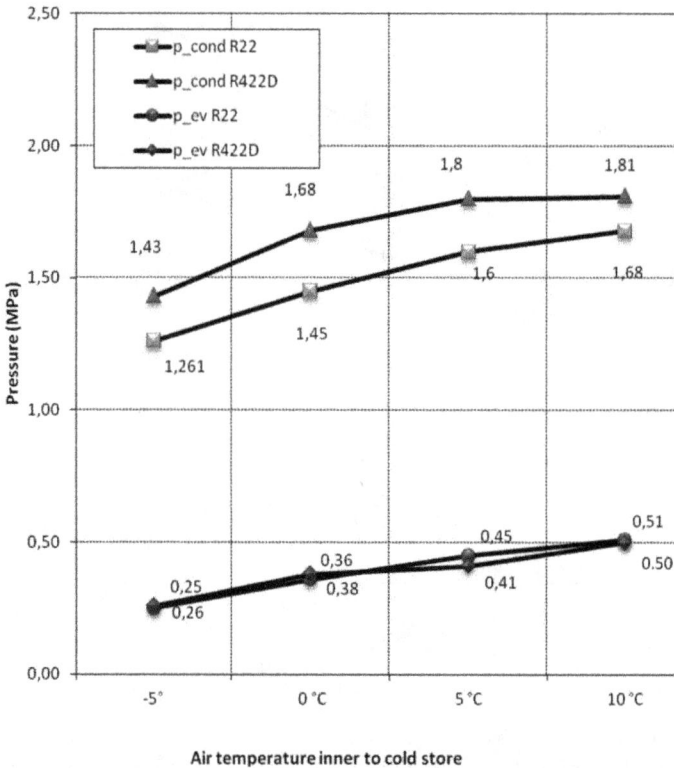

**Figure 23.** Condensing and evaporating pressure vs. air temperature inner to cold store.

To enhance the heat exchange at condenser two methods could be considered:

- if the blower have different operating speeds, select the highest speed available, otherwise, change the blower with another having higher volumetric capacity;
- replace the condenser with one having a larger surface.

It is important to underline that the first solution is better than second, since cheaper and technically easier, but it could lead to a larger absorption of electrical power by blowers, negatively affecting the COP. Based on theoretical considerations [32], it is possible to expect a 15% increase of COP for a 1.5 bar reduction of the condensing pressure.

When R422D is used as refrigerant, a further improvement of the energy performance could be obtained by installing an electronic expansion valve instead of the thermostatic one. As showed by Lazzarin and Noro [33], for any refrigerants the electronic expansion valve allows a lower condensation pressure in systems equipped with air-cooled condensers, thanks to the ability of monitoring the variations of theoutside air temperature. Consequently, they [33] indicated, on average, an 8% reducing of energy consumption for Mediterranean locations and a 15% for North-European locations.

Once completed the experimental investigation, we considered a new scenario still to evaluated a sensitivity analysis for the TEWI difference, now aimed to highlight the way to reduce the environmental impact of the R22 retrofit with R422D. For this purpose, we have reported in Fig. 24 the change of the TEWI difference percentage as a function of the leakage rate per year, while in Fig.25 we have done it as a function of the energy saving per year. For both figures, we have identified three scenarios:

- Scenario A: it represents the parameter domain, for which for every test conditions the environmental impact for R422D becomes higher than that for R22.
- Scenario B: it represents the parameter domain of transition, for which at least for one test condition the environmental impact for R422D becomes lower than that for R22.
- Scenario C (or eco-friendly scenario): it represents the parameter domain, for which for every test conditions the environmental impact for R422D becomes lower than that for R22.

Furthermore, we reported two dashed axes, whose intersection represents the breakeven point between the TEWI of R22 and that of R422D for all test conditions. Fig. 24 shows that for leakage rate per year lower than 5.7% the TEWI of R422D becomes lower than that of R22. The scenario B is very narrow and this allows achieving of the breakeven point almost simultaneously for all operating conditions. Considering 5% as technical limit for no-hermetical plant, it can see that the scenario C (Fig. 24) occurs for leakage rate per year values including in the narrowest range 5.0 ÷ 5.4 %, and it leads to a maximum $\Delta$TEWI (absolute value) included between -6.0 ÷ -2.0%. The scenario C is technically feasible but it leads to an increase of the management costs: leakage check more frequent could lead to reduce the leakage rate, as indicated in [34].

Considering the change of the energy saving (Fig. 25), it can be seen that the scenario B is very large and it starts for ~20% energy saving; the breakeven point is reachable for ~70% energy saving. These results are not reassuring, because a 70% energy saving should correspond to a plant efficiency equal to 30%, when usually, for actual plant operating under the same conditions here investigated, the efficiency can be equal to 25%. Differently, as above mentioned, 20% energy saving could be obtained as a result both of the heat exchange improvement and of the use of electronic expansion valves. Since both the strong reduction of leakage rate and the high energy saving lead to an increase of the total cost of the retrofitting operations, it is necessary to consider an overlay parameter change. For this purpose, we have developed another scenario considering the simultaneous change of both parameters.

In Fig. 26 we have reported four different charts, each referring to different test condition in terms of air temperature inner to cold store. For each chart, we have drawn a solid black line, which identifies the border between the scenario characterized by a negative TEWI difference (eco-friendly scenario) and that by a positive TEWI difference. This time the eco-friendly scenario is identified by means of a 2 D domain (leakage rate & energy saving): one has to select one couple of values for leakage rate and energy saving per year in order to obtain a reduction of TEWI consequently to the adoption of R422D. That the overlaying of the effects allows obtaining strong TEWI reduction by means of cheap operations: minor reductions of leakage rate and energy saving are required.

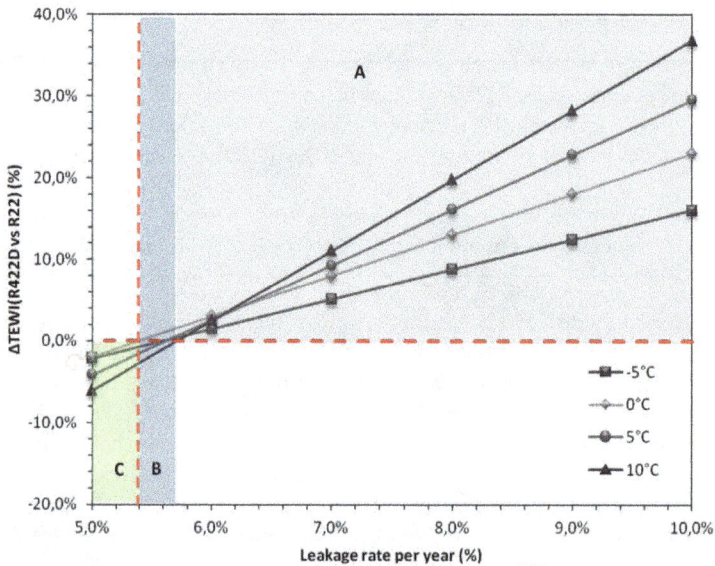

**Figure 24.** Difference of TEWI vs. leakage rate per year and operating scenarios.

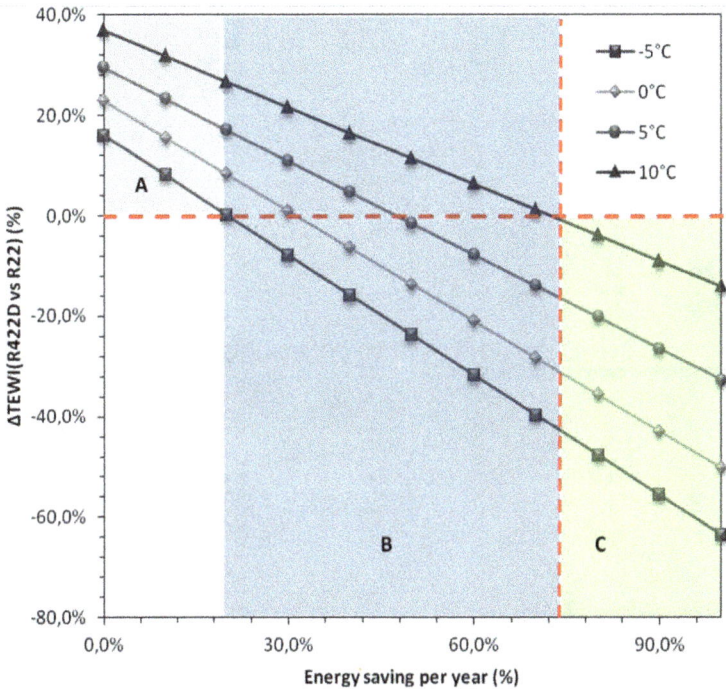

**Figure 25.** Difference of TEWI vs. energy saving per year and operating scenarios.

## 4.7. Concluding remarks

An experimental investigation has been carried out study the environmental impact of the R22 retrofit with R422D and to draw possible eco-friendly scenarios [35]. The experimental investigation consisted of two parts:

- Storage investigation, aimed at developing of TEWI analysis.
- Performance investigation, aimed at analyzing the behavior of the plant under steady state conditions.

For both investigations, we have considered four operating conditions in terms of the air temperature inner to cold store: -5, 0, 5, 10°C. To emulate actual operating conditions we choose Milan as reference locality.

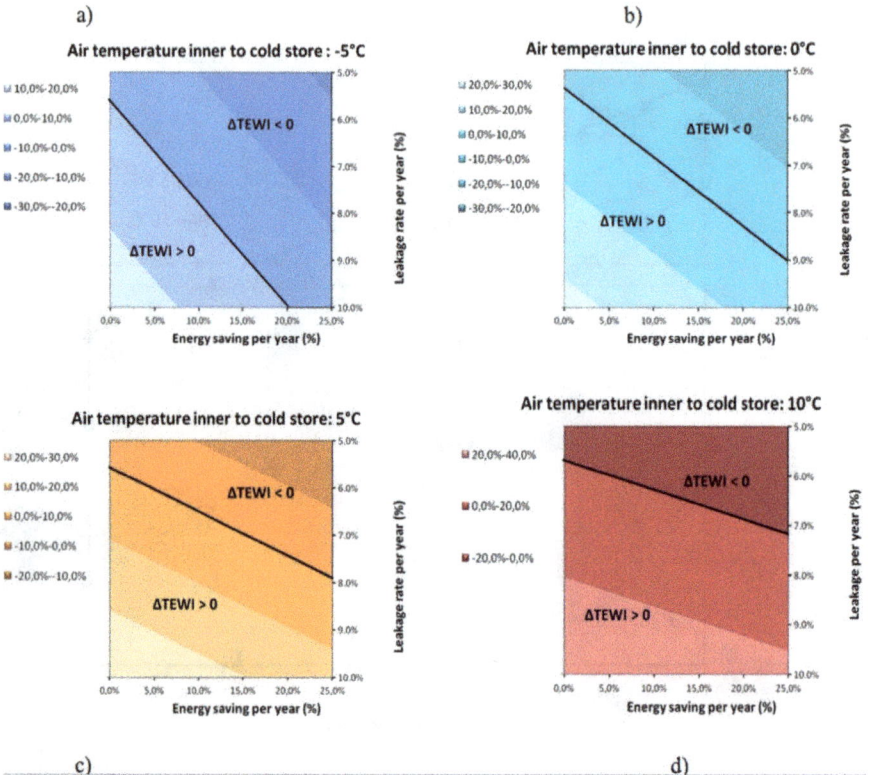

**Figure 26.** Operating scenario charts a) at -5°C, b) at 0 °C, c) 5°C and d) 10°C.

Subsequently, a scenario and sensitivity analysis for the TEWI difference has been introduced to study ways of reducing the environmental impact of the R22 retrofit with R422D. For this purpose leakage rate and improved efficiency have been considered as parameters and three parametric analyses have been developed: two have been carried out by changing only one parameter per time, while the other one by changing simultaneously both parameters. Based on our investigation, we can draw the following conclusions:

1.  The storage investigation have demonstrated that for each test conditions the R22 retrofit with R422D leads to an increase of the energy consumption up to 28.9% worsening $CO_2$ emissions.

2. Since the GWP of R422D is much higher than that of R22 and even if the charge of R422D is 8% lower than that of R22, the direct effect of the R422D is 42% higher than that of R22.
3. As a consequence of the R22 retrofit with R422D, the plant investigated has shown an increase of TEWI up to 36.8 %.
4. The performance investigation highlighted that the operation with R422D is less efficient than that with R22. In particular, the difference between the COP for R22 and that for R422D is, on average, 20%, and it grows with the raising of the air temperature of the inner cold store.
5. R22 retrofit with R422D leads to an increase of the condensing pressure, which indicates that the heat exchange surface of the condenser is insufficient to reject the thermal power, worsening the efficiency.
6. To improve the energy performance and then to reduce the indirect effect, we proposed two ways: improving of the heat exchange surface and adoption of electronic expansion valves. Based on theoretical considerations it is possible to obtain a 20% reduction of energy consumption.
7. The scenario and sensitivity analysis for the TEWI difference have demonstrated that for each test condition there are some operating eco-friendly scenarios. In particular, if the parameters change simultaneously, the eco-friendly scenario results technically feasible: both lower reductions of leakage rate and lower energy saving are required.

## 5. An experimental evaluation of the greenhouse effect in R134a substitution with R744

### 5.1. R134a substitution with R744

In the field of the mobile refrigeration systems, the European Parliament already set a regulation of F-Gases phase out that bans the use of refrigerants having GWPs exceeding 150. Such regulation is in effect since 2011.

R134a is an HFC with zero ODP and a GWP of 1300. According to the above mentioned European regulation on F-Gases, the use of R134a will be banned in mobile systems. R744 is a natural fluid ($CO_2$) therefore with no ODP and negligible direct contribution to global warming and can be a substitute of R134a. In this study, the impact of the substitution of R134a with R744 on global warming was studied through experimental evaluations of the TEWI index under different operating conditions.

The experimental tests discussed in this study compare a commercial R134a refrigeration plant subjected to a cold store and a prototype R744 system working as a classical spit-system to cool air in a trans-critical cycle.

Table 11 reports the parameters adopted for the TEWI evaluation. The annual operating hours in the TEWI simulation are 8760. These correspond to a commercial refrigerator cold store according to Dir.94/2/CE [36].

| Parameter | Value |
|:---------:|:-----:|
| H | 8760 h |
| PL | 5% year |
| PR | 95% |
| V | 10 years |
| $\alpha$ | 0.6 kg $CO_2$/kWh$_e$ |

**Table 11.** Parameters for TEWI evaluation.

## 5.2. Experimental equipment

### 5.2.1. Refrigeration plant working with carbon dioxide

Figure 27 shows a sketch of the experimental plant.

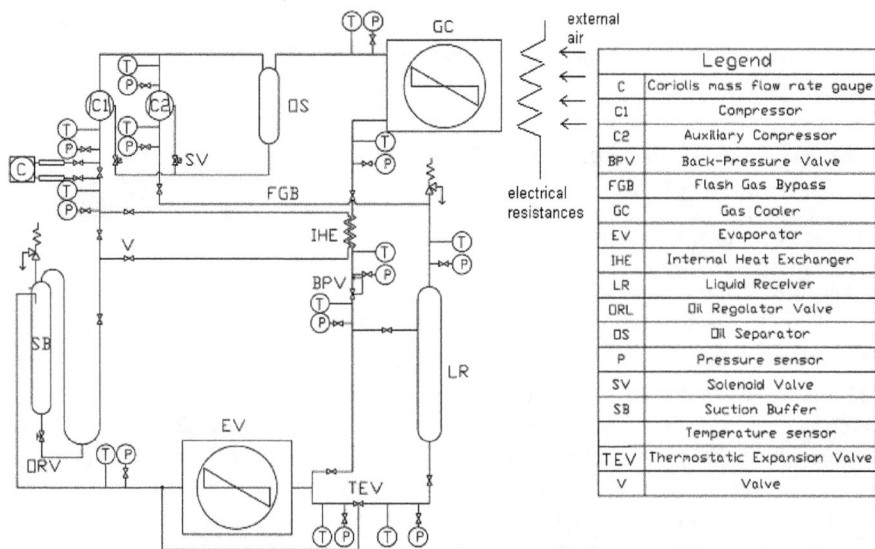

**Figure 27.** Sketch of the carbon dioxide experimental plant.

Basically, there are two single-stage hermetic reciprocating compressors, an oil separator, an air gas-cooler, a liquid capacity, an air evaporator, an electronic expansion valve (EEV) and an electronically-regulated back pressure valve (BPV). The main compressor is a semi-hermetic one. At evaporation temperatures of 5°C and of 30°C at the gas-cooler exit, when the pressure is 80 bar, the refrigerating power is about 3000W. An internal heat exchanger (IHX) between the refrigerant flow at the compressor suction and at the exit of the gas-cooler is provided. The lamination occurs thanks to the back-pressure valve and to the electronic expansion one. An auxiliary circuit can be used to by-pass the back-pressure valve, in order to vary the evaporation temperature. The air temperature on the condenser is regulated by an air-flow driven by a blower in a thermally insulated channel. Its temperature is modulated by some electrical resistances. This simulates variable external conditions, as well.

The plant is fully instrumented, in order to evaluate its performance as a whole, as well as that of each single component. The pressure and the temperature of the carbon dioxide are measured both at the inlet and at the outlet of each device. Mass flow rate is monitored at the main compressor suction (see Figure 27). Two watt transducers are used to measure the electrical power supplied to the compressors. Table 12 summarizes all the characteristics of the plant instrumentation.

| Transducers | Range | Uncertainty |
|---|---|---|
| **Coriolis effect flowmeter** | 0÷2 kg/min | ± 0.2% |
| Piezoelectric absolute pressure gauge | 1÷100 bar | ± 0.4% |
| RTD 100 4 wires | -100÷500 °C | ± 0.15°C |
| Wattmeter | 0.5÷6 kW | ± 0.2% |

**Table 12.** Transducers specifications.

The Coefficient of Performance is evaluated with eq. (4), with uncertainty of ± 3.8 % according to the procedure suggested by Moffat [37].

The uncertainty of the TEWI was calculated by applying the error propagation theory to Eq. (1). A balance is used to measure the refrigerant charge with an uncertainty of ± 0.2% in the range 0-100 kg. Sand et al. [18] suggested that a minimum of 20% uncertainty exists for the GWP values assigned to refrigerants by the Intergovernmental Panel on Climate Change (IPCC). These uncertainties, when combined with other estimates and assumptions of the analysis lead to a TEWI uncertainty of 10%.

### 5.2.2. Refrigeration plant working with R134a

The experimental vapour compression plant, subjected to a commercially available cold store and reported in Figure 28, is made up of a semi-hermetic reciprocating compressor.

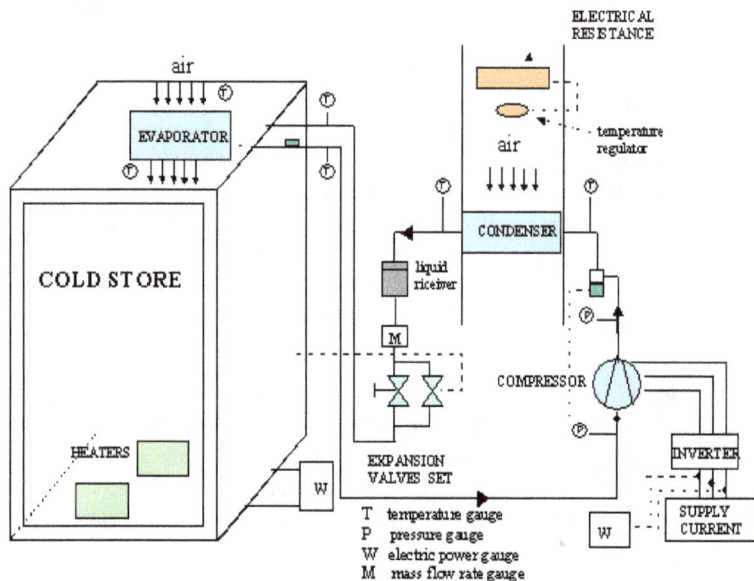

**Figure 28.** Sketch of the R134a experimental plant.

It was designed for the fluid R134a, according to the manufacturer specifications. The plant is supplied with a three-phase current (380 V phase-phase), an air condenser followed by a liquid receiver, a manifold with two expansion valves, a thermostatic one and a manual one mounted in parallel, to feed an air cooling evaporator inside the cold store.

In the evaporation temperature range -20 ÷10°C with a 35°C condensing temperature, working with the R134a at the nominal frequency of 50Hz, the compressor refrigerating capacity varies in the range 1.4÷4.4 kW. A blower provides an air-flow within a thermally insulated channel where some electrical resistances are located. Their power-modulation enables the control of the air temperature on the condenser and simulates different external conditions.

The refrigeration duty in the cold store is simulated by means of regulated electrical resistances. The electric power is measured by means of a watt transducer whose specifications are reported in Table 13.

Further specifications of the experimental plant were reported in a previous work [38-41]. The COP values calculated in accordance with (7), should be considered with an uncertainty less than ±0.5%. The smaller accuracy in the calculus of the COP pertaining to the R744 plant is due to a different accuracy of the pressure gauges used for the R744, as compared to those used for the R134. The lower accuracy of the pressure gauges for the R744 stems from the larger measurement range required.

| Transducers | Range | Uncertainty |
|---|---|---|
| **Coriolis effect flowmeter** | 0÷2 kg/min | ± 0.2% |
| RTD 100 4 wires | -100÷500 °C | ± 0.15°C |
| Piezoelectric absolute pressure gauge | 1÷10 bar | ± 0.2% |
| | 1÷30 bar | ± 0.5% F.S |
| Wattmeter | 0÷3 kW | ± 0.2% |

**Table 13.** Transducers specifications.

## 5.3. Results and discussion

The experimental, R744 plant was optimized in order to maximize its energetic performance.

The direct contribution to global warming was evaluated on the basis of the measured plant charge. The R744 charge is the optimal one, corresponding to 6.87 kg. That for R134a is of 2.45 kg. The direct contribution to the greenhouse effect pertaining to R134a during the plant useful life is always greater that that pertaining to R744. The direct contribution of R744 is negligible respect to the indirect one, whereas that of R134a is comparable.

System performances are compared for two evaporation temperatures of 0 and 5 °C, respectively, by varying the temperature of the external air over the gas-cooler and the condenser. Both plants develop a refrigerant power of 3000 W.

In the evaluation of the total contribution to greenhouse effect, three scenarios were considered: a commercial refrigerator cold store, a classical split-system and a mobile refrigeration system.

In the first scenario a commercial refrigeration cold store was considered. According to Dir.94/2/CE, it works for 8760 annual operating hours.

Figure 29 reports TEWI as a function of external air temperature for an evaporating temperature of 0 and 5 °C, respectively. Figure clearly shows that, for both refrigerant fluids, TEWI increases with the external air temperature because of the COP decrease.

Indeed, COP decreases due to the decrease of the refrigerant fluid enthalpy at the evaporator inlet because of the pressure increase at the condenser/gas-cooler. Its rise leads to a higher enthalpy at the compressor outlet, thus increasing the specific work of compression. The COP decrease for R744 is more marked than that of R134a.

The performance of both refrigerant fluids at fixed external air temperature decreases with evaporating temperature. This effect is more marked for the plant working with the transcritical cycle. In a typical transcritical cycle the expansion valve provides a dominant contribution to the overall energy loss [42,43]. The irreversibility of the expansion process increases by decreasing the evaporating pressure

Figure 29 clearly shows that, for a refrigerating system working as a commercial refrigerator cold store, the TEWI pertaining to R134a is always lower than that of R744 from a minimum of -22% to a maximum of -73%. The higher indirect contribution to global warming of R744 with respect to R134a always prevails.

The second scenario is a classical split-system, that according to Dir.29/1/2003 works with 500 annual operating hours.

Figure 30 reports TEWI as a function of external air temperature for an evaporating temperature of 0 and 5 °C.

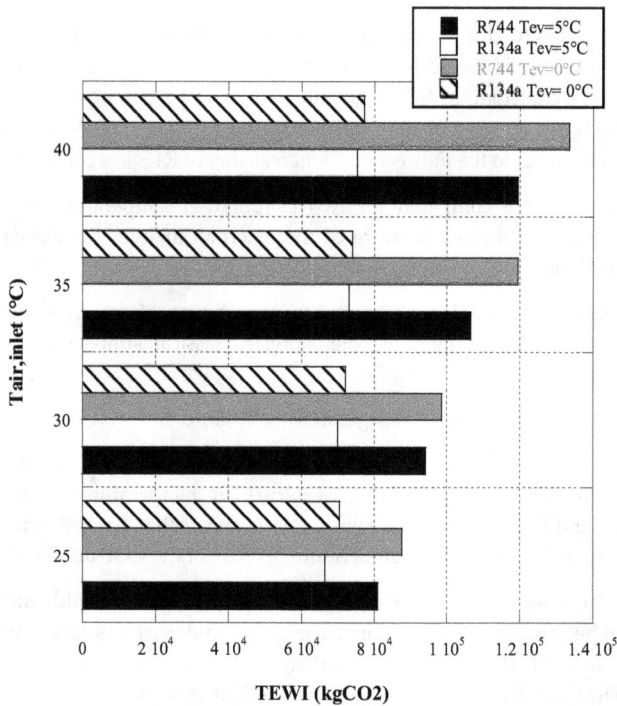

**Figure 29.** TEWI as a function of air temperature at the inlet of the condenser/gas-cooler, for two evaporating temperature, in a commercial refrigerator cold store.

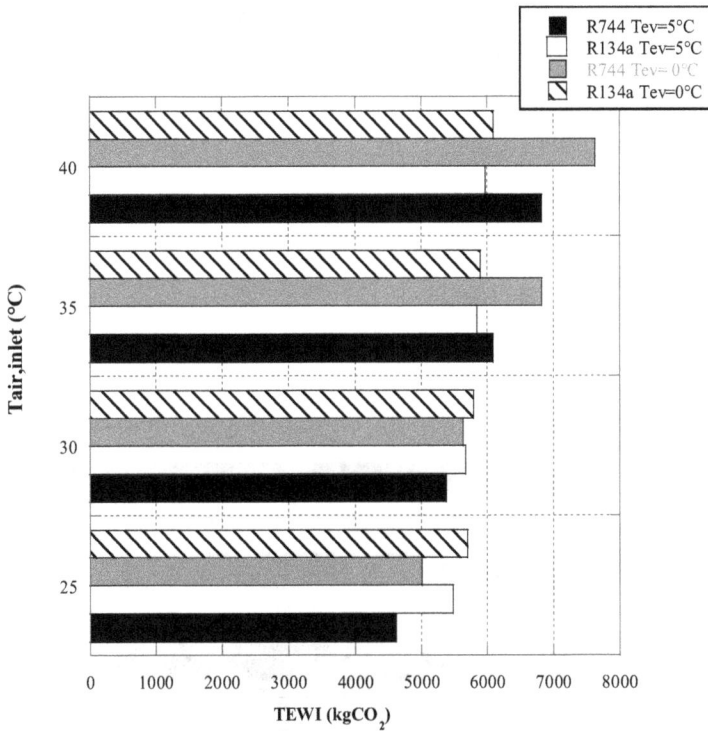

**Figure 30.** TEWI as a function of air temperature at the inlet of the condenser/gas-cooler, for two evaporating temperature, in a classical split-system.

In this case TEWI of R134a is higher than that of R744 for external air temperature of 25 and 30 °C (from a minimum of 3 to maximum of 16%). In these experimental conditions the system working with R744 has a lower global warming impact as compared to a system with R134a. This effect is due to the lower direct contribution of R744 that prevails on the indirect one. In the operating conditions corresponding to external air temperature of 35 and 40 °C, the TEWI of R134a is lower that that of R744 (from a minimum of -4 to maximum of -25%).

The third scenario is a mobile refrigeration system

In the following analysis, a 1000 kg small commercial car was considered. For this car, 98 g of $CO_2$ are released per km [46] in a typical urban cycle with a mean engine power of 5 kW. On this basis, the fraction of $CO_2$ emission produced by a 3 kW air conditioning device employing R134a was evaluated. The R134a charge is 800 g. That of $CO_2$ is 1.20 kg.

From these data and from the previous experimental results, it is possible to evaluate the TEWI for both R134a and R744.

In figure 31 TEWI is reported as a function of the refrigerant leakage rate per year.

The figure clearly shows that TEWI of R134a is lower than that of R744 (from a minimum of -4.7 to a maximum of -22%) in the yearly refrigerant leakage rate of 5 and 20%, respectively. At 25% the two fluids show the same TEWI. On the contrary, at 30% leakage, the R134a TEWI exceeds that of R744 of +4%.

In a typical mobile air conditioning device, the yearly refrigerant leakage rate is around 25%.

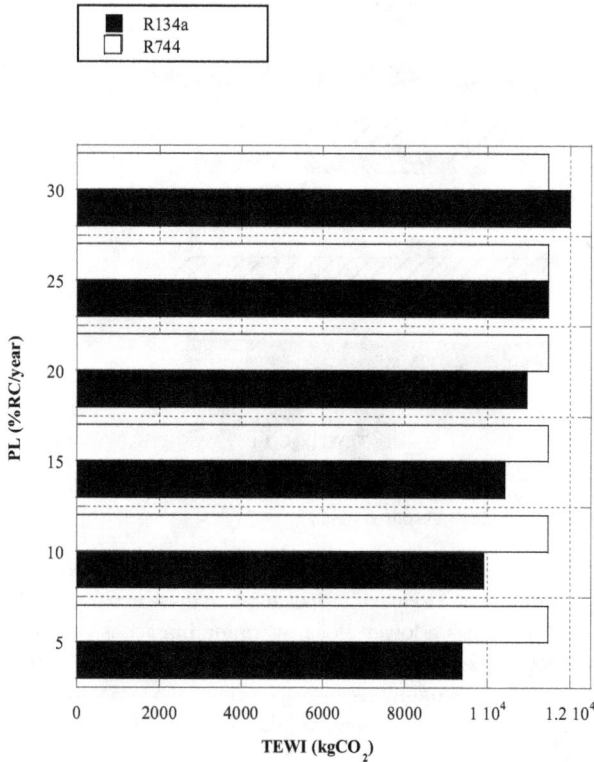

**Figure 31.** TEWI as a function of refrigerant leakage rate in a mobile refrigeration system.

## 5.4. Concluding remarks

The substitution of R134a was analyzed in terms of global warming effect. Indeed, R134a will be banned in mobile systems, according to the regulation on F-Gases. R134a has a relevant direct global warming effect stemming from its absorption power of long-wave radiations, that depends on its GWP and on the fraction of refrigerant charge released into the atmosphere. A possible substitute of R134a could be carbon dioxide, i.e. a natural fluid with negligible direct contribution to global warming.

An experimental comparison between R134a and R744 was performed in terms of the total equivalent warming impact (TEWI) that combines the effect of the direct emissions of refrigerants with the indirect effect of energy consumption.

The experiments compare a commercial R134a refrigeration plant subjected to a cold store and a prototype R744 system working in a trans-critical cycle. A preliminary analysis was carried out, in order to maximize the energetic performance of the trans-critical cycle.

In the comparison, different, common working conditions were considered by varying the external temperature and that of the evaporation. In all the test runs, the energetic efficiency of the trans-critical cycle is always lower than that of the plant working with R134a. This leads to a greater indirect global warming effect of the plant working with R744.

Three different scenarios were considered: a commercial refrigerator cold store, a classical split-system and a mobile refrigeration system.

Based on our investigation, the following conclusions can be drawn:

1) For a refrigerating system working as a commercial refrigerator cold store at fixed refrigerant leakage rate per year (5%), the TEWI pertaining to R134a is always lower than that of R744 by a minimum of -22% up to a maximum of -73%.

2) A refrigerating system working as a classical split (at fixed 5% refrigerant leakage rate per year) with R134a has a greater global warming impact as compared to a system with R134a for external air temperature of 25 and 30 °C (from a minimum of 3 to maximum of 16%). Whereas, in the operating conditions corresponding to external air temperature of 35 and 40 °C, the TEWI of R134a is lower that that of R744 (from a minimum of -4 to maximum of -25%).

3) In a mobile refrigeration system, the TEWI of R134a is lower than that of R744 (from a minimum of -4.7 to a maximum of -22%) for yearly refrigerant leakage rates between 5 and 20%. At 25% leakage, the two fluids show the same TEWI, whereas, at 30% leakage, the R134a TEWI exceeds that of R744 by +4%. In view of these results, R744 does not appear to be a suitable substitute for R134a. The latter, however, must be ruled out in mobile systems, in any case. Therefore, alternative fluids such as HFO (HydroFluoro-Olefin) should be considered.

## 6. Conclusions and recommendations

The impact of vapour compression refrigerant fluids on global warming has been analyzed by means of an experimental study. Vapour compression plants have both a direct and an indirect contribution to global warming. The Total Equivalent Warming Impact (TEWI) index takes into account both the direct and indirect contribution to global warming.

In this chapter the substitution of R22 and R134a has been studied. R407C and R422D are possible substitutes for R22. R22 is an HCFC, with an ODP of 0.05 and a GWP of 1700. Both refrigerant fluids are drop in substitutes: R407C is an HFC with zero ODP and a GWP 6% lower than that of R22, R422D is an HFC with zero ODP and a GWP 31% greater than that of R22. R134a is a single hydrofluorocarbon (HFC) compound. It has no chlorine content and therefore no ozone depletion potential, and a $GWP_{R134a}$ of 1300.

In the field of mobile refrigeration systems, the European Parliament (EU 842/2006) already set F-Gases phase out regulation that bans the use of refrigerants having GWPs in excess of 150. According to this regulation the use of R134a will be banned in mobile systems. Pure $CO_2$ (R744), with no direct contribution to global warming, can be a substitute of R134a. Experimental tests have been carried out on different vapour compression pilot plants for a range of operating conditions, the prototype R744 system works in a trans-critical cycle

From the experimental analysis the following general conclusions can be drawn:

- R22 direct contribution to the greenhouse effect is greater than that pertaining to R407C (+15%). The COP corresponding to R407C is 3.3-19% lower than that pertaining to R22.
- In the experimental tests corresponding to a condensation temperature in the range 43 / 47 °C and to an evaporation temperature in the range -12 / -7 °C, the TEWI pertaining to R407C exceeds that of R22 by about 11%. Therefore, the substitution of R22 with R407C should be unacceptable if specific reference is made to the greenhouse effect.
- In the experimental tests corresponding to a condensation temperature in the range 53/58 °C and to an evaporation temperature inlet range 2 / 10 °C, R22 and R407C behave in a similar way as regards the greenhouse effect.
- For high evaporation and condensation (over 50°C) temperatures, the TEWI of R407C is slightly lower than that of R22. As a consequence, the substitution of R22 with R407C is favourable, since no harm is produced to the ozone layer and no increase in $CO_2$ emission is made.
- GWP of R422D is much higher than that of R22 and even if the charge of R422D is 8% lower than that of R22, the direct effect of the R422D is 42% higher than that of R22.
- The plant working with R422D is less efficient than that with R22. In particular, the difference between the COP for R22 and that for R422D is, on average, 20%, and it grows with the raising of the air temperature of the inner cold store.
- For each test conditions the R22 retrofit with R422D leads to an increase of the energy consumption up to 28.9% , worsening $CO_2$ emissions, with an increase of TEWI up to 36.8 %. Therefore the substitution of R22 with R422D is always unacceptable from the point of view of greenhouse effect.
- In all the test runs, the energetic efficiency of the trans-critical cycle is always lower than that of the plant working with R134a. This leads to a greater indirect global warming effect of the plant working with R744.
- For a refrigerating system working as a commercial refrigerator cold store the substitution of R134a with R744 is always unacceptable because the TEWI pertaining to R134a is always lower than that of R744 (from -22% to -73%).

- A refrigerating system working as a classical split with R134a has a greater global warming impact as compared to a system with R744 for external air temperature of 25 and 30 °C (from a minimum of 3 to maximum of 16%). Whereas, in the operating conditions corresponding to external air temperature of 35 and 40 °C, the TEWI of R134a is lower that that of R744 (from a minimum of -4 to maximum of -25%).
- In a mobile refrigeration system, the TEWI of R134a is lower than that of R744 (from a minimum of -4.7 to a maximum of -22%) for yearly refrigerant leakage rates between 5 and 20%. At 25% leakage, the two fluids show the same TEWI, whereas, at 30% leakage, the R134a TEWI exceeds that of R744 by +4%. In view of these results, R744 does not appear to be a suitable substitute for R134a.

# 7. Nomenclature

## Symbols

$CO_{2,dir}$ = direct contribution to global warming ($kg_{CO2}$)
$CO_{2, indir}$ = indirect contribution to global warming ($kg_{CO2}$)
COP = Coefficient Of Performance
E = energy consumption (kWh)
GWP = Global Warming Potential ($kg_{CO2}$ $kg_{refrigerant}$)
H = annual operating hours (h/years)
h = enthalpy (kJ/kg)
$\dot{m}$ = refrigerant mass flow rate (kg/s)
ODP = Ozone Depletion Potential
p = pressure (Pa)
$P_L$ = accidental refrigerant leaks per year (% refrigerant charge/year)
$P_R$ = recycling rate (% refrigerant charge)
$Q_{ref}$ = refrigerant power (kW)
RC = refrigerant charge (kg)
T = temperature (°C, K)
TEWI = Total Equivalent Warming Impact ($kg_{CO2}$)
V = plant useful life (years)
$\dot{W}$ = compression power (kW)

## Greek symbols

$\alpha$ = $CO_2$ emission from power conversion ($kg_{CO2}$/kWhe)
$\beta$ = compression ratio
$\Delta TEWI$ = TEWI difference
$\varepsilon$ = efficiency

## Subscripts

air = air

amb = external ambient
co = condenser
cold = air inner to cold store
D  = daily
el= electrical
EV= evaporator
hot= external air ambient
i= i-th
in= inlet
m= monthly
MT = mean thermodynamic
out = outlet
ref = refrigerant
w = water
wg = water glycol mixture

## Author details

C. Aprea* and A. Maiorino
*Dipartimento di Ingegneria Industriale, Università di Salerno,*
*via Ponte Don Melillo, Fisciano, Salerno, Italia*

A. Greco
*DETEC, Università degli Studi di Napoli Federico II, P.le Tecchio, Napoli, Italia*

## 8. References

[1] M.J. Molina, F.S. Rowland, 1974, Stratospheric sink for chlorofluoromethanes: Chlorine atom catalyzed destruction of ozone, Nature n. 249, pp.810-812.
[2] C. Muller, 1993, Atmospheric ozone and greenhouse gases observation: an update, Proc. Int. Conf. Energy Efficiency Refrig. and Global Warming Impact, Commission B1/2 IIR, University of Ghent, Belgium, pp. 45-54.
[3] Montreal Protocol on Substances That Deplete the Ozone Layer, 1987. United Nations (UN), New York, NY, USA (1987 with subsequent amendments).
[4] UNEP, 2007a. Decisions Adopted by the Nineteenth Meeting of the Parties to the Montreal Protocol on Substances that Deplete the Ozone Layer. United Nations Environment Programme (UNEP) Ozone Secretariat, Nairobi, Kenya, (2007)
[5] S. Fisher, F. A. Creswick, 1988, Energy use impact of chlorofluorocarbon alternatives, ORNL/CON-273.
[6] S. Fisher, F. A. Creswick, 1988, How will CFC bans affect energy use, ASHRAE Journal, vol. 30, n. 11.

* Corresponding Author

[7] Brohan, P., Kennedy, J.J., Harris, I., Tett, S.F.B., Jones, P.D., Uncertainty estimates in regional and global observed temperature changes: a new dataset from 1850, J. of Geophysical Res. 111, (2006), D12106.

[8] Rayner, N.A., Brohan, P., Parker, D.E., Folland, C.K., Kennedy, J.J., Vanicek, M., Ansell, T.J., Tett, S.F.B, Improved analyses of changes and uncertainties in marine temperature measured in situ since the mid-nineteenth century: the HadSST2 dataset. Journal of Climate 19, (2006), 446–469.

[9] Kyoto Protocol to the United Nations Framework Convention on Climate Change, 1997. United Nations (UN), New York, NY, USA.

[10] Calm J. M., The next generation of refrigerants – Historical review, considerations, and outlook, Int. J. Ref. 31 (2008) 1123–1133.

[11] Horrocks, P.,EU F-gases Regulation and MAC Directive, ECCP-1 Review. European Commission Environment Directorate, Brussels, Belgium, (2006).

[12] Regulation (EC) No 842/2006 of the European Parliament and of the Council of 17 May 2006 on certain fluorinated greenhouse gases, Official J. of E.U., 49, (2006), L161.

[13] "Electrical Energy Reduction in Refrigeration and air Conditioning", Ken Landymore, Report Smartcool System Inc., 2007.

[14] J. M.Calm, 2002, Emissions and environmental impacts from air-conditioning and refrigeration systems, International Journal of Refrigeration, vol. 25, pp. 293-305.

[15] "Annual Energy Review 2010" Report DOE/EIA-0384 (2010), October 2011. US Department of Energy,USA.

[16] J. M.Calm, D.A. Didion, 1998, Trade-off in refrigerant selections: past, present, and future, International Journal of Refrigeration, vol. 21, pp. 308-321.

[17] J. M.Calm, 2006, Comparative efficiencies and implications for greenhouse gas emissions of chiller refrigerants, International Journal of Refrigeration, vol. 29, pp. 833-841.

[18] J.R. Sand, S.K. Fisher, V.D. Baxter, 1997, Energy and global warming impacts of HFC refrigerants and emerging technologies, Report sponsored by alternative fluorocarbons environmental acceptability study (AFEAS), U.S.Department of Energy, www.afeas.org.

[19] J.R. Sand, Fisher S.K., Baxter V.D., 1999, TEWI analysis: its utility, its shortcomings, and its results. Proceedings of the Taipei international conference on atmospheric protection.

[20] Third assessment report, Integovernative Panel on Climate Change, 2001, http://www.ipcc.ch/.

[21] Guideline method of calculating TEWI, 2006, British Refrigeration Association, Table 3.

[22] R. Maykot, G.C. Weber, R.A. Maciel, 2004, Using the TEWI methodology to evaluate alternative refrigerants technologies, International Refrigeration and Air Conditioning Conference at Purdue, July 12 – 15.

[23] S. Malla, 2009, $CO_2$ emissions from electricity generation in seven Asia-Pacific and North America countries: a decomposition analysis, Energy Policy, vol. 35, pp.5938-5952.

[24] R. Quadrelli, S. Peterson, 2007, The energy-climate challenge: recent trends in $CO_2$ emissions from fuel combustion, Energy Policy, vol. 35, pp. 5938-5952.

[25] C. Weber, Koyama M., S. Kraines, 2006, $CO_2$ emissions reductions potential and costs of decentralized energy system for providing electricity, cooling and heating in an office-building in Tokyo, Energy vol. 31, pp. 3041-3061.

[26] C.S. Psomopoulos, I. Skoula, C. Karras, A. Chatzimpiros, M. Chionidis, 2010, Electricitysavings and $CO_2$ emissions reduction in buildings sector: how important the network losses are I the calculation?, Energy, vol. 35, pp. 485-490.

[27] Jabaraj D.B., Narendran A., Mohan Lal D., Renganarayanan S., Evolving an optimal composition of HFC407C/HC290/HC600a mixture as an alternative to HCFC22 in window air conditioners, Int. J. of Th. Sci. 46 (2007) 276–283.

[28] Sarkar J., Cycle parameter optimization of vortex tube expansion transcritical CO2 system, Int. J. of Th. Sci. 48 (2009) 1823–1828.

[29] C. Aprea, A. Greco, 1998, An experimental evaluation of the greenhouse effect in R22 substitution, Energy Conversion & Management, vol.39, n.9, pp.877-887.

[30] Greco A., Mastrullo R., Palombo A., 1997, R407C as an alternative to R22 in vapour compression plant: an experimental study, Int. Journ. of Energy Resaerch, vol. 21, pp.1087-1098.

[31]DuPont, Retrofit guidelines for DuPontTM Isceon® M029  (R422D) Refrigerant, Techincal Information art.46, http://www.refrigerants.dupont.com (2007)

[32]Arora A., Sachdev H.L., Thermodynamics analysis of R422 series refrigerants as alternative refrigerants to HCFC22 in a vapor compression refrigeration system; 33 (8), (2009) 753-65.

[33]Lazzarin R., Noro M., Experimental comparison of electronic and thermostatic expansion valves performances in an air conditioning plant, Int. J. Ref., 31,(2008),113-118.

[34]Jabaraj D.B., Narendran A., Mohan Lal D., Renganarayanan S., Evolving an optimal composition of HFC407C/HC290/HC600a mixture as an alternative to HCFC22 in window air conditioners, Int. J. of Th. Sci. 2007; 46:276–283

[35] C. Aprea, A. Maiorino, 2011, An experimental investigation of the global environmental impact of the R22 retrofit with R422D, Energy, vol. 36, pp.1161-1170.

[36] Commission Directive 94/2/EC of 21 January 1994 implementing Council Directive 92/75/EEC with regard to energy labeling of household electric refrigerators, freezers and their combinations.

[37] R.J. Moffat,1985, Using uncertainty analysis in the planning of an experiment, Trans. ASME, J. Fluids Eng., Vol. 107, pp. 173-178.

[38] C. Aprea, A. Maiorino, 2009, Transcritical CO2 refrigerator and sub-critical R134a refrigerator: a comparison of the experimental results, Int. Jour. of Energy Research, vol.33, pp.1040-1047.

[39] P. Nekså, 2002, $CO_2$ heat pump systems, Int. J. Refrigeration, vol. 25, pp.421-427.

[40] C. Aprea, A. Maiorino, 2008, An experimental evaluation of the transcritical $CO_2$ refrigerator performances using an internal heat exchanger, Int. Journal of Refrigeration, vol. 31, pp. 1006-1011.

[41] C. Aprea, A. Maiorino, 2009, Heat rejection pressure optimization for a carbon dioxide split system: an experimental study, Applied Energy, vol. 861, pp. 2373-2380.

[44] A. Fartaj, D.S.K. Ting, W.Yang, 2004, Second law analysis of the transcritical $CO_2$ refrigeration cycle, Energy Conversion & management, Vol.45, pp.2269-2281.

[45] Y.B. Tao, Y..L. He, W.Q. Tao, 2010, Exergetic analysis of a transcritical residential air-conditioning system based on experimental data, Applied Energy, vo. 87, pp.3065-3072.

[46 ] Regulation (EC) 1999/100/CE.

# Chapter 2

# A New Perspective for Labeling the Carbon Footprint Against Climate Change

Juan Cagiao Villar, Sebastián Labella Hidalgo,
Adolfo Carballo Penela and Breixo Gómez Meijide

Additional information is available at the end of the chapter

## 1. Introduction

Irrespective of the current social and economic problems, the fact is that hurricane-force winds hover over our current way of life, and ultimately over our very civilization. Progressive deforestation, water shortages, loss of biodiversity, the scarcity of natural resources exposed to their own ecological limits. The result of all of this is the relentless generation of waste, emissions and discharges into an increasingly limited absorptive capacity of the planet.

The economic debt, in any form, whether it is consumer-related, national or foreign, which we hear about every day on the news, is insignificant compared to the ecological debt we are acquiring. In 1997 a study by the team of Robert Costanza, specialist in environmental economics, estimated the average value of the global ecosystem services to be around the 33 billion dollars annually. That same year the global GDP was only 18 billion. For example, the Global Footprint Network (GFN) calculations of April 2011 showed that Spain entered an "ecological debt" situation, having consumed by that time the total annual budget in terms of natural resources.

It is possible to adapt an economic model, to fix it, and replace it, but trying to expand the planet is simply utopian. Like it or not, our planet is finite and a finite system is incompatible with a subsystem (economic) whose paradigm is based on continuous and unlimited growth. Somehow we have to reconcile growth and sustainability, and to do so, our companies need to access transparent and comparable information to be able to make the best decisions so as not to compromise either their growth or the impact on the planetary ecosystem.

Obviously, growth and better living conditions have to reach developing countries where per capita income is less than a dollar a month, but it doesn't seem consistent to raise growth based on production patterns that are supported by 'dirty' technologies in developed countries. Identifying sinks in a critical absorption situation and ecosystems with a falling supply in natural resources, on which we base our economy, are critical to our survival.

One of the most critical impacts identified during the last century was the likely failure of the absorption capacity of our atmosphere to operate as a sink for so-called greenhouse gases (GHGs) without producing drastic changes in climatic conditions. These gases are named for their characteristic ability to pass short wavelength radiation from the sun and retain heat from the earth in the form of long wavelength radiation, which leads to the greenhouse effect.

Reports issued by the Intergovernmental Panel on Climate Change (IPCC), which includes the largest community of experts, are warning us that, like everything in life, a little bit of everything is good but too much of one thing can be lethal.

One of the main problems is the extraordinarily high rate of GHG emissions which our society has been generating for more than 100 years. This inhibits any reaction from the flora and fauna as well as the human race, which is encountering an increasingly unpredictable system from a climatic point of view. The planet will absorb these greenhouse gases without any problem, but the species that inhabit it will have enormous difficulties in adapting to new conditions. The scenario painted by the experts could not be more daunting, and urgent warnings for action must be sent out to the general public, businesses and individuals.

In answer to this impending scenario, Carbonfeel has been designed with a core mission: to organize information and knowledge on the carbon footprint, making it universally useful and accessible to all society. In short, the point is to provide companies with the best available techniques for calculation and exchange of information within the processes of inventory, management, reduction and offsetting of GHG emissions generated by their own activities.

This information will allow companies to participate actively in improving their behavior, without having any effect on their business. Quite the contrary; their activities will start to focus on production patterns based on eco-efficiency and eco-design, and therefore lead to a reduction in costs. Moreover, customers will recognize a continuous improvement effort based on a credible label supported by many different certifiers, consultants, companies, associations, universities and others.

The message is very clear to society. Various organizations have joined together to facilitate the expansion of a responsible economy to help businesses generate goods and services in a friendly environment, avoiding the wide variety of labels and certificates with a commercial purpose only. We understand from Carbonfeel that making business compatible with and respectful of the environment is not an option, rather it is the only valid way for modern business. Whether we recognize this or simply look away depends on the conscience of each and every individual.

Carbonfeel provides the public and private world a true environmental accounting system based on the universal indicator, the carbon footprint (CF), a scorecard that will help them choose the best practices in their processes and procure less intensive goods and services, all tending towards a low carbon culture.

## 2. The Carbonfeel project

### 2.1. Why Carbonfeel? The initiative

Carbonfeel (http://www.carbonfeel.org) is a collaborative initiative promoted by the Environmental Forum Foundation (http://www.forumambiental.org), the Interdisciplinary University Group Carbon Footprint and the technology company Atos (http://atos.net). The project provides procedural solutions, methodological and technological processes of calculation, verification, certification and labeling of the carbon footprint both at the corporate level and in terms of products and services.

Any organization that has in its principles of corporate social responsibility the fight against climate change as a priority, is invited to participate within the profile appropriate to their interests, either actively collaborating in the dissemination of calculation and verification projects, or simply calculating their footprint. Through this network of collaboration we have a carbon footprint that is truly accessible, transparent and comparable.

Carbonfeel starts out from a methodological basis proposed by the Compound Method based on Financial Accounts (MC3), inherited from the ecological footprint concept that has been extended worldwide by its creators William Rees and Mathis Wackernagel (http://www.footprintnetwork.org). The project takes advantage of other emerging methodological trends such as GHG Protocol, PAS 2050 or ISO 14064 standards and the future ISO 14067 and 14069, in order to get an approximation of the real calculation.

Supporting an integrated approach, the incorporation of information technologies makes Carbonfeel an innovative project that has burst into the market to completely change the focus of the classic studies of life cycle analysis, whose drawbacks in cost and study time had already been reported by different analysts. This also became evident after the announcement in January 2012 of the multinational company Tesco (a pioneer in carbon footprint labeling), which, after five years of activities in projects of calculation, abandoned its initial plan to label all their products with their carbon footprint, blaming the fact that "a minimum of several months of work" would be necessary to calculate the footprint of each product and the lack of collaboration and monitoring of suppliers and other retailers.

The Guardian previously reported that Tesco would take centuries to fulfill his promise, as the supermarket adds labels at speeds of 125 products a year. A Tesco spokesman expressed their expectations to new ways of undertaking the calculations "We are fully committed to the carbon footprint and to helping our customers make greener choices. No final decision has been taken and we are always on the lookout to find ways to better communicate the carbon footprint of products in a way that informs and enriches our clients".

Other corporations that have undertaken calculation at the corporate level express their disappointment at not being able to assume scope 3 (the footprint inherited from their suppliers) because of the lack of standardization and collaboration in the supply chain, which makes the inclusion of the suppliers' footprints in this puzzle completely unreliable.

The great paradox of the Carbonfeel method is that companies get a carbon footprint at the corporate level and the life cycle of all products and services without any restriction on the scopes, with the information provided in great detail. Moreover this information is more extensive and of a higher quality as it is based on primary data (real footprint of its suppliers), and all at a cost and a time frame fully accessible to any corporation.

The telematic assembly technique provides an entire life cycle, where each corporation analyzes its own emissions (scope 1 and land use) on an autonomous basis for calculating the indirect footprint or inherited from its suppliers by the telematic assembly.

This report shows step by step how it is possible to have more and better quality information to help companies transform their patterns of production and consumption habits towards a low carbon culture, and all this in a way that is totally accessible to the entire business community, from micro-businesses to SMEs and large corporations.

## 2.2. Mission and objectives

The network of actors involved in the initiative offers our society a way of working with a clearly defined mission and objectives:

Mission

- Organize information and knowledge about the carbon footprint, making it universally useful and accessible to all society.
- Promote new patterns of production in organizations and a real transformation of consumer habits in society, both directed towards a low carbon culture.

Objectives

- Standardization of a methodology for the calculation of the carbon footprint based on an integrated approach (organization and product/service), always in strict compliance with the existing international standards in use, both at the corporate level (ISO 14064, GHG Protocol and future ISO 14069) and product level (PAS 2050 and future ISO 14067).
- Standardization and automation of the verification and certification processes of the carbon footprint.
- Make available to the general public an accesible, transparent and comparable labeling process of the carbon footprint.
- Incorporation of all the above points in the information society through the use of the new technologies required in the initiative.

As mentioned previously, countless labels and certifications are saturating the market. Some of these are based on calculation methods that have been accommodated to certain interests of the contracting company, a fact which only serves to undermine the credibility of the different studies. This type of dynamics is being used by companies interested in 'greenwashing' their products and actions. This sometimes leads to an unfair scenario in which companies that are truly committed to the environmental improvement of their products are put in a situation where their clients can not appreciate the goodness of their acts.

Carbonfeel emerged as a proposal that incorporates a common language based on consensus to the vast network of actors involved in the calculation of the carbon footprint. It is based on information technologies which allow data exchange to flow quickly and reliably, providing accounting and labeling processes that are renewed annually.

Carbonfeel seeks the incorporation of all types of businesses into the process of calculation and certification. It is no longer a marketing tool only affordable to large corporations, but has become a basic environmental accounting tool for the future assessment and analysis of improvement actions. Thus, even the smallest company will be eligible for certification. Moreover transparency is ensured under the rules and calculation methods accepted by all, without any problems related to subjectivities or cut-off criteria in the delimitation of the calculation, and thereby obtaining comparability as a source of competitiveness.

The reasons why a project like Carbonfeel has arisen and keeps on growing daily, fall under four different perspectives: social, economic, environmental and institutional.

- **Social perspective**: introducing concepts such as the carbon footprint and eco-labeling, which today are still unknown (in 2010, only 23% of Spanish consumers, compared to 94% of British or 97% of Japanese, had heard about the carbon footprint, according to studies conducted by TNS).
- **Economic outlook**: making it easier for companies that actually opt for an alternative "green" production style so that they can have a favorable commercial scenario and, thereby, facilitating their growth.
- **Environmental perspective**: promoting a real change in production patterns in organizations and a real transformation of consumer habits towards a low carbon culture.
- **Institutional perspective**: providing consensual solutions that may allow the homogenization of the many initiatives of institutions at national, regional and even local levels, who want to inventory, monitor and promote attitudes and sustainable practises in businesses and citizens.

## 2.3. Holding the roof

If we ask ourselves what kind of results a carbon accounting method should provide, the most appropriate answer is to help reduce emissions of greenhouse gases. Any other purpose would seem banal. Isn't the carbon accounting technique supposed to combat climate change? Because if the idea is to use it as a tool for promoting green products and the corporate image, then there are better marketing tools without having to pervert a method that was created for a very clear purpose.

Therefore, disregarding other objectives related to the current economic situation, we must pay attention to an overview of the results in the attempt to find a working method for calculation, verification and labeling which will be truly useful in the fight against global climate change. Indeed, there is nothing more useful for companies than to provide information that facilitates the reduction of emissions in relative terms (per unit of production of a product or service, and emissive intensity), but also in absolute terms (for the whole corporation). It's of little use if we lower emissions relatively while, on the other hand, corporate emissions grow due to other actions that are not within the scope of the current study.

## The roof

Imagine that what we want to do is build a house with a roof called "to lower emissions" and the aim is that the roof should be as large as possible, so that the larger the size of the roof, the greater our success in fighting climate change.

But we cannot put a roof over nothing; we need a structure that supports it. What are the requirements, given that the greater the support, the greater the roof "to lower emissions" will be?

## The beams

To lower carbon emissions there are few roads to choose from. A simple but subtly devastating vision of the problem indicates that we can do basically three things:

- Change our patterns of production, either by identifying processes for improvement, identifying good product design that is more environmentally friendly in the vector of climate change.
- Identify measures of eco-efficiency in the consumption of energy and materials in our business and production processes.
- Change our habits, both from the standpoint of providing information to the final consumer (B2C) of our products and services, and to provide ourselves with information from our network of suppliers (B2B) in order to help us to inherit the smallest footprint possible of products and services that feed our production system.

These are the three basic and essential beams required to sustain our roof, and if they are well-managed they will allow us to reduce emissions from our corporate activities.

Note that all three require the processing of data quickly and reliably. Let's explore this point that will lead us to the following levels of support for our house.

## The columns

How can we change our production and consumption patterns, and at the same time identify eco-efficient measures in our activities to help us cut emissions?

There is a clear answer to this question ... we must have reliable and quality data, and this information must have three properties that ensure the stability of our house:

- Accessibility
- Transparency
- Comparability

Let's consider each of these three points together because all three are intimately intertwined.

We refer to **accessibility** as the option for all businesses, from the smallest company to the large corporations, to make a claim for a carbon footprint certificate for their products, according to the prices and time frames of the projects adapted to the size of the contracting company, without omitting, in any case, the other two pillars: transparency and comparability. The incorporation process of calculating all the business is necessary; a global problem requires the involvement of all.

At present, the size of projects based on calculation techniques using the classic life cycle analysis, in which a link in the chain (the company who wants to calculate the footprint of its product) bears the burden of the whole calculation effort by drawing up complete process maps for the product and its life cycle. However, owing to the sheer size of these projects, both financially and operationally, they cannot be assumed by the majority of small and medium-sized enterprises.

Negotiating the scope of the studies, a common practice adopted by many companies to reduce costs, isn't the solution because it prevents reliable management and threatens the basic column of the home we propose to build: the **comparability** of results.

Therefore, the method needs to be accessible to all companies so that they will have a chance to show their carbon intensity, and thus, to improve themselves using benchmarking techniques supported by the comparability of results.

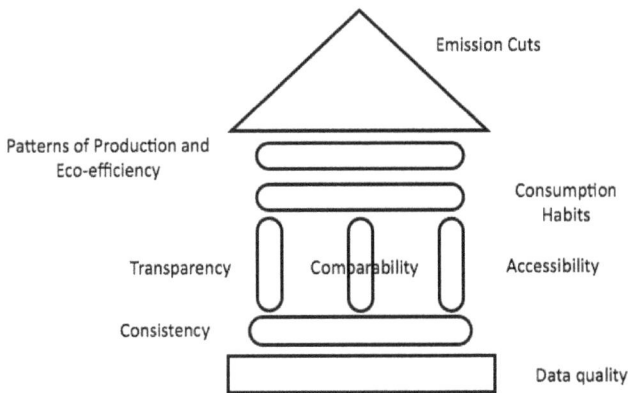

**Figure 1.** Requirements for a useful carbon accounting method in the fight against climate change

The existence of a spreadsheet calculation scheme that ensures the same scope for any project provides credibility and confidence to companies who want to 'play' on a scenario with identical rules and conditions. Thus, each company will be sure that their calculations have been carried out in the same way as their competitors.

This assumption is necessary to obtain an exemplary and transparent certificate, and the only way to do this is by a reporting method in line with a clear and objective scheme of calculation. How many times have we read about the total emission compensation for a given organization, where it is impossible to compare the study limits, calculation schemes and data sources that underlie the study?

Indeed, many of them are just "green-washing" strategies that confuse consumers and prolong a scene truly unfair to companies that are committed to an environmental strategy for its activities.

Transparency will provide confidence to all stakeholders and will eliminate from the carbon footprint market opportunistic corporations with marketing labels that try to displace corporations that are truly committed to sustainability and fighting climate change. Under these conditions of non-transparency, the proposed house will have little chance of supporting the roof that today's world demands of us.

Finally we will describe below the column that will provide definitive support to the structure.

Comparability is one of the most wanted features in a carbon footprint labeling process. It is essential to boost competitiveness in favor of corporate environmental improvement. Without comparability, the carbon footprint has no meaning and becomes just another environmental label.

A green purchasing policy, public or private, means including the carbon footprint as a standard for the environmental certification of products and services. The lack of comparability is one of the main excuses given by certain business sectors not to accept or promote green purchasing policies based on the concept of the carbon footprint.

Corporations seeking solutions that may allow them to flood the market with products and services with a lower carbon load need to identify improvements. Without comparable references in the market, these companies cannot carry out their mission; they cannot buy less carbon intensive goods in the market.

## The foundation

These columns, representing  accessible, transparent and comparable information, require a foundation to support them. This is what gives strength to the structure:

- Data consistency
- Data quality

To understand what consistency means in this context, it is necessary to explain the great paradox of the carbon footprint, which, in the words of Juan Luis Doménech, chief ideologue of the MC3 methodology, clearly shows how inefficient it is to maintain separate approaches to the corporation and products.

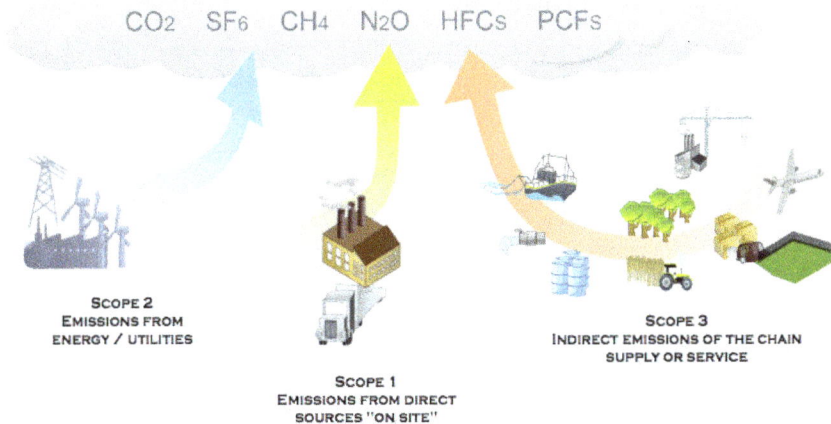

CO₂ SF₆ CH₄ N₂O HFCs PCFs

SCOPE 2
EMISSIONS FROM
ENERGY / UTILITIES

SCOPE 3
INDIRECT EMISSIONS OF THE CHAIN
SUPPLY OR SERVICE

SCOPE 1
EMISSIONS FROM DIRECT
SOURCES "ON SITE"

**Figure 2.** Classification of activities subject to emission rights (scope 1 and 2 are adopted)

"The methods of the classic life cycle analysis or methods focused on processes (ACV-P, PAS 2050) are not easy to implement as they require the participation of several companies on the value chain. Data acquisition based on the "most relevant processes" varies according to the analyst, and the "cutoff criteria" (as the value chain could be infinite) seriously compromise the comparability between products.

On the other hand, methods focused on the organization (such as ISO 14064 and GHG Protocol) are partial; they allow emissions called "scope 3" (materials, services, contracts, travel, construction, waste, etc.) to be voluntary and may vary from company to company. This also undermines comparability, at least for now, unless future editions correct this situation. On top of this, they are free to choose the calculation method of the actual carbon footprint and the emission factors. The latter should only come from reliable sources."

Carbonfeel is committed to an integrated approach, in which, as in any cost accounting method, partial studies are abandoned and a global vision of the company as a GHG emitter is undertaken in order to enter the gases emitted into the company's accountability in all the products and services generated.

This is the consistency which we are referring to. In the economic sphere, any accountant generally applies an integrated approach. Any other alternative with a partial character would not be accepted by a financial department. The corporate carbon footprint and the footprint of products and services that have no consistency will never be able to offer a scenario based on comparability, and therefore we will be seriously damaging a fundamental column of our building.

Finally, there is a basic foundation that supports the entire building: the quality of the survey data. To address this point, we must first distinguish the subtle difference between primary data and secondary data.

Primary data are obtained from a source through direct measurements, or provided by the same supplier that certifies that measurement to us in the case of an inherited footprint. In some ways it is a fact that closely reflects the local situation under study.

When primary data are not accessible due to the high cost in obtaining this information or simply because the provider does not provide it, we turn to what is called secondary data, provided by reliable sources. Conversion factors, databases or simulation tools give us a valid approximation to the data.

Logically, it is desirable that the calculations are supported by primary data to have a better approach to the real data. However, in the current state of the art, this is not true, and there is a lucrative business to be had in providing companies with secondary data to support their calculations.

Somehow these secondary sources, which are needed in the current state of the art, indicate, for example, that 100 gr. of sugar has a given carbon footprint load based on a life cycle study carried out under certain conditions. This figure is only a simplification that causes almost all companies to end up giving the same results for their studies due to the fact that all of them are based on the same reference data, rather than relying on the myriad scenarios that make up the current sugar production situation. It is not the same to have a local supplier, in this sugar production process case, than one that is 10,000 km away.

Thus, when discussing data quality, we refer to the fact that the proposed working method should be oriented towards the development and distribution of raw data (actual data derived from measurements provided by the supplier), and not to the business of secondary data. The role of secondary data in a calculation methodology is necessary, but as an alternative, not as an end.

## 2.4. State of the art

Before getting into the working method of Carbonfeel, we must make some comments on the state of the art which will help provide some insight into the advantages offered by the proposed method.

In a study commissioned by the European Commission in 2010, a total of 80 corporate CF calculation methodologies were identified, and 62 in the approach to products or services, each with countless variations and sectorial "sensitivities".

It could be said that once the calculation has been made, even within the same methodology, the results can be quite different depending on the analyst conducting the study, the collaboration of the chain and data sources used.

Obviously, there are a wide variety of reasons why companies choose to use a corporate approach or a product approach. Corporate carbon footprint studies are mainly undertaken when the company's activity is subject to emission permits and therefore it is mandatory to do so. In addition to this, a company may undertake these studies to communicate a green image to third parties and finally, these studies are undertaken to identify possible sources of inefficiencies that result in cost savings in energy and resources.

In the product approach, interests have nothing to do with the emission permits, but lie closer to the promotion of products by associating them with a green image, to meet the requirements of international customers, and even to identify improvements in the eco-efficiency of the production process and the use and disposal of the products and services under study.

As shown, they are all partial interests. If a company undertakes annual corporate carbon footprint studies in order to contribute to the fight against climate change, these partial interests will automatically be fullfilled as they will have a real environmental accounting method with relevant information to manage and communicate as they see fit.

Carbonfeel proposes a method that gets more and better information at a lower cost, which favors the annual monitoring of this type of accounting, as in the field of economics.

We cannot imagine a company doing accounting processes only every two or three years, so why is it acceptable in the environmental field?

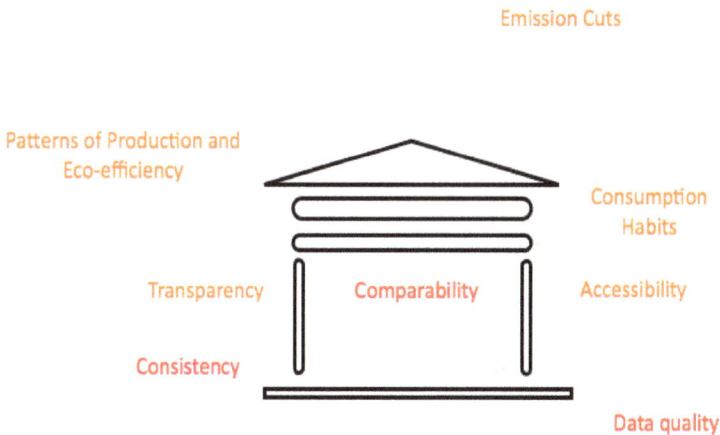

**Figure 3.** What does the state of art tell us?

The answer is obvious, because there are no available data to avoid the great effort in terms of time and economics that a company must exert if it wants to undertake this type of environmental accounting.

While we have outlined the difficulties of conducting studies with partial approaches (corporate vs product/service level), let us do a brief inventory of the current situation and its impact on the construction of this building whose roof we call "emission cuts".

The calculation process at a corporate level faces the following obstacles:

- Great difficulty reaching scope 3. Collecting the supplier's indirect footprint is an impossible mission for many corporations. In addition to the procedural difficulty involved in "forcing" providers to do the calculation, it is based on a totally non-standard assembly process in which each provider chooses the method to calculate the footprint of their products. This creates great distortion and the results lack credibility.
- Voluntary choice of the calculation method, and the scope and the emission factors as long as they come from 'reliable sources'. This leaves the spreadsheet open.
- Inconsistency with the footprint of products or services when these are calculated.
- Legislation compliance ($CO_2$ emission rights) rather than searching for scenarios of competitiveness among enterprises.
- Risk of outsourcing scope 3. Indeed, if it is decided not to calculate the footprint, then all that is needed is to outsource the activities (eg. transportation) so as not to include the footprint in the studies as they are not part of scopes 1 and 2.
- Risk of dispersion of the network. This is perhaps the most serious drawback. The corporate carbon footprint, despite all the potential it has to do a complete analysis of the corporation's resource consumption, may become a mere bureaucratic procedure.
- It is not possible to compare emission intensity. The basic indicator that informs us about $CO_2$ emissions per monetary income of a corporation is disabled by not including all ranges in the calculation studies.

Regarding the approach to products based on a life cycle analysis, the following are identified:

- Great difficulty in project development since the participation of many companies is required. Projects become a repetitive search for information within a company network usually with little willingness to cooperate, either because they are not interested, or because it is hard work getting the required information. This causes unaffordable time and costs for many corporations.
- Accessibility based on the negotiation of the scope of studies. As it is virtually impossible to face the whole cycle with all its ramifications, the cutoff criteria may be capable of being negotiated subjectively, simply according to economic criteria.
- It calculates potential impacts, not real impacts. By eliminating corporate carbon accounting, LCA studies face process maps with theoretical material, according to data provided by companies on inputs and outputs, data which are based on patterns of behavior often very far from the business reality under study.
- Risk of "tailor made" labeling. One of the biggest risks is the profusion of non-transparent product labeling, based on studies whose sole purpose is to bring carbon footprint labeled products to the market but with hardly any verifiable indications.

- No application criteria of secondary sources. The method is mostly based on secondary data support, on which there is no consensus either locally based or sectorial, causing distortion in the calculations and avoiding possible scenarios for comparison.
- High subjectivity of the analyst and the contracting firm on the calculation specifications.
- Comparability is ruled out due to problems arising from the above points. Comparability can not be assumed.
- Indirect carbon charges are dismissed. Studies based on process maps rule out carbon loads from 'non relevant' processes for the corporate character. The relevance or non relevance of these processes is not regulated or is difficult to verify.
- Focused on the business of secondary data. In order to rule out the primary character of the data due problems with availability, secondary data bases, which are hard to upgrade, are promoted. These data provide estimations, but in no case can they calculate a carbon load close to the real business performance.
- Inconsistency of product-level calculations from the corporate perspective. Life cycle assessments lose business perspectives. They focus on the product in search of patterns of behavior, leaving aside the real carbon loads of the corporations which belong to the chain under study.

## 2.5. The integrated approach: The key question

Once the open points of the approaches are detected, we shall see in this section the calculation of both the corporate and products/services footprint. The method not only closes many of these points, but reinforces consistency, transparency and finally ensures the comparability of results opening up a spectrum of possibilities for action in the business world to encourage changes in production patterns, eco-efficiency and consumption habits.

Paradoxically, a calculation based on an integrated approach is both more economical and more complete. It includes all ranges. The company stops worrying about the tracking of emissions that are out of view (scope 3 in the approach to corporate and upstream approach to product based on LCAs) and focuses exclusively on the part of their responsibility, the direct emissions and the organization's land use. Therefore, time of calculation is drastically reduced, making it assumable to all the business.

There is no doubt that if a corporation is seriously facing a study of carbon footprint with high quality information in order to improve emissions intensity of their activities (Kg of $CO_2$ emitted by € of income), an integrated approach has to be undertaken.

If we talk about accounting, there is no general manager that takes more seriously into consideration the importance of data quality and consistency than a CFO. This CFO, when performing cost accounting, never conducts an 'upstream' research on costs which has impacted the income statement of its products in the life cycle. Obviously it is impossible to assume such studies because no longer are they economically sustainable and results are useless due to the uncertainty that they cause. He simply counts the costs of the organization and then splits them between the actual production, which gives a true picture of the corporate cost accounting of each of their products and services.

So, why does a Director of Environment face ACVs projects with a high level of economic demand for accounting the carbon footprint of its products, when it only provides potential emission values as it misses the whole business perspective? We can list many reasons, but, from the technical point of view, we would say that it is impossible to assess the actual cost or carbon footprint per functional unit of each good or service purchased by the organization. This is the key difference between why a CFO carries out an integrated accounting and why a Director of Environment cannot perform it.

But, if this should be solved, if somehow someone had a method that moved all purchases, usually in monetary value, to carbon footprint, the problem would be solved, at least in part.

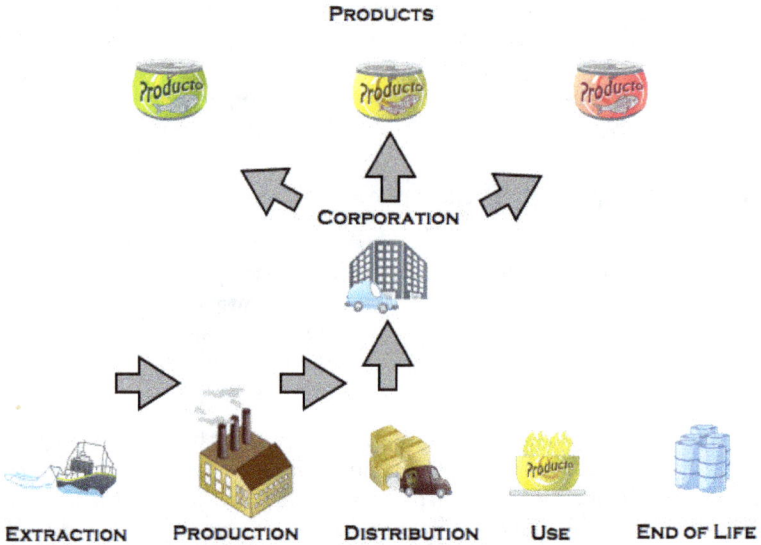

**Figure 4.** The integrated approach

## 2.6. The Compound method based on Financial Accounts (MC3)

The Compound Method based on Financial Accounts (MC3) has two different uses. Firstly, the MC3 provides an inventory of materials, goods, services and generated wastes transformed into a common unit, EqtCO$_2$. This information is useful to elaborate environmental policies and corrective measures based on the CF at an organizational level.

Secondly, the footprint of a company can be assigned to the produced goods. In this case, the organizational footprints assess the product's CF across the supply chain, identifying the footprint at every phase of the life cycle. The distribution of the footprint of every organization among the produced goods requires unitary footprints expressed in Gha/t and/or tCO$_2$/t. When a firm purchases a good, the acquiring company will use unitary footprints to estimate his organizational footprint.

## Organisation level

The MC3 was developed by Doménech [1,2]. Initially the method assessed the CF of companies and organizations. Nowadays, the method also estimates the CF of goods and services throughout the supply chain.

In both cases, the starting point of the MC3 is the estimation of the CF of organizations. This chapter briefly describes the method at this level. A more detailed explanation can be found in [1-4].

The origin of MC3 can be found in the concept of household footprint [5]. In this way, based on the matrix of consumptions versus land present in the spreadsheet for the calculation of households' footprint [5], Doménech [1] prepares a similar consumption land-use matrix (CLUM) (see table 2), which contains the consumptions of the main categories of products needed by a company. The land-use matrix also includes sections for the wastes generated and the use of land. These consumptions/wastes will be transformed into land units and greenhouse emissions [6]. Carbonfeel initiative has improved this CLUM matrix, including new categories of products, emissions and conversion factors (MC3.V.2).

The needed information to estimate CF using MC3 is mainly obtained from accounting documents such as the balance sheet and the income statement, which clearly state the activities that are associated with every entity: MC3 estimates the footprint of all the goods and services considering information from financial accounts. Wastes generated and built-up surface by all the facilities of the company are also included. Further information from other company departments with specific data about certain sections (waste generation, use of land by the organization's facilities, among others) may also be necessary in case this information is excluded from the financial accounts. The footprint is calculated in a spreadsheet, which also works as the CLUM matrix.

The rows of this CLUM matrix show the footprint of each category of product/service consumed. The columns present, among other elements, different land-use categories for CF, into which the footprint is divided. Columns are divided into five groups. The first one (see column 1) corresponds to the description of the different categories of consumable products. These are classified into 9 major categories showed in Table 1. One can include as many products as desired within each category.

The second group (columns 2–6) shows each product's consumption, expressed in specific units. The units in the first column of the group are related to product's characteristics (e.g., electricity consumption, in kWh). The second column indicates the value of the consumptions in monetary units, while the third shows consumptions in tonnes. The fifth column reveals energy corresponding to each consumption expressed in gigajoules (GJ),

obtained by multiplying tonnes of product (third column) by the quantity of energy used by
tonne in its production (GJ/t) (fourth column) [7].

| Consumption sections | Consumption categories |
|---|---|
| 1. Direct emissions | 1.1. Fuels |
| | 1.2. Other direct emissions |
| 2. Indirect emissions | 2.1. Electricity |
| | 2.2. "Other indirect emissions" |
| 3. Materials | 3.1. Flow materials (merchandise) |
| | 3.2. Non-redeemable materials |
| | 3.3. Redeemable materials (generic) |
| | 3.4. Redeemable materials (construction) |
| | 3.5. Use of public infrastructures |
| 4. Services and contracts | 4.1. Low mobility services |
| | 4.2. High mobility services |
| | 4.3. Passenger transport services |
| | 4.4. Merchandise transport service |
| | 4.5. Use of public infrastructures |
| 5. Agricultural and fishing resources | 5.1. Clothing and manufactured products |
| | 5.2. Agricultural products |
| | 5.3. Restaurant services |
| 6. Forestry resources | |
| 7. Water footprint | 7.1. Consumption of drinking water |
| | 7.2. Consumption of non-potable water |
| 8. Land use | 8.1. On land |
| | 8.2. On water |
| 9. Waste, discharges and emissions | 9.1. Non-hazardous waste |
| | 9.2. Hazardous waste |
| | 9.3. Radioactive waste |
| | 9.4. Discharges in effluents |
| | 9.5. Emissions |
| | 9.5.1. GEG Gases Kyoto Protocol |
| | 9.5.2. Other GEG or precursors |
| | 9.5.3. Other atmospheric emissions |

**Table 1.** Sources of emissions considered in the carbon footprint (MC3.V.2)

Energy intensity factors comprise the amount of energy used in the production of every
product included in the CLUM matrix, considering an average supply chain. At this
moment, they are mainly obtained from the European Commission [8-10], Simmons [11],
Wackernagel [5,12] and different public institutions such as Spanish Office for Climate
Change (OECC) and The Institute for Energy Diversification and Saving (IDAE) and
Intergovernmental Panel on Climate Change IPCC [13]. The third group of columns

(columns 7 and 8) show emission factors for every category of product. Emission factors are mainly obtained from the European Commission [8-10,14,15], IPCC [13], OECC and IDAE.

The fourth group contains six columns (9–14) showing the distribution of the footprint among different categories of land. These are the same as that used for the countries' ecological footprint ($CO_2$ absorption, cropland, pastures, forests, built-up land, and fishing grounds).

Finally, MC3 estimates the organizations' counter footprint. The counter footprint concept starts from the positive regard for the companies' availability of natural capital, despite the desirable reduction of their footprint by being more efficient and by curbing consumption. Therefore, investments in this kind of productive space reduce their footprint. In this way, this indicator could encourage the private sector's involvement in the preservation of natural spaces [2], as which is positive in terms of sustainability [6].

*Product level*

Since the year 2005 a team of researchers from the Universities of Oviedo, Cantabria, Valencia, Cádiz, Santiago de Compostela and La Coruña, coordinated by Juan Luís Domenech, have been developing MC3 at an organisation level. A member of this group, Adolfo Carballo-Penela, of the University of Santiago de Compostela, has broadened the scope of the method to products, and has developed the theoretical and practical knowledge needed to determine how they should be ecolabelled [3].

Information from products considering supply chains is useful for both, companies and final consumers. Companies can reorganize their existing processes, obtaining environmental improvements and reductions of costs. Ecolabelling processes based on CF allow consumer's purchase decisions to have a positive influence in achieving a more sustainable world.

From the MC3's perspective, CF throughout the life of a good or service considers those land and emissions required/generated by each of the companies involved in its production, from the phase of raw materials up to the retail point. Every company itself is a phase of the supply chain.

Figure 5 shows an example of this way of proceeding. In this case, the supply chain is composed by four companies which produce canned tuna fish: a fishing company, a preserves company, a carrier and a restaurant. If the customer of the restaurant applies for lower CF products, the restaurant must reduce its footprint to meet this demand. Actions like reducing consumption of goods and waste generation, recycling activities, or technologies that are more efficient would be effective in this case. The purchase of goods with a lower footprint is also a valid option, replacing present suppliers for other lower-footprint providers. Asking present suppliers to reduce their CF and, therefore, their product's footprint is a possible recommendation as well. The demand for lower footprint products can be extended to all the participants of the considered lifecycle and to all the goods of the economy.

| Product category | Annual consumption | | | | | Emission factor | | Footprint by productive space | | | | | | GF | Counter footprint |
|---|---|---|---|---|---|---|---|---|---|---|---|---|---|---|---|
| | Consumption units [unit/year] | €, $, …. without VAT [€/year] | Tones [t/year] | Energy intensity [GJ/t] | GJ [GJ/year] | $[tec\ CO_2/t]$ | $[t\ CO_2/GJ]$ | CO2 absorption land | Cropland | Pastures | Forests | Built-up land | Fisheries ground | | |
| 1.- Direct emissions | | | | | | | | | | | | | | | |
| 2.- Indirect emissions | | | | | | | | | | | | | | | |
| 3.- Materials | | | | | | | | | | | | | | | |
| 4.- Services and contracts | | | | | | | | | | | | | | | |
| 5.- Agriculture and fishing resources | | | | | | | | | | | | | | | |
| 6.- Forestry ressources | | | | | | | | | | | | | | | |
| 7.- Water footprint | | | | | | | | | | | | | | | |
| 8.- Soil use | | | | | | | | | | | | | | | |
| 9.- Waste, discharges and emissions | | | | | | | | | | | | | | | |

**Table 2.** Structure of the spreadsheet showing the CF CLUM matrix.
t: tonnes; VAT (value added tax)

The adoption of MC3's supply chain approach requires establishing links among the CF of the different companies of the supply chain. When each of the participants in the lifecycle of a product acquires different goods from the company situated in the previous phase, they are also acquiring the CF incorporated in that good. If every participant communicates the unitary footprints of the goods and services that produces (e.g. eqtCO2/t of product) to the following phase of the supply chain, the needed connection is made. Footprints per tonne of product (unitary footprints) are obtained dividing the total footprint of every company by its production. Table 3 collects an example of this way of proceeding.

This case is similar to that shown in Figure 5. In this case, a retailer replaces the carrier. This example assumes that each participating company produces only one tonne of one product, the canned tuna fish, which is purchased by the next company in the supply. Every company also acquires 1 ton of the rest of the used products. Information of the CF is shown in Table 3.

| Company | CF |
|---|---|
| Fishing company (EqtCO$_2$/t of product) | 8.0 |
| Fuel (EqtCO$_2$/t of product) | 2.0 |
| Bait fish (eqtCO$_2$/t of product) | 6.0 |
| ... | |
| Preserves company | 15.0 |
| Tuna fish (EqtCO$_2$/t of product) | 8.0 |
| Machinery (EqtCO$_2$/t of product) | 7.0 |
| ... | |
| Retailer | 17.5 |
| Tuna fish (EqtCO$_2$/t of product) | 15.0 |
| Fuel (EqtCO$_2$/t of product) | 2.5 |
| ... | |
| Restaurant | 21.0 |
| Tuna fish (EqtCO$_2$/t of product) | 17.5 |
| Electricity (EqtCO$_2$/t of product) | 3.5 |

**Table 3.** An example of unitary footprints application in the lifecycle of canned tuna fish

The fishing company would estimate its footprint using the unitary footprints of the acquired goods, in this example, fuels (2.0 EqtCO$_2$/t of product) and bait fish (6.0 EqtCO$_2$/t of product). Considering these values, the CF of one tonne of tuna fish at this phase of the supply chain is 8.0 EqtCO$_2$/t of product. The preserves company acquires a tonne of tuna fish, which means, 8.0 EqtCO$_2$/t of product. This company adds footprint from the consumption of one ton of machinery (7.0 EqtCO$_2$/t of product), being its total footprint of 15.0 EqtCO$_2$/t of product, the only commercialized product of the preserves company.

In this example, the retailer's purchase of fuel generates 2.5 EqtCO$_2$/t of product. In addition, this firm acquires 1 tonne of tuna fish (15 EqtCO$_2$/t of product) from the preserves company.

This means a total footprint of 17.5 EqtCO$_2$/t of product of tuna fish, sold to the restaurant. This company also adds 3.5 EqtCO$_2$/t of product, from the electricity used in its activities, which implies a CF of 21.0 EqtCO$_2$/t of product of tuna fish at the end of the supply chain. This value would be showed in an ecolabel that collects the CF of this preserved fish tuna.

We want to remark that the total footprint of the tuna fish is not estimated as the sum of the footprint of all the companies involved in the supply chain (61.5 EqtCO$_2$/t of product). By doing this, the footprint of tuna is multiple-counted since every company includes the fish's footprint of the previous phase. The tuna fish's CF is estimated considering the added footprint in every stage of the supply chain.

*Starting of the method*

The use of MC3 to estimate the CF of products needs of unitary footprints for each of the categories of products collected in the CLUM matrix. These unitary footprints come from secondary data from pilot studies. The pilot studies are based on the energy intensities and emission factors usually used by the MC3, besides results from other supply chain studies that estimate the emissions from primary data.

The transmission of CF across the supply chain and its use as an ecolabel will depend on the will of the participants in the supply chain to estimate their footprint. The success of the adopted approach depends on the organizations' awareness of the advantages of estimating the footprint of their products. Environmental marketing differentiation and savings related to a more efficient use of materials and energy along the supply chain are relevant questions that should be considered [16]. However, Carbonfeel initiative will provide involved companies with enough information to estimate the footprint of the products they purchase.

The support of national or regional governments seems to encourage companies' participation in countries like the United Kingdom, where DEFRA and the Carbon Trust have developed a key role to accelerate the process. In the absence of public sector participation, interested companies should encourage customers and providers to estimate their CF and communicate them along the supply chain.

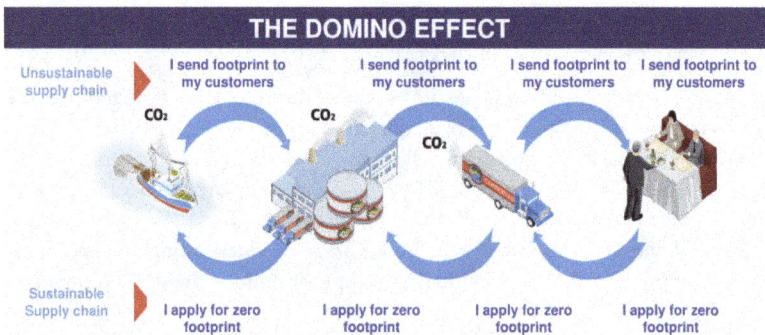

**Figure 5.** An example of supply chain according to MC3: tuna fish in preserve [2]

*Boundaries of the analysis*

MC3 is based on the cradle-to-gate life cycles. This means that MC3 assesses CF from the raw materials phase to the retailing phase, by including all the activities required to extract the raw materials for the product, manufacture the product, and ship the product to the point of purchase. MC3 does not consider footprints from the use and disposal of goods. MC3's footprints collect the demand of land/emissions of $CO_2$ of all the goods and services acquired by every company, the generated wastes, and the built-up land in each of the phases of the lifecycle.

*Transmission of the information across the supply chain and ecolabelling process*

In case of goods for final consumption and services, the information about unitary footprints ($tCO_2/t$) should be incorporated in the common price labels, tickets and similar documents. Invoices, delivery notes, contracts, budgets or any other documents containing prices should add CF information at the intermediate phases of the lifecycle. This is the way that the CF information is available during the entire life cycle and transmitted. Once Carbonfeel begins to work, information technologies will simplify the process of communicating footprints among companies.

When a company acquires a product, the purchase documents should include the unitary footprint accumulated until that moment and making possible to use that information for estimating its organizational footprint. If a supplier does not provide information on a product, Carbonfeel database will supply this information. This database includes the unitary footprints on standard lifecycles for the main categories of products included in the CLUM matrix. They are obtained from pilot studies. These unitary footprints show information from the different stages of the supply chain. Considering the case of the tuna fish (Table 3) different unitary footprints for "Tuna fish", "Preserved tuna fish", "Preserved tuna fish: retailer" and Preserved tuna fish: restaurant" should be available for the MC3 users. At this moment, the Carbonfeel database is under development. Obtaining detailed information about more goods and services requires an increase in the number of pilot studies.

*Assessment of the exposed method*

Similar to the other methodological approaches, the MC3 has some strengths and limitations, summarized in the following sections.

In previous articles, authors have stated that the MC3 is a complete, transparent and technically feasible method based on Wackernagel and Rees compound method. Working with MC3 does not require extensive expert staff inputs and everybody working with spreadsheets will be able to calculate CF. MC3 is also a flexible and complete method. MC3 can be adapted to the characteristics of different types of companies, collecting the footprint from all the products consumed and wastes generated by a company [7].

The fact that the information comes from accessible financial documents, and that every company covers a complete phase of the lifecycle implies lower economic and time costs, besides delimiting clearly the products and activities that are under analysis. This ensures comparability among products.

The theoretical presentation of the method requires determining participants in the supply chain. In practice, every company gets the environmental information of the purchased products from their suppliers or from the Carbonfeel database. According to the European Commission [17], the market could become a powerful force for delivering environmental improvement. The role of markets as the main source of environmental information on products, thereby absorbing environmental performance as a competitive issue, is an important strength of the method. The identification between a corporation and the supply chain phase also favours the collection of information, obtained from every company [6].

The way of estimating the CF avoids double counting problems with some intermediate inputs, a relevant question in this context [18,19]. Organisational footprints are useful in terms of making decisions on improving environmental performance of organizations but never in terms of aggregating environmental impacts. This aggregation is only possible in terms of the products.

This analysis is less detailed than conventional process-based life cycle assessment. The organization's activities are not divided into detailed simple processes that show the amount of energy, and materials consumed in every stage of the production. Instead of doing this, MC3 includes all the goods, services and wastes consumed/generated for the organizations in a period. The use of unitary footprints or energy intensities and other aggregated information allow MC3 to estimate CF.

## Benefits of the integrated approach

Recalling the three basic pillars necessary for a carbon accounting method to be useful to the company in its fight against climate change, we note that an integrated approach, like MC3, provides a number of benefits that can solve many of the open issues identified in the approaches focused on the organization and on an individual product.

Transparency

- All calculations are based on reliable sources of recognized standing and free access.
- There are neither subjective criteria of the study design limits nor cutoff criteria, since the scope is complete.
- As a result, customers and consumers are well aware that the Carbonfeel label guarantees studies that have been conducted on an equal basis in all participating organizations. A company facing a Carbonfeel project can communicate this to the interested parties, who will accept and trust in these studies.
- The information is not just potential, it closely reflects the true business reality of the organization and provides critical indicators of emission intensity which, with the inclusion of all scopes, provides an idea of the company's situation in terms of carbon accounting.

Accessibility

- The information is found within the company; it isn't necessary to to get it from the network of suppliers. The calculation is completely autonomous and does not depend on other organizations.
- As a result, the study times are speeded up exponentially. This process will be optimized over time once the elements involved in automating the calculation, exchange and assembly of information have been identified.
- Moreover, the project cost drops dramatically by not requiring mapping processes and the subsequent investigation in the whole supply chain.

Comparability

- By not having to develop cut off criteria, studies ensure full comparability.
- In the near future it will be possible to design carbon footprint labels type III of sectorial goods and services as long as comparability of results is guaranteed.

Added to these benefits, the integrated approach provides a foundation which ensures that the three columns will support the building. The consistency of the results, defined as the consistency between the Corporate Carbon Footprint and the Carbon Footprint of products and services.

## 2.7. The pending issue

Note that an integrated approach can be improved by adding a foundation to provide greater stability to the building to be constructed, which results in more transparency, comparability and accessibility. It is therefore more likely to transform our patterns of production and consumption.

As mentioned, virtually all methods of calculation are oriented towards the use of secondary data when incorporating emissions from the lifecycle or footprint of our suppliers. Multiple databases with commercial or free access grow asynchronously, which adds a new point of controversy to the calculations, leading to a lack of comparability of the results.

MC3 provides the factors to estimate the carbon footprint based on sources and conversion factors that continue to be a secondary database such as, for example, used energy intensities.

We understand that a working method of carbon accounting should be aimed at facilitating the integration of primary data, i.e. the actual footprint of goods or services which are acquired or participate in a given life cycle.

The integrated approach favors this. If somehow we could operate like a CFO and get the cost of what you buy on each bill, i.e. the actual carbon footprint per functional unit that a Director of Environment has to charge to their accounts and then multiply it by the real consumption, we will be laying a vital foundation: data quality.

The 'green coin' can become a reality if we pay attention to the technological factor faced by the carbon footprint as a problem of information exchange.

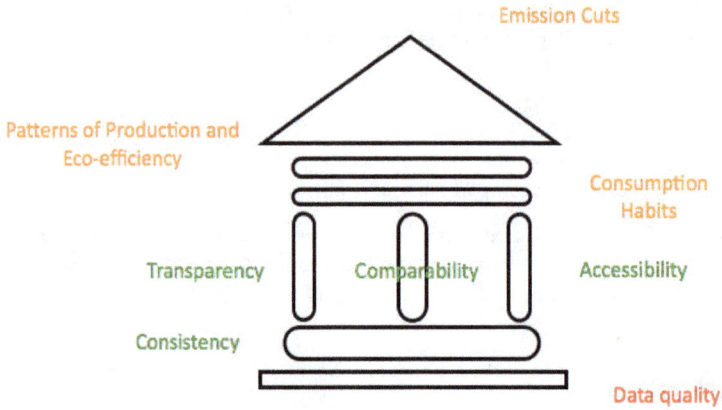

**Figure 6.** What does the integrated approach report?

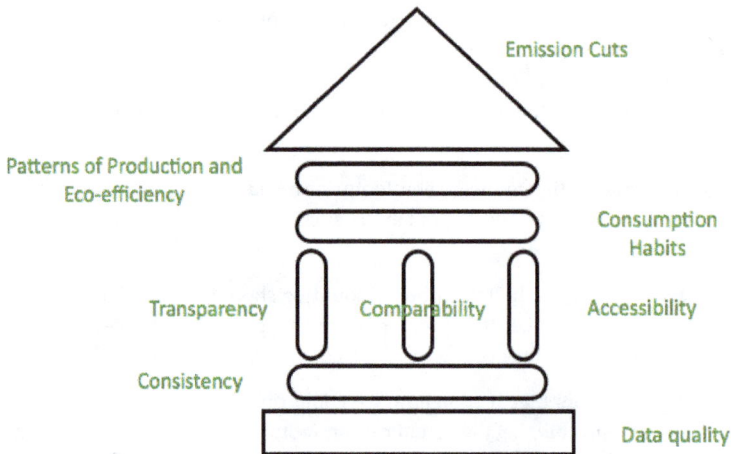

**Figure 7.** What does Carbonfeel provide?

PRODUCTS

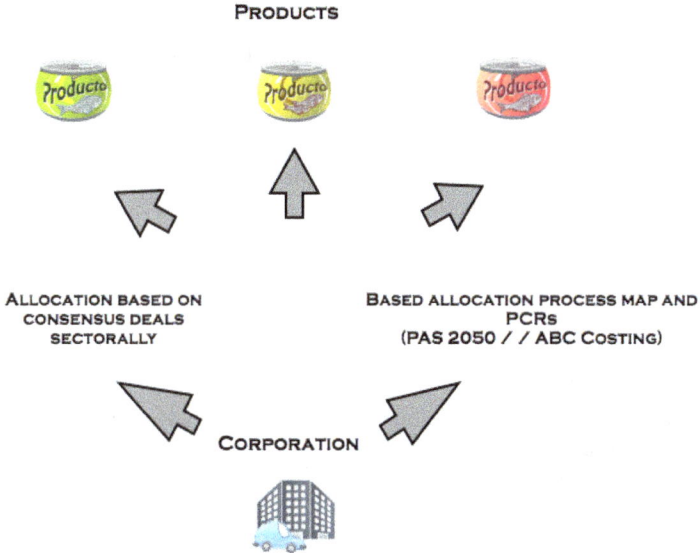

ALLOCATION BASED ON
CONSENSUS DEALS
SECTORALLY

BASED ALLOCATION PROCESS MAP AND
PCRs
(PAS 2050 / / ABC COSTING)

CORPORATION

**Figure 8.** The integrated approach: moving from the corporation to the products.

## 2.8. Connecting the network: The role of information technology

Carbonfeel relies on information technology in order to provide the benefits of an integrated approach, the foundation related to data quality, i.e. obtaining the real carbon footprint of each good or service consumed.

This solves one of the great challenges of the technical studies related to life cycle analysis, which is nothing more than having the ability to 'assemble' the "real" footprint or primary data from each of links in the chain involved in the processes of the product life cycle to be calculated.

From the viewpoint of a computer analyst, this problem, faced from the perspective of the information exchange between various partners, requires only two things:

- Consensus on the semantics of computation
- Cooperation of the parties

Carbonfeel has a Committee of technical experts familiarized with MC3, input-output analysis, life cycle analysis such as PAS 2050, and others, which take the best solutions provided by each one to achieve an integrated approach. All this work is related to adopting some form of calculation. The semantic analyst's job is to compile these agreements into electronic dictionaries that provide the rules for computer analysts and databases so they can develop software able to calculate the carbon footprint based on these rules, and more importantly, to exchange information between different actors.

*Assembling the life cycle*

As discussed earlier, the mission of Carbonfeel is to organize information and knowledge about the carbon footprint, making it universally useful and accessible to all society. Translating this purpose to a practical language, we can say that, based on an integrated approach and the best available techniques, Carbonfeel determines how to calculate a carbon footprint on a neutral level (valid for any type of company), and also how to calculate the footprint for a particular sector. Sectorial standards will be created to be able to apply the rules to all economic activities sectors.

Once these standards, rules and calculation schemes are stablished, always in strict compliance with existing ISOs and future ISO 14064, ISO 14067 and ISO 14069, it will be possible to develop a software able to calculate, and what is more important, to exchange information between different actors.

Once the corporate footprint has been calculated, the deployment to products and services of the corporation is carried out primarily by means of two basic techniques which are, curiously enough, the same techniques that a CFO normally uses:

- Distribution of carbon loads directly to the products and services according to agreed sectorial schemes. This scheme is recommended for small and medium enterprises or corporations with little variety of products and services.
- Distribution of loads on a map of processes and activities. In fact, it is very similar to an ABC Costing study, well known in the accounting field. This method is ideal for the identification of inefficient processes and activities and is recommended for large corporations with complex process maps.

In the second case there is a clear connection with calculation techniques based on Life Cycle Assessment, already introduced into the market as PAS 2050 and Product Categories Rules PCRs. These may acquire a new dimension in the benefits they provide when focusing on an integrated approach.

With these raw elements, it is possible to consider (based on an integrated approach) combining the worlds of economics and the environment by implementing the carbon accounting system in exactly the same way that any organization does its financial accounting.

The idea is as simple as it is powerful. Each one of the goods or services purchased must be assessed as a debit in the footprint of the branch company. The products involved, goods or services sold generate the cumulative footprint passed on to the next link in the chain once the allocation of corporate footprint of the goods and services produced by the organization has been made.

Does this Carbonfeel footprint calculation represent a product life cycle as promoted by the standards of the ISO 14040 series? In fact, it is a life cycle from cradle to the gate, ready to be assembled in the following link (customer buying the product or service), but with a substantial difference as compared to a classic project. While in the latter the footprint has been calculated for a single organization (which has commissioned the study), with the assembly method, every link has estimated its own part of the whole life cycle, independently and based on actual footprints of its first level providers.

The question arises: what if the suppliers are not in the network of calculation and do not provide their footprint? This is the point where Carbonfeel resorts to the secondary data to come as close as possible to the reality of the study.

It is important to note that the integrated Carbonfeel approach is oriented towards a telematics assembly of "real" primary carbon footprint data. The secondary data cease to be the only possible data to take on an alternative role. The technology exists. Developing the semantics and required software is only a matter of time. The benefits are for everyone, in both the B2B and B2C environment. The entire network is benefited thanks to the accessibility provided by information technology. Government, businesses and citizens will have quality, consistent, transparent, comparable and accessible information. The building will have a strength that will help us fight climate change with better weapons.

A Carbonfeel project offers companies a real environmental accounting method based on a universal indicator such as the carbon footprint, which analyzes the corporation and each of the products and services generated.

## TO THE GREEN COIN

THE MISSION IS TO FIND THAT 'PURCHASE PRICE', CERTIFY AND EXCHANGE.

**Figure 9.** Moving towards the green coin

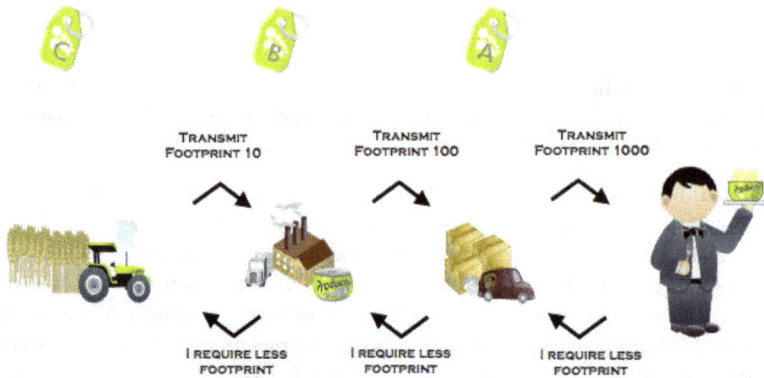

**Figure 10.** Global labeling (B2B and B2C)

## 3. Application of MC3 methodology and Carbonfeel philosophy in a case study: Calculation of the corporate carbon footprint of a cement industry in Spain

The climate change is one of the biggest problems the humanity copes with nowadays. Therefore, reducing the $CO_2$ emissions of sectors such as the cement industry, whose emissions account for roughly 5% of the total $CO_2$ emissions worldwide [20], is a primary goal in order to comply with the objectives laid down in the Kyoto protocol.

Throughout the next pages it will be presented the application of the organization-product-based-life-cycle assessment (hereafter OP-LCA) methodology (MC3) to three types of cement facilities in order to calculate the corporate carbon footprint of the cement manufactured in three different ways.

Our goal is not only to determine the footprint calculated in this way but also to demonstrate that the comparability between different brands and products is totally possible, thus providing a serious alternative to process-product-based-life-cycle assessment (P-LCA) methodologies. Moreover, as a result of the analysis, it will be also possible to identify the best ways forward to achieve the lowest possible footprint.

As mentioned, this case study was carried out with three potential scenarios in mind: Case A pertaining to a conventional integral plant which we will call "current", Case B which refers to a grinding plant and Case C, an integral plant which has been subject to the best available techniques (BAT).

The three scenarios were modeled with the same productivity of 1,000,000 t/year in order to simplify the comparability between them. Their differences as far as operability is concern can be drawn from the next descriptions of each one.

## Case A – "Current integral plant"

In general terms, this type of plant includes a line that consists of the following processes:

After the crude cement has been ground and dried, the resulting product is a powder which is 80% calcareous, 19% clay and 1% iron corrector. The moisture is roughly 8%.

Next it goes through the four-step cyclone exchanger. There, heat is transferred from the gases to the crude cement; the residual moisture is dried; the water making up the clay is lost; and decarbonation begins.

After this operation, the crude product is placed into the rotating kiln at approximately 900ºC while the decarbonation, fusion and clinkerization reactions are completed. The fuel used is coke, which raises the temperature inside the kiln to nearly 2000ºC.

The newly formed clinker leaves the kiln at around 1500ºC and is then released onto a grate cooler, where it is cooled to 100ºC by means of air exchange. The gases from both the cyclone exchanger and the kiln are filtered before being released into the atmosphere.

Finally, the cement is ground along with the additives in a tubular ball mill until a particle size of roughly microns is achieved. The product is then packaged and stored until it is shipped.

## Case B – "Grinding plant"

This plant receives the clinker from outside sources, generally imported from China and/or Turkey. Therefore, this plant simply grinds the outsourced clinker with the additives and ships the cement obtained.

At the present time, the great majority of the plants use closed circuit tubular ball mills with a highly efficient turbo-separator which allows the fineness of the cement to be controlled by means of centrifugal force. The dust is removed from the mill and the turboseparador by sleeve filters. Since the clinker may contain some moisture, the grinding facility usually has a hot gas oven. It does not generally have a drying chamber. Instead this process is carried out in the first grinding chamber. The final product, the cement, has a particle size of around 30 microns. It is transported by means of aerogliders to rubber belt bucket elevators and then to the cement stock silos. Finally the product is packaged and stored until it is shipped.

Since clinker is imported, the plants are forced to manufacture cement with the minimum percentage of this material, in order to cut costs. Therefore it is logical to use the maximum amount of subproducts as additives. However, we must remember that in Spain independent companies do not have access to materials like slag from blast furnaces or fly ash, which are monopolized by integral cement factories, which have exclusive agreements with the producers of these materials. Therefore, there are two possible options: either the manufacture of cement is carried out with limestone additives and a high content of clinker or with the use of materials such as bottom ash which may be classified as natural calcined pozzolana, since Standard EN 197-1 does not currently allow its commercialization as power station ash.

## Case C – "BAT integral plant"

The operating process is the same as case A, but here the factory uses all the technologies and principles associated with the concept of the "Best Available Techniques" of the IPPC, at the current level. Below is a description of the latest innovations as compared to the previous reference.

After the crude product has been ground and dried, the dry powder undergoes homogenization, which produces a uniform product that facilitates even heating. This is carried out by means of a controlled flow system of the different layers of material as they enter the silo consisting of a mixture of compressed air at the exit of the filter, with or without a separate chamber. The stock of powder is the equivalent of around three days of kiln time.

Next the powder enters the cyclone exchanger, this time, during five successive steps, where the heat is transferred from the gases to the crude cement; the residual moisture is dried; the water making up the clay is lost; and decarbonation begins.

At the base of the exchanger, part of the total fuel is injected into the kiln, in the precalcination system, with the help of the combustion air supply from the head of the kiln through an ad hoc tertiary air duct. Fuel injection is carried out in several steps to reduce the emission of NOx.

According to the principle of BAT, the percentage of total fuel injected into the precalcinator is around 70%, with 30% being injected into the head burner (as opposed to 25% and 75% respectively in case A). Secondary fuels or waste materials such as used oils, paints, solvents and a certain amount of biomass are used.

The clinker leaves the kiln at approximately 1500°C, and is then released onto a high-efficiency grate cooler with air injection control grate plates, where it is cooled to 100°C by means of air exchange. Some of this air enters the kiln by the flue effect as secondary combustion air, and some goes to the tertiary air duct while the remainder is released into the atmosphere after being purified through the pertinent filter.

Similar to case A, the clinker is mixed with the additives and is ground, but this time, in a vertical mill with a highly efficient turboseparator that allows the fineness of the cement to be controlled by centrifugal force. Just like the other two cases, the final size of the particles is roughly 30 microns. Finally the product is packaged and stored until it is shipped.

## Results and discussion

The results obtained with the computation tool are summarized in Table 4 to make them easier to understand and to be able to establish comparative criteria between the different scenarios.

| CONSUMPTION CATEGORIES | A - Current integral plant | B - Grinding plant | C - BAT integral plant |
|---|---|---|---|
| Direct emissions | 756.005,3 | 6.653,5 | 608.998,0 |
| Indirect emissions | 39.040,0 | 18.369,3 | 33.829,8 |
| Material footprint | 15.810,6 | 677.435,7 | 15.710,4 |
| Footprint of services and contracts | 2.818,3 | 131.728,7 | 2.825,1 |
| Agricultural and fishing footprint | 0,0 | 0,0 | 0,0 |
| Forestry footprint | 8.746,2 | 8.840,7 | 8.734,7 |
| Water footprint | 659,9 | 61,2 | 192,0 |
| Soil use footprint | 58,2 | 9,6 | 56,8 |
| Footprint of wastes, emissions and discharges | 180.428,3 | 59.344,6 | 119.943,1 |
| Soil use counter footprint | 11,6 | 2,5 | 11,6 |
| | | | |
| TOTAL FOOTPRINT | 1.003.566,8 | 902.443,3 | 790.289,9 |
| TOTAL COUNTER FOOTPRINT | 11,6 | 2,5 | 11,6 |
| **NET FOOTPRINT** | **1.003.555,2** | **902.440,8** | **790.278,3** |

**Table 4.** $CO_2$ emissions from different "inputs" (in $tCO_2$/year)

Based on the results obtained and reported in the previous section, the following observations can be drawn:

(1) By dividing the total footprint (1,003,555.2 $tCO_2$/year for case A, 902,440.8 $tCO_2$/year for case B and 790,278.3 $tCO_2$/year for case C) by the productivity (1,000,000 t/year), it is possible to obtain the amount of $CO_2$ that must be released to manufacture a ton of cement. These values are 1.00 $tCO_2$/tcement for case A, 0.90 $tCO_2$/tcement for case B and 0.79 $tCO_2$/tcement for case C. Therefore, the carbon footprint of cement may be considered high in theory. We are well aware of the enormous effort the sector has made over the years in attempting to reduce their $CO_2$ emissions by applying a number of different techniques, but there are still areas left to be explored and completed. In view of the results, some guidelines aimed at reducing $CO_2$ emissions in cement plants can be obtained:

**Fuels and electric energy**: Both direct and indirect emissions, easily account for the greatest part of the total footprint (75.33%+3.89% in case A and 77.06%+4.28% in case C). In case B, they seem to be lower (only 0.73%+2.02%) but that is just because a grinding plant does not need to use the kilns but it will do incorporate the intrinsic footprint of the purchased clinker (included into the materials category). It is also clear how the use of BAT allows for the reduction of the sum of direct and indirect emissions from 795,045.3 $tCO_2$/year (case A) to 642,827.8 $tCO_2$/year (case C); i.e., a reduction of 152,217.5 t $CO_2$/year (nearly 20%). Some possible measures would be the use of secondary fuels and sustainable energy sources or the reduction of the percentage of clinker in cements.

**Materials**: the footprint produced by materials is not too large in cases A (15,810.6 $tCO_2$/year) and C (15,710.4 $tCO_2$/year) accounting for 1.58% and 1.99% of their total footprints, respectively. However, owing to the enormous, but unavoidable amount of outsourced clinker purchased by plant B, along with the fact that this material has a large intrinsic footprint, in the case of the grinding plant, it is the materials category that accounts, without a doubt, for the greatest part of the total footprint (74.66%), reaching an absolute value of 677,435.7 $tCO_2$/year. For this reason, the lines of action taken to reduce the footprint should be based on more sustainable constructions and the optimization of the use of aggregates and minerals in general, but above all, in case B, it is primordial to reduce the amount of clinker in cement.

**Services and contracts**: it is not a highly significant category in cases A (2,818.3 $tCO_2$/year, the 0.28%) and C (2,825.1 $tCO_2$/year, the 0.36%) but it really is in case B, where the fact that the clinker must be imported with its consequent transport by ship, raises case B $CO_2$ emissions to 131,728.7 $tCO_2$/year (14.52% of its total footprint). In general, this footprint can be reduced by contracting the services of the most efficient companies in environmental terms. Another category of importance is the contracting of "office" services with a high added value, whose carbon footprint can be reduced mainly by saving on energy. In case B, it is clear that the best way forward is to find more efficient and sustainable means of transport for the imported clinker as well as closer suppliers, what would minimize the necessary traveled distances (despite the fact that the China is currently the largest export market owing to its low prices).

**The agricultural and fishing footprints**: this category has not been introduced into the analysis. This category is usually the first to be omitted for reasons of discipline although it can take on great importance in certain businesses and multinational companies, owing to the expenses incurred from travel, and the resulting cost in sustenance (not to mention company dinners, social events and invitations).

**Forestry footprint**: this footprint does not have very high values in relation to the total footprint (8,746.2 $tCO_2$/year in case A, 8,840.7 $tCO_2$/year in case B and 8,734.7 $tCO_2$/year in case C) with incidence percentages of 0.87%, 0.98% and 1.11% respectively. However, it should be controlled by making sure that the wood is certified and that it comes from forests managed under sustainable development programs and by demanding cellulose and wood products from suppliers with a small footprint or with plans to reduce their footprint. The forestry footprint that cannot be reduced should be offset by the counter footprint. At the present time there are companies that invest in creating forests, parks, pastures, etc. in order to increase their counter footprint, thereby decreasing their total net footprint. Therefore, the investment of natural capital in non-company owned land or even distant pieces of land should not be ruled out.

**Water footprint**: What is striking in this case is the great reduction achieved by using the best available techniques, decreasing the consumption from 269,725.0 $m^3$/year in case A to 78,462.0 $m^3$/year in case C (which is equal to a reduction of 659.9 $tCO_2$/year in case A to 192.0 $tCO_2$/year of case C, i.e., a reduction of 71%). Not only the new production techniques were important in this case, but also BAT are now being used for the selective collection of water whereby consumption is reduced through recycling techniques, collection of rainwater, greywater and others. In general terms, potable water (which requires a distribution network, pumping facilities, potablization processes, etc. and unnecessarily adding an enormous footprint to the product) should never be used for industrial processes.

**Soil use**: plants type A and C require much more soil than plant type B. Therefore, while the footprints of cases A and C (58.2 and 56.8 $tCO_2$/year respectively), that of case B is 9.6 $tCO_2$/year. Generally speaking, the footprint corresponding to this concept is small, which is no reason not to optimize this occupied space to the maximum. Moreover, the footprint can be reduced by means of counter footprint, for instance, new green zones and garden areas that form part of the property where the cement plants are located and which also serve as a screen of vegetation to combat contamination.

**Waste materials**: this category accounts for the 17.98% in case A, 6.54% in case B and 15.18% in case C, so they are all substantially high values, which would indicate that any influence exerted on them would lead to considerable savings in the total footprint, what was demonstrated by the achieved reduction of the waste material footprint (33.52%) which the use of BAT technologies got.

(2) Comparing the results A and C, it can be said that while the best available techniques are a complete solution, they allowed an important reduction of 213,276.9 $tCO_2$/year (21.25% of total emissions, from 1,003,555.2 $tCO_2$/year in case A to 790,278.3 $tCO_2$/year in case C). Summarized in Table 5 we can see the reductions depending on the consumption category.

| CONSUMPTION CATEGORIES | Reduction (%) |
| --- | --- |
| Footprint of direct emissions | 19.45 |
| Footprint of indirect emissions | 13.35 |
| Footprint of materials | 0.63 |
| Footprint of services and contracts | -0.24 |
| Agricultural and fishing footprint | 0.00 |
| Forestry footprint | 0.13 |
| Water footprint | 70.90 |
| Soil use footprint | 2.41 |
| Footprint from wastes | 33.52 |
| Soil use counter footprint | 0.00 |

**Table 5.** Reduction of the percentage of the footprint according to consumption categories, thanks to the use of BATs

On the one hand, it is clear, even at first glance, how important are these techniques, above all, as far as the consumption of fuel, energy and water as well as waste emissions (particularly solid particles) are concern. On the other hand, they do not reduce other considerable footprints, such as the footprint of materials or that of services and contracts which even got worse. This singular effect is explained by the fact that the BAT need tend to involve more complicated processes, the human factor of control, security, projects and planning, etc. become more necessary, and this results in an increase of the footprint.

(3) As mentioned, case B does not include some important processes, such as pre-heating, burning in the kiln or cooling of the clinker. This way, it presents less consumption of fuel and electricity by saving on these high energy consumption operations. Moreover, it also has a lower demand for soil (since these are smaller plants) and a smaller quantity of wastes (for instance, solid particles present in the exhaust gases from combustion in the clinker kiln). However, it presents a greater consumption of raw materials since imported clinker must be purchased and transported from distant points of manufacture (generally from China). So the smaller footprint of this option is only due to that fact that the intrinsic footprint included in the imported clinker comes only from the energy footprint and not the total footprint.

Methodologies like MC3 allow for the inclusion of the intrinsic footprint of the materials consumed by means of their energy intensity or their embodied energy, but there are other factors such as the consumption of water needed to manufacture them, services and contracts involved in their manufacture, the demand for soil, forestry resources, etc. that the methodology is unable to process [21] and for this reason a slightly lower value was obtained. In fact if the plant that manufactures the clinker in China were similar to that plant, for example in case A, the final footprint that should be obtained in case B, would be the one pertaining to case A, along with the added footprint of transporting the clinker from China (since in case A this process is not carried out).

With this, what we are attempting to clarify is that by building grinding plants instead of integral plants, we are far from solving the problem of $CO_2$ emissions, instead we are transferring the problem from one country to another (generally to the less developed countries), for reasons related to the cost of energy, labor, etc. Therefore, if we analyze the situation in global terms, the problem is aggravated even more, since the technologies of these countries are generally less efficient and the long chains of merchandise supply and transport become a necessity.

## 4. Conclusion

Like it or not, our planet is finite and a finite system is incompatible with an economic subsystem whose paradigm is based on continuous and unlimited growth. Somehow we have to reconcile growth and sustainability, and to do so, our companies need to access transparent and comparable information to be able to make the best decisions so as not to compromise either their growth or the impact on the planetary ecosystem.

In answer to this, Carbonfeel has been designed with a core mission: to organize information and knowledge on the carbon footprint, making it universally useful and accessible to all society. In short, the point is to provide companies with the best available techniques for calculation and exchange of information within the processes of inventory, management, reduction and offsetting of GHG emissions generated by their own activities.

This information will allow companies to participate actively in improving their behavior, without having any effect on their business. Quite the contrary; their activities will start to focus on production patterns based on eco-efficiency and eco-design, and therefore lead to a reduction in costs. Moreover, customers will recognize a continuous improvement effort based on a credible label supported by many different certifiers, consultants, companies, associations, universities and others.

Carbonfeel starts out from a methodological basis proposed by the Compound Method based on Financial Accounts (MC3), inherited from the ecological footprint concept that has been extended worldwide by its creators William Rees and Mathis Wackernagel. The project takes advantage of other emerging methodological trends such as GHG Protocol, PAS 2050 or ISO 14064 standards and the future ISO 14067 and 14069, in order to get an approximation of the real calculation.

Supporting an integrated approach, the incorporation of information technologies makes Carbonfeel an innovative project that has burst into the market to completely change the focus of the classic studies of life cycle analysis, whose drawbacks in cost and study time had already been reported by different analysts.

The great paradox of the Carbonfeel method is that companies get a carbon footprint at the corporate level and the life cycle of all products and services without any restriction on the scopes, with the information provided in great detail. Moreover, this information is more extensive and of a higher quality as it is based on primary data (real footprint of its suppliers), and all at a cost and a time frame fully accessible to any corporation.

The telematic assembly technique provides an entire life cycle, where each corporation analyzes its own emissions (scope 1 and land use) on an autonomous basis for calculating the indirect footprint or inherited from its suppliers by the telematic assembly. The company stops worrying about the tracking of emissions that are out of view (scope 3 in the approach to corporate and upstream approach to product based on LCAs) and focuses exclusively on the part of their responsibility, the direct emissions and the organization's land use. Therefore, time of calculation is drastically reduced, making it assumable to all the business.

The 'green coin' can become a reality if we pay attention to the technological factor faced by the carbon footprint as a problem of information exchange.

A Carbonfeel project offers companies a real environmental accounting method based on a universal indicator such as the carbon footprint, which analyzes the corporation and each of the products and services generated.

Finally, the application of the organization-product-based-life-cycle assessment methodology (MC3) to three types of cement facilities in Spain (case A pertaining to a conventional integral plant, case B which refers to a grinding plant and case C, an integral plant which has been subject to the best available techniques BAT) shows that if we compare results A and C, the best available techniques allow an important reduction of 213,276.9 $tCO_2$/year (21.25% of total emissions, from 1,003,555.2 $tCO_2$/year in case A to 790,278.3 $tCO_2$/year in case C).

## Author details

Juan Cagiao Villar and Breixo Gómez Meijide
*Department of Mathematical Methods and Representation, Civil Engineering School, University of A Coruña, A Coruña, Spain*

Sebastián Labella Hidalgo
*Atos Consulting and Technology Services. Atos Spain S.A. Barcelona. Spain*

Adolfo Carballo Penela
*Department of Business Management and Commerce, University of Santiago de Compostela, Santiago de Compostela, Spain*

## 5. References

[1] Doménech, J.L. (2004) Huella ecológica portuaria y desarrollo sostenible. Puertos 2004; 114: 26-31.

[2] Doménech, J.L. (2007) Huella ecológica y desarrollo sostenible. Madrid: AENOR Ediciones. 398 p.

[3] Carballo-Penela A., (2010) Ecoetiquetado de bienes y servicios para un desarrollo sostenible. Madrid: AENOR Ediciones. 360 p.

[4] Carballo-Penela A., Doménech J.L. (2010) Managing the carbon footprint of products: the contribution of the method composed of financial statements (MC3). International Journal of Life Cycle Assessment. 15: 962–969.

[5] Wackernagel M., Dholakia R., Deumling D., Richardson D. Redefining Progress, Assess your Household's Ecological Footprint 2.0, March 2000. [cited 5 September 2006]. Available: http://greatchange.org/ng-footprint-ef_household_evaluation.xls.

[6] Carballo-Penela A., Mateo-Mantecón I., Doménech, J.L., Coto-Millán P. (2012) From the motorways of the sea to the green corridors' carbon footprint: the case of a port in Spain. Journal of Environmental Planning and Management (in press). Available: http://www.tandfonline.com/action/showAxaArticles?journalCode=cjep20

[7] Carballo-Penela A., García-Negro MC, Doménech J.L. (2009) A methodological proposal for the corporate carbon footprint: an application to a wine producer company in Galicia (Spain). Sustainability Journal. 1: 302-318.

[8] European Comission (2007a Wheel-to-wheels Analysis of Future Automotive Fuels and Powertrains in the European Context. Versión 2c, March 2007. Available: http://www. ies. jrc.ec.europa.eu/wtw.html

[9] European Comission (2007b) Libro verde: adaptación al cambio climático en Europa: opciones de actuación para la UE. COM (2007) 354 final, 29-06-2007. Available: http://eur-lex.europa.eu/LexUriServ/LexUriServ.do?uri=COM:2007:0354:FIN:ES:PDF

[10] European Comission (2007c) Creación de una alianza mundial para hacer frente al cambio climático entre la UE y los países en desarrollo pobre más vulnerables al cambio climático.COM (2007) 540 final, 18-09-2007. Available: http://eur-lex.europa.eu/ LexUriServ/LexUriServ.do?uri=COM:2007:0540:FIN:ES:PDF

[11] Simmons C, Lewis K, Barrett J (2006) Two feet-two approaches: a component-based model of ecological footprinting. Ecological Economics. 32: 375-380.

[12] Wackernagel M. The Ecological footprint of Italia: calculation spreadsheet. USA: ICLEI; 1998 [cited 30 June 2005]. Available: http://www.iclei.org/ICLEI/ef-ita.xls

[13] International Panel on Climate Change (IPCC). IPCC Fourth Assessment Report (AR4) Changes in Atmospheric Constituents and in Radiative Forcing. UK: IPCC; 2007 [cited 13 October 2008]. Available: http://ipcc-wg1.ucar.edu/wg1/Report/AR4WG1_Print_Ch02.pdf

[14] European Comission (2008a) La lucha contra el cambio climático. La Unión Europea lidera el camino. Available: http://ec.europa.eu/publications/booklets/move/75/es.pdf .

[15] European Comission (2008b) Description and detailed energy and GHG balance of individual pathways. Available: http://www.ies.jrc.ec.europa.eu/wtw.html

[16] Wiedmann T, Lenzen M (2009) Unravelling the impacts of supply chains. A new Triple-Bottom-Line Accounting Approach. In: Schaltegger S, Bennett M, Burrit R, Jasch C, Editors. Environmental Management Accounting for Cleaner Production. Amsterdam: Springer Netherlands. pp. 65-90

[17] European Comission (2006) Making product information work for the environment. Brussels: Final Report of the Integrated Product Policy Working Group on Product Information;2006[cited3April2009].Available:http://ec.europa.eu/environment/ipp/pdf/2 0070115_report.pdf

[18] Global Footprint Network (GFN). Ecological footprint standards 2006. Oakland: Global Footprint Network; 2006.. Available: http://www.footprintnetwork.org

[19] Global Footprint Network (GFN). Ecological Footprint Standards 2009. Oakland: Global Footprint Network; 2009 [cited 11 July 2009]. Available: http://www.footprintnetwork. org

[20] Humphreys, K., Mahasenan, M., 2002. Towards a Sustainable Cement Industry – Substudy 8: Climate Change. World Business Council for Sustainable Developrment: Cement Sustainability Initiative.

[21] Cagiao, J., Gómez, B., Doménech, J.L., Gutiérrez, S., Gutiérrez, H.  (2011) Calculation of the corporate carbon footprint of the cement industry by the application of MC3 methodology. Ecological Indicators, 11 (2011): 1526-1540.

# Chapter 3

# Study of Impacts of Global Warming on Climate Change: Rise in Sea Level and Disaster Frequency

Bharat Raj Singh and Onkar Singh

Additional information is available at the end of the chapter

## 1. Introduction

Global warming and climate change refer to an increase in average global temperatures. Natural events and human activities are believed to be main contributors to such increases in average global temperatures. The climate change, caused by rising emissions of carbon dioxide from vehicles, factories and power stations, will not only effects the atmosphere and the sea but also will alter the geology of the Earth. Emissions of carbon dioxide due to our use of fossil energy will change the climate and the temperature is estimated to increase by 2 to 6° Celsius within year 2100, which is a tremendous increase from our current average temperature of 1.7° Celsius as predicted by IPCC. This may cause huge changes to our civilization, both positive and negative, but the total impact on our society is currently very uncertain. Forecasts indicate that major storms could devastate New York City in next decade whereas Gulf countries will get affected badly well before.

Global warming primarily caused by increases in "greenhouse" gases such as Carbon Dioxide ($CO_2$), Nitrous oxide ($NO_x$), Sulphur dioxide ($SO_2$), Hydrogen etc., . A warming planet thus leads to climate changes which can adversely affect weather in different ways. Some of the prominent indicators for a global warming are detailed below:

i.   Temperature over land
ii.  Snow cover on Hills
iii. Glaciers on Hills
iv.  Ocean Heat content
v.   Sea Ice
vi.  Sea level
vii. Sea surface temperature

viii. Temperature Over Ocean

ix.   Humidity

x.    Tropospheric Temperature

Past decade, according to Scientists in 48 Countries, it was recorded warmest time phase during meeting of *National Oceanic and Atmospheric Administration* (NOAA), on July 28, 2010. Although since decades, scientists and environmentalists have been warning that the *way* we are using Earth's resources is not sustainable. Alternative technologies have been called for repeatedly, seemingly falling upon deaf ears or, cynically, upon those who don't want to make substantial changes as it challenge their bottom line and reduces their current profits.

Global warming in today's scenario is threat to the survival of mankind. In 1956, an US based Chief consultant and oil geologist *Marion King Hubert, (1956)* predicted that if oil is consumed with high rate, US oil production may peak in 1970 and thereafter it will decline. He also described that other countries may attain peak oil day within 20-30 years and many more may suffer with oil crises within 40 years, when oil wells are going to dry. He illustrated the projection with a bell shaped *Hubert Curve* based on the availability and its consumptions of the fossil fuel. Large fields are discovered first, small ones later. After exploration and initial growth in output, production plateaus and eventually declines to zero.

Crude oil, coal and gas are the main resources for world energy supply. The size of fossil fuel reserves and the dilemma that when non-renewable energy will be diminished, is a fundamental and doubtful question that needs to be answered. A new formula for calculating, when fossil fuel reserves are likely to be depleted, is presented along with an econometrics model to demonstrate the relationship between fossil fuel reserves and some main variables (Shahriar Shafiee et.al. 2009). The new formula is modified from the Klass model and thus assumes a continuous compound rate and computes fossil fuel reserve depletion times for oil, coal and gas of approximately 35, 107 and 37 years, respectively. This means that coal reserves are available up to 2112, and will be the only fossil fuel remaining after 2042.

In India, vehicular pollution is estimated to have increased eight times over the last two decades. This source alone is estimated to contribute about 70 per cent to the total air pollution. With 243.3 million tons of carbon released from the consumption and combustion of fossil fuels in 1999, India is ranked fifth in the world behind the U.S., China, Russia and Japan. India's contribution to world carbon emissions is expected to increase in the coming years due to the rapid pace of urbanization, shift from non-commercial to commercial fuels, increased vehicular usage and continued use of older and more inefficient coal-fired and fuel power-plants (*Singh, BR, et al., 2010*).

Thus, peak oil year may be the turning point for mankind which may lead to the end of 100 year of easy growth, if self-sufficiently and sustainability of energy is not maintained on priority. This chapter describes the efforts being made to explore non-conventional energy resources such as: solar energy, wind energy, bio-mass and bio-gas, hydrogen, bio-diesel which may help for the sustainable fossil fuel reserves and reduce the tail pipe emission and other pollutants like: $CO_2$, $NO_x$ etc.. The special emphasis is also given for the storage of energy such as compressed air stored from solar, wind and or other resources like: climatic energy to maintain energy sustainability of $21^{st}$ century. This may also leads to environmentally and ecologically better future.

## 2. Weather watch: Byron's view of the glaciers

In September 1816, Lord Byron set off from Geneva with his friend Hob house, and kept a journal for his half-sister Augusta. Lodged at the Curate's, set out to see the Valley; heard an Avalanche fall, like thunder; saw Glacier – enormous. Storm came on, thunder, lightning, hail; all in perfection, and beautiful (**Fig 1**).

**Figure 1.** Byron described glaciers in Geneva as "neither mist nor water" in September 1816. (Photograph: John Mcconnico/AP)

He said that he was on horseback; Guide wanted to carry his cane; he was going to give it to him, when he recollected that it was a Swordstick, and he thought lightning might be attracted towards him; kept it himself; a good deal encumbered with it, and his cloak, as it was too heavy for a whip, and the horse was stupid, and stood still with every other peal," he records in Byron: Selections from Poetry, Letters & Journals (Nonesuch Press.)

Got in, not very wet; the cloak being staunch, H wet though. H took refuge in a cottage; sent man, umbrella and cloak (from the Curate's when he arrived) after him. He sees a torrent like the tail of a white horse streaming in the wind, such as it might be conceived would be that of the 'pale horse' on which Death is mounted in the Apocalypse. It is neither mist nor water, but something between both; its immense height (nine hundred feet) gives it a wave, a curve, a spreading here, a condensation there, wonderful and indescribable. He looks again the next day: the Sun upon it forming a *rainbow* of the lower part of all colours, but principally purple and gold; the bow moving as you move; he never saw anything like this.

## 3. Global warming issues

### 3.1. Effect of global warming

With increases in the Earth's global mean temperature i.e., global warming, the various effects on climate change pose risks that increases. The IPCC (*2001d and 2007d*) has organized many of these risks into five "reasons for concern:

- Threats to endangered species and unique systems,
- Damages from extreme climate events,
- Effects that fall most heavily on developing countries and
- The poor within countries, global aggregate impacts (i.e., various measurements of total social, economic and ecological impacts), and large-scale high-impact events.

The effects, or impacts, of climate change may be physical, ecological, social or economic. Evidence of observed climate change includes the instrumental temperature record, rising sea levels, and decreased snow cover in the Northern Hemisphere. According to the Intergovernmental Panel on Climate Change (*IPCC, 2007a:10*), "[most] of the observed increase in global average temperatures since the mid-20th century is *very likely* due to the observed increase in [human greenhouse gas] concentrations". It is predicted that future climate changes will include further global warming (i.e., an upward trend in global mean temperature), sea level rise, and a probable increase in the frequency of some extreme weather events. United Nations Framework Convention on Climate Change has agreed to implement policies designed to reduce their emissions of greenhouse gases.

### 3.2. Effect of climate change

The phrase climate change is used to describe a change in the climate, measured in terms of its statistical properties, e.g., the global mean surface temperature. In this context, climate is taken to mean the average weather. Climate can change over period of time ranging from months to thousands or millions of years. The classical time period is 30 years, as defined by the World Meteorological Organization. The climate change referred to may be due to natural causes, e.g., changes in the sun's output, or due to human activities, e.g., changing the composition of the atmosphere. Any human-induced changes in climate will occur against the background of natural climatic variations.

Climate change reflects a change in the energy balance of the climate system, i.e. changes the relative balance between incoming solar radiation and outgoing infrared radiation from Earth. When this balance changes it is called "radiative forcing", and the calculation and measurement of radiative forcing is one aspect of the science of climatology. The processes that cause such changes are called "forcing mechanisms". Forcing mechanisms can be either "internal" or "external". Internal forcing mechanisms are natural processes within the climate system itself, e.g., the meridional turnover. External forcing mechanisms can be either natural (e.g., changes in solar output) or anthropogenic (e.g., increased emissions of greenhouse gases).

Whether the initial forcing mechanism is internal or external, the response of the climate system might be fast (e.g., a sudden cooling due to airborne volcanic ash reflecting sunlight), slow (e.g. thermal expansion of warming ocean water), or a combination (e.g., sudden loss of albedo in the arctic ocean as sea ice melts, followed by more gradual thermal expansion of the water). Therefore, the climate system can respond abruptly, but the full response to forcing mechanisms might not be fully developed for centuries or even longer.

The most general definition of *climate change* is a change in the statistical properties of the climate system when considered over long periods of time, regardless of cause, *whereas* Global warming" refers to the change in the Earth's global average surface temperature. Measurements show a global temperature increase of 1.4 °F (0.78 °C) between the years 1900 and 2005. Global warming is closely associated with a broad spectrum of other climate changes, such as:

- Increases in the frequency of intense rainfall,
- Decreases in snow cover and sea ice,
- More frequent and intense heat waves,
- Rising sea levels, and
- Widespread ocean acidification.

## 3.2.1. Risk of intense rainfall

There are two studied made here to elaborate the risk of intense rain fall one by United States and other one by United Kingdom. They have warned that these risks are due to extreme climate change, thus we have to curb the global warming issues in phases. The summaries of study are given below:

i.   *Two 500-Year Floods in Just 15 Years:* In the United States, The Great Flood of 1993—devastating communities along the Mississippi River and its tributaries in nine Midwestern states—was one of the most costly disasters. Thousands of Americans were displaced from their homes and forced to leave their lives behind, hundreds of levees failed, and damages soared to an estimated $12 to 16 billion. A mere 15 years later, history is repeating itself in the Midwest as the rainswollen Cedar, Illinois, Missouri and Mississippi Rivers and their tributaries top their banks and levees, leaving hundreds of thousands of people displaced. With rainfall in May-June 2008 about two to three times greater than the long-term average, soybean planting is behind schedule and some crops may have to be replanted. This remarkably quick return of such severe flooding is not anticipated by currently used out-of-date methodologies, but is what we should expect as global warming leads to more frequent and intense severe storms. Inadequate floodplain management is also responsible for the extent of damages from both floods, especially over-reliance on levees and the false sense of security they provide to those who live behind them. About 28 percent of the total new development in the seven states over the past 15 years has been in areas within the 1993 flood events. The National Wildlife Federation says that to limit the magnitude of changes to the climate and the impacts on communities and wildlife, we must curb global warming pollution. The National Wildlife Federation recommends that policy makers, industry, and individuals take steps to reduce global warming pollution from today's levels by 80 percent by 2050. That's a reduction of 20 percent per decade or just 2 percent per year. Science tells us that this is the only way to hold warming in the next century to no more than 2°F. This target is achievable with technologies either available or under development, but we need to start taking action now to avoid the worst impacts (See: www.nwf.org/globalwarming).

ii.   *Extreme rainfall and flood risk in the UK:* Multi-day rainfall events are an important cause of recent severe flooding in the UK and any change in the magnitude of such events may have severe impacts upon urban structures such as dams, urban drainage systems and flood defences and cause failures to occur. Regional pooling of 1-, 2-, 5- and 10-day annual maxima for 1961 to 2000 from 204 sites across the UK is used in a standard regional frequency analysis to produce GEV growth curves for long return-period rainfall events for each of nine defined climatological regions. Temporal changes in 1-, 2-, 5- and 10-day annual maxima are examined with L-moments using both a 10-year moving window and fixed decades from 1961-70, 1971-80, 1981-90 and 1991-2000. A bootstrap technique is then used to assess uncertainty in the fitted decadal growth curves and to identify significant trends in both distribution parameters and quantile estimates.

There has been a two-part change in extreme rainfall event occurrence across the UK from 1961-2000. Little change is observed at 1- and 2-day duration, but significant decadal level changes are seen in 5- and 10-day events in many regions. In the south of the UK, growth curves have flattened and 5- and 10-day annual maxima have decreased during the 1990s. However, in the north, the 10-day growth curve has steepened and annual maxima have risen during the 1990s. This is particularly evident in Scotland. The 50-year event in Scotland during 1961-1990 has become an 8-, 11- and 25-year event in the Eastern, Southern and Northern Scotland pooling regions respectively during the 1990s. In northern England the average recurrence interval has also halved. This may have severe implications for design and planning practices in flood control.

Increasing flood risk is now recognised as the most important sectoral threat from climate change in most parts of the world, with recent repeated severe flooding in the UK and Europe causing major loss of property and life, and causing the insurance industry to threaten the withdrawal of flood insurance cover from millions of UK households. This has prompted public debate on the apparent increased frequency of extremes and focussed attention in particular on perceived increases in rainfall intensities. Climate model integrations predict increases in both the frequency and intensity of heavy rainfall in the high latitudes under enhanced greenhouse conditions. These projections are consistent with recent increases in rainfall intensity seen in the UK and worldwide.

## 3.2.2. Decreases in snow cover and sea ice

*Decreasing snow cover and land-ice extent continue to be positively correlated with increasing land-surface temperatures:* Satellite data show that it is quite likely to have been decreases of about 10% in the extent of snow cover since the late 1960s. There is a highly significant correlation between increases in Northern Hemisphere land temperatures and the decreases. There is now ample evidence to support a major retreat of alpine and continental glaciers in response to 20th century warming. In a few maritime regions, increases in precipitation due to regional atmospheric circulation changes have overshadowed increases in temperature in the past two decades, and glaciers have re-advanced. Over the past 100 to 150 years, ground-based observations show that there is possibility of a reduction of about two weeks in the annual duration of lake and river ice in the mid- to high latitudes of the Northern Hemisphere (**Fig.2**).

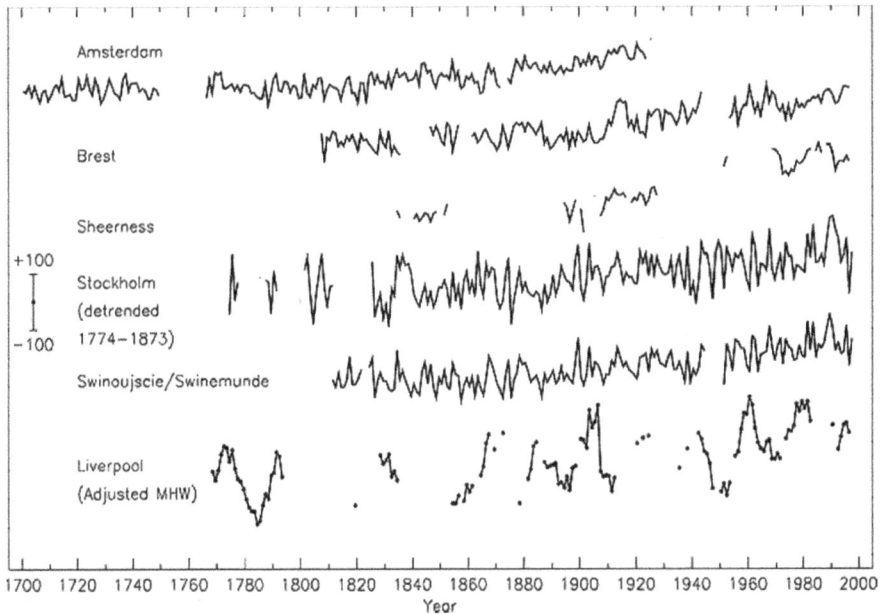

**Figure 2.** Time-series of relative sea level for the past 300 years from Northern Europe: Amsterdam, Netherlands; Brest, France; Sheerness, UK; Stockholm, Sweden (detrended over the period 1774 to 1873 to remove to first order the contribution of post-glacial rebound); Swinoujscie, Poland (formerly Swinemunde, Germany); and Liverpool, UK. Data for the latter are of "Adjusted Mean High Water" rather than Mean Sea Level and include a nodal (18.6 year) term. The scale bar indicates ±100 mm.

*Northern Hemisphere sea-ice amounts are decreasing, but no significant trends in Antarctic sea-ice extent are apparent:* A retreat of sea-ice extent in the Arctic spring and summer of 10 to 15% since the 1950s is consistent with an increase in spring temperatures and, to a lesser extent, summer temperatures in the high latitudes. There is little indication of reduced Arctic sea-ice extent during winter when temperatures have increased in the surrounding region. By contrast, there is no readily apparent relationship between decadal changes of Antarctic temperatures and sea-ice extent since 1973. After an initial decrease in the mid-1970s, Antarctic sea-ice extent has remained stable, or even slightly increased.

*New data indicate that there likely has been an approximately 40% decline in Arctic sea-ice thickness in late summer to early autumn between the period of 1958 to 1976 and the mid-1990s, and a substantially smaller decline in winter*: The relatively short record length and incomplete sampling limit the interpretation of these data. Interannual variability and inter-decadal variability could be influencing these changes.

## 3.2.3. More frequent and intense heat waves

A recent study shows that an increase in heat-absorbing greenhouse gases intensifies an unusual atmospheric circulation pattern already observed during heat waves in Europe and North America. As the pattern becomes more pronounced, severe heat waves occur in the Mediterranean region and the southern and western United States. Other parts of France, Germany and the Balkans also become more susceptible to severe heat waves. "Extreme weather events will have some of the most severe impacts on human society as climate changes, "says Meehl.

Heat waves can kill more people in a shorter time than almost any other climate event. According to records, 739 people died as a result of Chicago's July, 1995, heat wave. Fifteen thousand Parisians are estimated to have died from heat in August, 2003, along with thousands of farm animals. For the study, Meehl and Tebaldi compared present (1961-1990) and future (2080-2099) decades to determine how greenhouse gases and sulfate aerosols might affect future climate in Europe and the United States, focusing on Paris and Chicago. They assumed little policy intervention to slow the buildup of greenhouse gases. During the Paris and Chicago heat waves, atmospheric pressure rose to values higher than usual over Lake Michigan and Paris, producing clear skies and prolonged heat. In the model, atmospheric pressure increases even more during heat waves in both regions as carbon dioxide accumulates in the atmosphere.

Heat wave is based on the concept of exceeding specific thresholds, thus allowing analyses of heat wave duration and frequency. Three criteria were used to define heat waves in this way, which relied on two location-specific thresholds for maximum temperatures. Threshold 1 (T1) was defined as the 97.5th percentile of the distribution of maximum temperatures in the observations and in the simulated present-day climate (seasonal climatology at the given location), and T2 was defined as the 81st percentile. A heat wave was then defined as the longest period of consecutive days satisfying the following three conditions:

i.   The daily maximum temperature must be above T1 for at least 3 days,
ii.  The average daily maximum temperature must be above T1 for the entire period, and
iii. The daily maximum temperature must be above T2 for every day of the entire period.

Because the Chicago heat wave of 1995 and the Paris heat wave of 2003 had particularly severe impacts, we chose grid points from the model that were close to those two locations to illustrate heat wave characteristics. This choice was subjective and illustrative given that there are, of course, other well-known heat waves from other locations. Also, we are not suggesting that a model grid point is similar to a particular weather station; we picked these grid points because they represent heat wave conditions for regions representative of Illinois and France in the model, and therefore they can help identify processes that contribute to changes in heat waves in the future climate in those regions. We chose comparable grid points from the National Centers for Environmental Prediction (NCEP)/NCAR reanalyses that used assimilated observational data for comparison to the model results.

Heat waves in Chicago, Paris, and elsewhere in North America and Europe will become more intense, more frequent and longer lasting in the 21st century, according to a new modeling study by two scientists at the National Center for Atmospheric Research (NCAR) in Boulder, Colo. In the United States, heat waves will become most severe in the West and South. The findings appear in the August 13(2004) issue of the journal Science. Gerald Meehl and Claudia Tebaldi, both of NCAR, examined Earth's future climate using the Parallel Climate Model, developed by NCAR and the U.S. Department of Energy (DOE).

During the 1995 Chicago heat wave, the most severe health impacts resulted from the lack of cooling relief several nights in a row, according to health experts. In the model, the western and southern United States and the Mediterranean region of Europe experience a rise in nighttime minimum temperatures of more than 3 degrees Celsius (5.4 degrees Fahrenheit) three nights in a row. They will occur more often: The average number of heat waves in the Chicago area increases in the coming century by 25 percent, from "Heat Waves of the 21st Century: More Intense, More Frequent and Longer Lasting." (Source: PHYSorg.com. 12 Aug 2004, http://phys.org/news806.html Page 1/21.66 per year to 2.08).

In Paris, the average number increases 31percent, from 1.64 per year to 2.15. They will last longer: Chicago's present heat waves last from 5.39 to 8.85 days; future events increase to between 8.5 and 9.24 days. In Paris, present-day heat waves persist from 8.33 to 12.69 days; they stretch to between 11.39 and 17.04 days in future decades.(Source: National Science Foundation)

## 3.2.4. Observed changes in sea level

Based on tide gauge data, the rate of global mean sea level rise during the 20th century is in the range 1.0 to 2.0 mm/yr, with a central value of 1.5 mm/yr (the central value should not be interpreted as a best estimate.

i.   *The causes for change of the sea level:* At the shoreline it is determined by many factors in the global environment that operate on a great range of time-scales, from hours (tidal) to millions of years (ocean basin changes due to tectonics and sedimentation). On the time-scale of decades to centuries, some of the largest influences on the average levels of the sea are linked to climate and climate change processes.

Firstly, as ocean water warms, it expands. On the basis of observations of ocean temperatures and model results, thermal expansion is believed to be one of the major contributors to historical sea level changes. Further, thermal expansion is expected to contribute the largest component to sea level rise over the next hundred years. Deep ocean temperatures change only slowly; therefore, thermal expansion would continue for many centuries even if the atmospheric concentrations of greenhouse gases were to stabilise.

The amount of warming and the depth of water affected vary with location. In addition, warmer water expands more than colder water for a given change in temperature. The geographical distribution of sea level change results from the geographical variation of thermal expansion, changes in salinity, winds, and ocean circulation. The range of regional variation is substantial compared with the global average sea level rise.

ii.   *Rise in sea Level:* Sea level also changes when the mass of water in the ocean increases or decreases. This occurs when ocean water is exchanged with the water stored on land. The major land store is the water frozen in glaciers or ice sheets. Indeed, the main reason for the lower sea level during the last glacial period was the amount of water stored in the large extension of the ice sheets on the continents of the Northern Hemisphere. After thermal expansion, the melting of mountain glaciers and ice caps is expected to make the largest contribution to the rise of sea level over the next hundred years. These glaciers and ice caps make up only a few per cent of the world's land-ice area, but they are more sensitive to climate change than the larger ice sheets in Greenland and Antarctica, because the ice sheets are in colder climates with low precipitation and low melting rates. Consequently, the large ice sheets are expected to make only a small net contribution to sea level change in the coming decades.

## 3.2.5. Widespread ocean acidification

*A new study says the seas are acidifying ten times faster today than 55 million years ago when a mass extinction of marine species occurred. And, the study concludes, current changes in ocean chemistry due to the burning of fossil fuels may portend a new wave of die-offs.* In other words, the vast clouds of shelled creatures in the deep oceans had virtually disappeared. Many scientists now agree that this change was caused by a drastic drop of the ocean's pH level. The seawater became so corrosive that it ate away at the shells, along with other species with calcium carbonate in their bodies. It took hundreds of thousands of years for the oceans to recover from this crisis, and for the sea floor to turn from red back to white. The clay that the crew of the *JOIDES Resolution* dredged up may be an ominous warning of what the future has in store. By spewing carbon dioxide into the air, we are now once again making the oceans more acidic.

## 3.3. Historical impacts of climate change

Approximately one millennium after the 7 Ka (32$^{nd}$ Century BCE) slowing of sea-level rise, many coastal urban centers rose to prominence around the world (Day, John W., et al. 2007). It has been hypothesized that this is correlated with the development of stable coastal environments and ecosystems and an increase in marine productivity (also related to an increase in temperatures), which would provide a food source for hierarchical urban societies.

The last written records of the Norse Greenlanders are from a 1408 marriage in the church of Hvalsey — today the best-preserved of the Norse ruins. Climate change has been associated with the historical collapse of civilizations, cities and dynasties. Notable examples of this include the Anasazi (Demenocal, P. B. 2001), Classic Maya (Hodell, David A., 1995), the Harappa, the Hittites, and Ancient Egypt (Jonathan Cowie, 2007). Other, smaller communities such as the Viking settlement of Greenland (transl. with introd. by Magnus Magnusson, 1983), have also suffered collapse with climate change being a suggested contributory factor (Diamond, Jared, 2005).

There are two proposed methods of Classic Maya collapse: environmental and non-environmental. The environmental approach uses paleoclimatic evidence to show that movements in the intertropical convergence zone likely caused severe, extended droughts during a few time periods at the end of the archaeological record for the classic Maya (Haug, Gh, et al., 2003). The non-environmental approach suggests that the collapse could be due to increasing class tensions associated with the building of monumental architecture and the corresponding decline of agriculture (Hosler D, et al., 1977), increased disease (Santley, Robert S.,et al., 1986) and increased internal warfare (Foias, Antonia E., et al.,1997). The Harappa and Indus civilizations were affected by drought 4,500–3,500 years ago. A decline in rainfall in the Middle East and Northern India 3,800–2,500 is likely to have affected the Hittites and Ancient Egypt.

Notable periods of climate change in recorded history include the medieval warm period and the little ice age. In the case of the Norse, the medieval warm period was associated with the Norse age of exploration and arctic colonization, and the later colder periods led to the decline of those colonies (Patterson, W.P., et al., 2007). Climate change in the recent past may be detected by corresponding changes in settlement and agricultural patterns. Archaeological evidence, oral history and historical documents can offer insights into past changes in the climate. Climate change effects have been linked to the collapse of various civilizations.

## 3.4. Global warming impacts of climate change

According to different levels of future global warming, impacts of climate has been used in the IPCC's Assessment Reports on climate change (*Schneider DH, et al., 2007*). The instrumental temperature record shows global warming of around 0.6 °C over the entire 20th century (*IPCC 2007d.1*). The future level of global warming is uncertain, but a wide range of estimates (projections) have been made (*Fisher, BS et al., 2007*). The IPCC's "SRES" scenarios have been frequently used to make projections of future climate change (*Karl, 2009*). Climate models using the six SRES "marker" scenarios suggest future warming of 1.1 to 6.4 °C by the end of the 21st century (above average global temperatures over the 1980 to 1999 time period as shown in **Fig.3**) (*IPCC 2007d.3*). The projected rate of warming under these scenarios would very likely be without precedent during at least the last 10,000 years (*IPCC 2001-SPM*). The most recent warm period comparable to these projections was the mid-Pliocene, around 3 million years ago (*Stern N., 2008*). At that time, models suggest that mean global temperatures were about 2–3 °C warmer than pre-industrial temperatures (*Jansen E., et al., 2007*).

## Global Land–Ocean Temperature Index

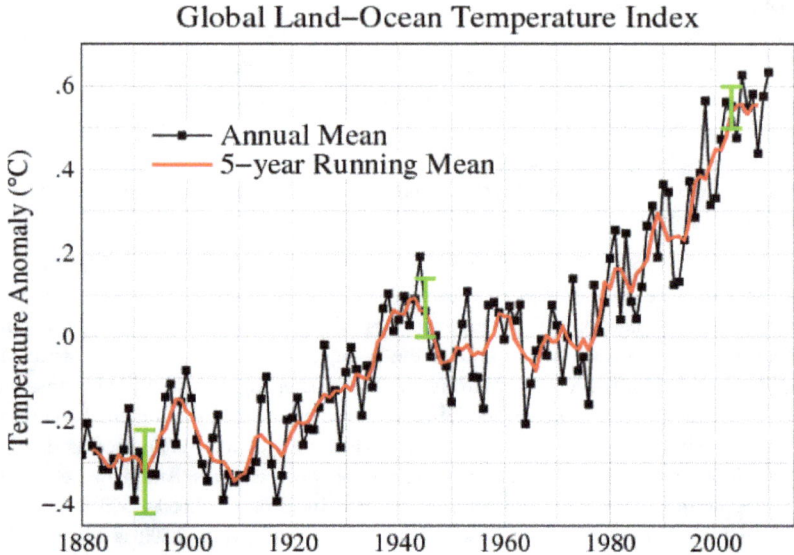

**Figure 3.** Global Land-Ocean, mean surface temperature difference from the average for 1880–2009 (Courtesy: Wikipedia.com)

The most recent report IPCC projected that during the 21st century the global surface temperature is likely to rise a further1.1 to 2.9 °C (2 to 5.2 °F) for the lowest emissions scenario used in the report and 2.4 to 6.4 °C (4.3 to 11.5 °F) for the highest (**Fig.4**).

### 3.5. Physical impacts of climate change

Working Group I's contribution to the IPCC Fourth Assessment Report, published in 2007, concluded that warming of the climate system was "unequivocal" (*Solomon S, 2007a*). This was based on the consistency of evidence across a range of observed changes, including increases in global average air and ocean temperatures, widespread melting of snow and ice, and rising global average sea level(*Solomon S, 2007b*).

Human activities have contributed to a number of the observed changes in climate (*Hegerl GC, et. al., 2007*). This contribution has principally been through the burning of fossil fuels, which has led to an increase in the concentration of GHGs in the atmosphere. This increase in GHG concentrations has caused a radiative forcing of the climate in the direction of warming. Human-induced forcing of the climate has likely to contributed to a number of observed changes, including sea level rise, changes in climate extremes (such as warm and cold days), declines in Arctic sea ice extent, and to glacier retreat (**Fig.5 & 6**).

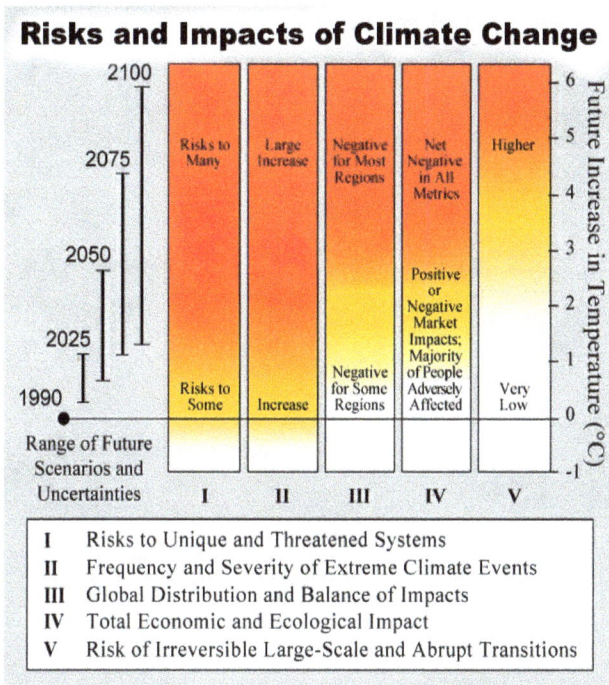

**Figure 4.** Projected Global Temperature Rise 1.1 to 6.4 °C during 21st century (Courtesy: Wikipedia.com)

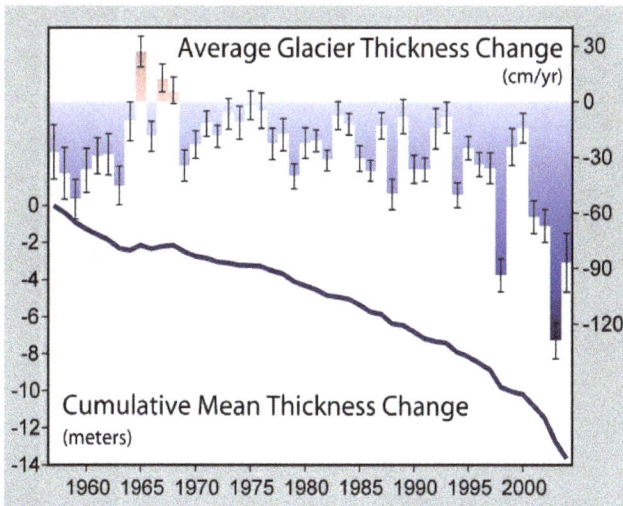

**Figure 5.** Decline in thickness of glaciers worldwide over the past half-century

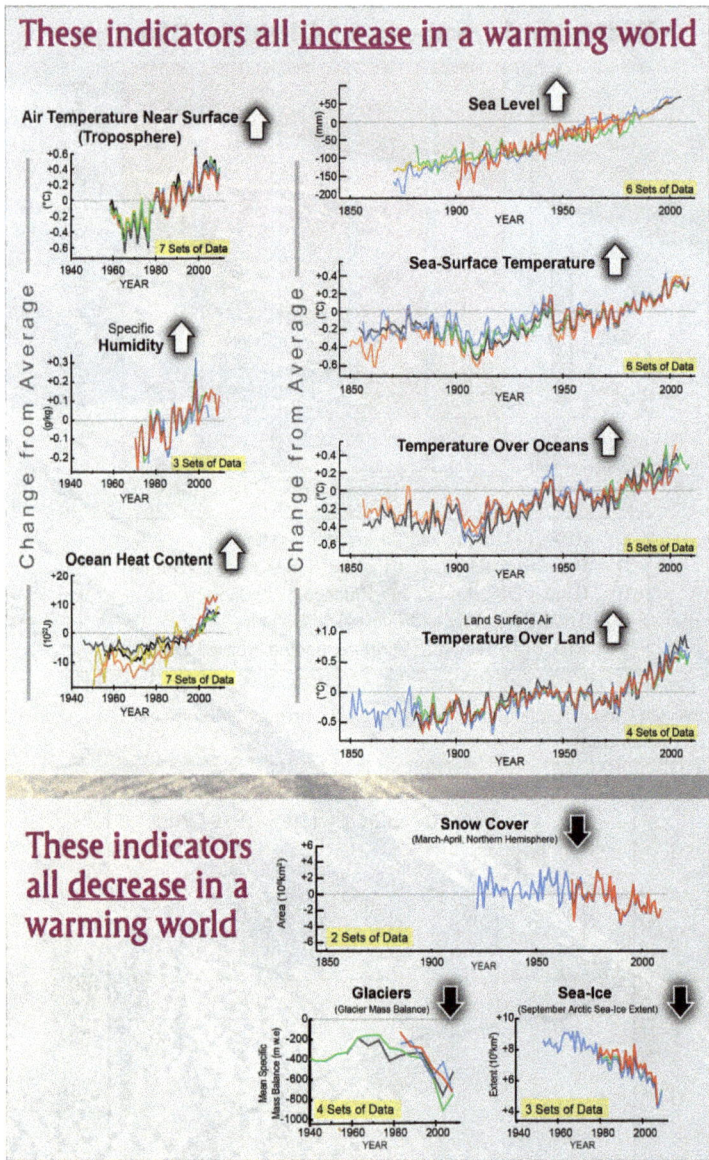

**Figure 6.** Key climate indicators that show global warming (Courtesy: Wikipedia.com)

Human-induced warming could potentially lead to some impacts that are abrupt or irreversible. The probability of warming having unforeseen consequences increases with the rate, magnitude, and duration of climate change.

## 3.6. Climate change effects on weather

Observations show that there have been changes in weather (*Le Treut H, et. al., 2007*). As climate changes, the probabilities of certain types of weather events are affected. Changes have been observed in the amount, intensity, frequency, and type of precipitation. Widespread increases in heavy precipitation have occurred, even in places where total rain amounts have decreased. IPCC (2007d) concluded that human influences had, more likely than not (greater than 50% probability, based on expert judgment), contributed to an increase in the frequency of heavy precipitation events. Projections of future changes in precipitation show overall increases in the global average, but with substantial shifts in where and how precipitation falls. Climate models tend to project increasing precipitation at high latitudes and in the tropics (e.g., the south-east monsoon region and over the tropical Pacific) and decreasing precipitation in the sub-tropics (e.g., over much of North Africa and the northern Sahara).

Evidence suggests that, since the 1970s, there have been substantial increases in the intensity and duration of tropical storms and hurricanes. Models project a general tendency for more intense but fewer storms outside the tropics.

## 3.7. Extreme weather, tropical cyclone, and list of atlantics hurricane records

Since the late 20th century, changes have been observed in the trends of some extreme weather and climate events, e.g., heat waves. Human activities have, with varying degrees of confidence, contributed to some of these observed trends. Projections for the 21st century suggest continuing changes in trends for some extreme events (**Fig.7**). Solomon *et al.* (2007), for example, projected the following likely (greater than 66% probability, based on expert judgment) changes:

- an increase in the areas affected by drought;
- increased tropical cyclone activity; and
- increased incidence of extreme high sea level (excluding tsunamis).

Projected changes in extreme events will have predominantly adverse impacts on ecosystems and human society.

### 3.7.1. Triggering earthquakes, tsunamis, avalanches and volcanic eruptions

Scientists are to outline dramatic evidence that global warming threatens the planet in a new and unexpected way – by triggering earthquakes, tsunamis, avalanches and volcanic eruptions. It is assessed that the Melting glaciers will set off avalanches, floods and mud flows in the Alps and other mountain ranges; torrential rainfall in the UK is likely to cause widespread erosion; while disappearing Greenland and Antarctic ice sheets threaten to let loose underwater landslides, triggering tsunamis that could even strike the seas around Britain.

At the same time the disappearance of ice caps will change the pressures acting on the Earth's crust and set off volcanic eruptions across the globe. Life on Earth faces a warm future – and a fiery one.

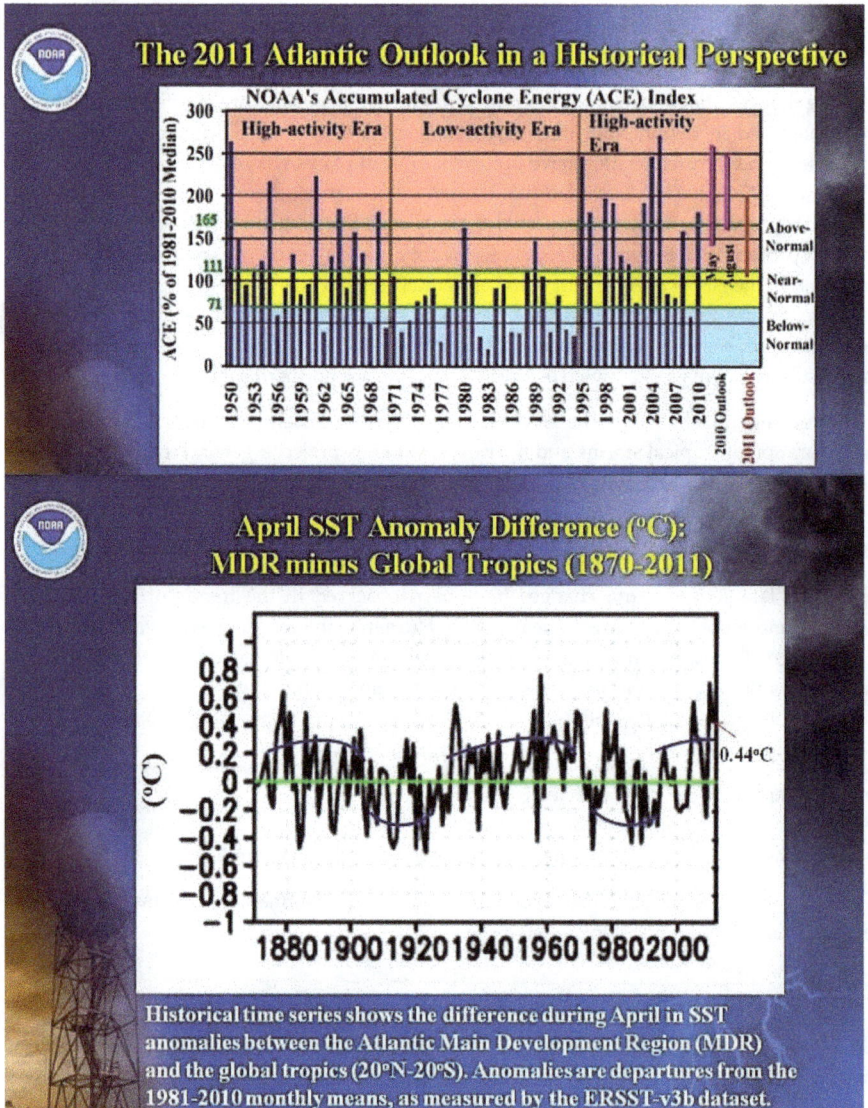

**Figure 7.** Accumulated cyclone energy in the Atlantic Ocean and the sea surface temperature difference which influences such, measured by the U.S. NOAA.

Not only are the oceans and atmosphere conspiring against us, bringing baking temperatures, more powerful storms and floods, but the crust beneath our feet seems likely to join in too, said Professor Bill McGuire, director of the Benfield Hazard Research Centre, at University College London (UCL).

Maybe the Earth is trying to tell us something, added McGuire, who is one of the organisers of UCL's Climate Forcing of Geological Hazards. Some of the key evidence will come from studies of past volcanic activity. These indicate that when ice sheets disappear the number of eruptions increases, said Professor David Pyle, of Oxford University's earth sciences department.

The last ice age came to an end between 12,000 to 15,000 years ago and the ice sheets that once covered central Europe shrank dramatically, added Pyle. The impact on the continent's geology can be measured by the jump in volcanic activity that occurred at this time.

In the Eiffel region of western Germany a huge eruption created a vast caldera, or basin-shaped crater, 12,900 years ago, for example. This has since flooded to form the Laacher See, near Koblenz. Scientists are now studying volcanic regions in Chile and Alaska – where glaciers and ice sheets are shrinking rapidly as the planet heats up – in an effort to anticipate the eruptions that might be set off.

Recently scientists from Northern Arizona University reported in the journal *Science* that temperatures in the Arctic were now higher than at any time in the past 2,000 years. Ice sheets are disappearing at a dramatic rate – and these could have other, unexpected impacts on the planet's geology.

According to Professor Mark Maslin of UCL, one is likely to be the release of the planet's methane hydrate deposits. These ice-like deposits are found on the seabed and in the permafrost regions of Siberia and the far north. These permafrost deposits are now melting and releasing their methane, said Maslin. You can see the methane bubbling out of lakes in Siberia. And that is a concern, for the impact of methane in the atmosphere is considerable. It is 25 times more powerful than carbon dioxide as a greenhouse gas.

**Figure 8.** Earthquake

A build-up of permafrost methane in the atmosphere would produce a further jump in global warming and accelerate the process of climate change. Even more worrying, however, is the impact of rising sea temperatures on the far greater reserves of methane hydrates that are found on the sea floor. It was not just the warming of the sea that was the problem, added Maslin. As the ice around Greenland and Antarctica melted, sediments would pour off land masses and cliffs would crumble, triggering underwater landslides that would break open more hydrate reserves on the sea-bed. Again there would be a jump in global warming. These are key issues that we will have to investigate over the next few years, he said.

There is also a danger of earthquakes, triggered by disintegrating glaciers, causing tsunamis off Chile, New Zealand and Newfoundland in Canada, NASA scientist Tony Song said recently (**Fig.8**). The last on this list could even send a tsunami across the Atlantic, one that might reach British shores. From other experts, it is said that the risk posed by melting ice in mountain regions, which would pose significant dangers to local people and tourists. The Alps, in particular, face a worryingly uncertain future, said Jasper Knight of Exeter University. Rock walls resting against glaciers will become unstable as the ice disappears and so set off avalanches. In addition, increasing melt-waters will trigger more floods and mud flows.

For the Alps this is a serious problem. Tourism is growing there, while the region's population is rising. Managing and protecting these people was now an issue that needed to be addressed as a matter of urgency, Knight said. "Global warming is not just a matter of warmer weather, more floods or stronger hurricanes. It is a wake-up call to Terra Firma," McGuire said.

## 3.7.2. Major storms could submerge New York city in next decade

Sea-level rise due to climate change could cripple the city in Irene-like storm scenarios, new climate report claims Irene-like storms of the future would put a third of New York City streets under water and flood many of the tunnels leading into Manhattan in under an hour because of climate change, a new state government report warns Wednesday 16th Nov' 2011 (Fig.9).

Sea level rise due to climate change would leave lower Manhattan dangerously exposed to flood surges during major storms, the report, which looks at the impact of climate change across the entire state of New York, warns. The risks and the impacts are huge, said Art deGaetano, a climate scientist at Cornell University and lead author of the ClimAID study. Clearly areas of the city that are currently inhabited will be uninhabitable with the rising of the sea.

Factor in storm surges, and the scenario becomes even more frightening, he said. Subway tunnels get affected, airports - both LaGuardia and Kennedy sit right at sea level - and when you are talking about the lowest areas of the city you are talking about the business districts. The report, commissioned by the New York State Energy Research and Development

Authority, said the effects of sea level rise and changing weather patterns would be felt as early as the next decade.

**Figure 9.** The Manhattan skyline as Hurricane Irene approached

By the mid-2020s, sea level rise around Manhattan and Long Island could be up to 10 inches, assuming the rapid melting of polar sea ice continues. By 2050, sea-rise could reach 2.5ft and more than 4.5ft by 2080 under the same conditions. In such a scenario, many of the tunnels - subway, highway, and rail - crossing into the Bronx beneath the Harlem River, and under the East River would be flooded within the hour, the report said. Some transport systems could be out of operation for up to a month.

The report, which was two years in the making, was intended to help the New York state government take steps now to get people out of harm's way - and factor climate change into long-term planning to protect transport, water and sewage systems. New York mayor Michael Bloomberg was so concerned that he went on to commission an even more detailed study of the city after receiving early briefings on the report. That makes him an outlier among his fellow Republicans, who blocked funds for creating a new climate service in budget negotiations in Congress this week.

DeGaetano said climate change would force governments to begin rethinking infrastructure. Most of New York City's power plants, water treatment plants, and sewage systems are right at sea level. City planners are also going to have to help people adapt. More than half a million people live in the New York flood plain, and, as the report noted, a significant portion of them are African American and Latinos. And floods are not the only potential danger of climate change. The report notes that New York could face average annual temperature rises of up to 5 degrees Fahrenheit by the middle of this century and by as much as 9 degrees by 2080.

In summer months, this could subject New Yorkers to power shortages and the risk of black-outs because of the extra need for air conditioning. Those without air conditioning - or who cannot afford the higher electricity bills - would be at greater risk of heat stroke. Those hotter conditions would have effects right across the state, playing havoc with New York State's wine and agricultural industries. Spruce and Fir trees would disappear from the Catskills and West Hudson River Valley, dairy cows would suffer heat stress, and popular apple varieties would decline, the report said.

## 4. IPCC expected to confirm link between climate change and extreme weather

Climate change is likely to cause more storms, floods, droughts, heatwaves and other extreme weather events, according to the most authoritative review yet of the effects of global warming. Report likely to conclude that man-made emissions are increasing the frequency of storms, floods and droughts on Thursday- 17 November 2011 16.32 GMT from New York. The Intergovernmental Panel on Climate Change will publish on 18 November 2011, its first special report on extreme weather, and its relationship to rising greenhouse gas emissions. The final details are being fought over by governments, as the "summary for policymakers" of the report has to be agreed in full by every nation that chooses to be involved. But the conclusions are expected to be that emissions from human activities are increasing the frequency of extreme weather events. In particular, there are likely to be many more heatwaves, droughts and changes in rainfall patterns.

Jake Schmidt of the US-based Natural Resources Defense Council said: This report should be a wake-up call to those that believe that climate change is some distant issue that might impact someone else. The report documents that extreme weather is happening now and that global warming will bring very dangerous events in the future. From the report you can see that extreme weather will impact everyone in one way or another. This is a window into the future if our political response doesn't change quickly.

This special report - one of only two that the IPCC is publishing before its 2014 comprehensive assessment of the state of climate change science - is particularly controversial as it deals with the relationship between man-made climate change and damaging events such as storms, floods and droughts. Some climate change skeptics and scientists cast doubt on whether the observed increase in extreme weather events can be attributed directly to human actions, or whether much of it is due to natural variability in the weather (**Fig. 10**).

The IPCC, a body of the world's leading climate scientists convened by the United Nations, is likely to conclude that extreme weather can be linked to man-made climate change, but that individual weather events can at present only rarely be linked directly to global warming.

The Red Cross warned that disaster agencies were already dealing with the effects of climate change in vulnerable countries across the world. "The findings of this report certainly tally with what the Red Cross Movement is seeing, which is a rise in the number of weather-related emergencies around the world," said Maarten van Aalst, director of the Red Cross / Red Crescent Climate Centre and coordinating lead author of the IPCC report. "We are committed to responding to disasters whenever and wherever they happen, but we have to recognise that if the number of disasters continues to increase, the current model we have for responding to them is simply impossible to sustain."

**Figure 10.** A Pakistani mother carries her children through flood waters in 2010. The IPCC report deals with the relationship between man-made climate change and extreme weather. (Photograph: K.M.Chaudary/AP)

Insurers are also worried. Mark Way, of the insurance giant Swiss Re, told the Guardian that the massive increase in insurance claims was causing serious concern. He said that between 1970 and 1989, the insurance industry globally had paid out an average of $5bn a year in weather-related claims, but that this had increased enormously to $27bn a year. Although not all of this was attributable to climate change - increasing population, urbanisation and prosperity also play a major part - he said insurers wanted governments to get to grips with the effects of climate change in order to prepare for likely damage and tackle the causes of global warming.

Mike Hulme at the Tyndall Centre said it would be dangerous for governments to use this report in order to justify directing overseas aid only to those countries that could be proved to be suffering from climate change, rather than other problems. In that scenario, he said: "Funding will no longer go to those who are most at risk from climate-impacts and with low adaptive capacity, but will go to those who are lucky enough to live in regions of the world where weather extremes happen to be most attributable by climate models to human agency. These regions tend to be in mid-to-high latitudes, with lots of good weather data and well calibrated models. So, goodbye Africa."

## 5. Conclusion

From the various studies and reports, it is evident that the with the current rate of carbon dioxide release in the atmosphere there would not only be the increase in the global temperature, but it will also cause rise in sea, level and increase the frequency of disasters. The following major challenges are noticed from the above study:

- Emissions from human activities are increasing the frequency of extreme weather events. In particular, there are likely to be many more heatwaves, droughts and changes in rainfall patterns.
- The temperature is estimated to increase by 2 to $6°$ Celsius within year 2100, which is a tremendous increase from our current average temperature of $1.7°$ Celsius (IPCC).
- By the mid-2020s, sea level rise around Manhattan and Long Island could be up to 10 inches, assuming the rapid melting of polar sea ice continues. By 2050, sea-rise could reach 2.5ft and more than 4.5ft by 2080 under the same conditions.
- Global warming threatens the planet in a new and unexpected way – by triggering earthquakes, tsunamis, avalanches and volcanic eruptions.
- Irene-like storms of the future would put a third of New York City streets under water and flood many of the tunnels leading into Manhattan in under an hour because of climate change.

These are few glimpses of future suspects; there may be much more bad implications of evils of climate change globally and humanity will be at high risk, developments will get shattered and rescue efforts will gain higher priorities

## Author details

Bharat Raj Singh*
*School of Management Sciences, Technical Campus, Lucknow, Uttar Pradesh, India*

Onkar Singh
*Harcourt Butler Technological Institute, Kanpur, Uttar Pradesh, India*

## Acknowledgement

Authors indebted to extend their thanks to the School of Management Sciences, Technical Campus, Lucknow and Harcourt Butler Technological Institute, Kanpur for providing the support of Library.

---

* Corresponding Author

# 6. References

Crowley, T. J.; North, G. R. (May 1988), "Abrupt Climate Change and Extinction Events in Earth History", Science 240 (4855): 996–1002, Bib code: 1988Sci...240...996C, doi: 10.1126/science.240.4855.996, PMID 17731712.

Day, John W., et al. (2007). "Emergence of Complex Societies After Sea Level Stabilized". Eos Transactions American Geophysical Union 88 (15): 169. Bibcode 2007EOSTr..88..169D. doi:10.1029/2007EO150001.

Demenocal, P. B. (2001). "Cultural Responses to Climate Change During the Late Holocene". Science 292 (5517): 667–673. Bibcode 2001Sci...292..667D. doi:10.1126/science.1059827. PMID 11303088.

Desanker, P., et al., "Executive summary", In McCarthy 2001, Chapter 10: Africa, Retrieved 2011-06-20.

Diamond, Jared (2005), Collapse: How Societies Choose to Fail or Succeed. Viking Adult. ISBN 0670033375.

Easterling, WE, et al., "5.4.1 Primary effects and interactions", In Parry 2007, Chapter 5: Food, Fibre, and Forest Products, pp. 282, Retrieved 2011-06-25.

Fisher, BS, et al (2007). "3.1 Emissions scenarios", In B Metz, et al. Issues related to mitigation in the long term context. Climate Change 2007: Mitigation, Contribution of Working Group III to the Fourth Assessment Report of the Intergovernmental Panel on Climate Change, Cambridge, UK & New York, NY, USA; CH: Cambridge University Press; IPCC, Retrieved: 2011-05-04.

Foias, Antonia E.; et al. (1997), "Changing Ceramic Production and Exchange in the Petexbatun Region, Guatemala: Reconsidering the Classic Maya Collapse". Ancient Mesoamerica 8 (2): 275. doi:10.1017/S0956536100001735.

Gille, Sarah T (February 15, 2002). "Warming of the Southern Ocean Since the 1950s", Science 295 (5558): 1275–7, Bibcode 2002Sci.295.1275G, doi:10.1126/science.1065863, PMID 11847337.

Goldemberg, 2006 J. Goldemberg, The promise of clean energy. Energy Policy, 34 (2006), pp. 2185–2190.

Haug, Gh., et al. (Mar 2003), "Climate and the collapse of Maya civilization". Science 299 (5613): 1731–5. Bibcode 2003Sci...299.1731H. doi:10.1126/science.1080444. ISSN 0036-8075. PMID 12637744.

Hodell, David A., et al. (1995), "Possible role of climate in the collapse of Classic Maya civilization". Nature 375 (6530): 391. Bibcode 1995Natur.375..391H. doi:10.1038/375391a0.

Hosler D., et al. (1977), "Simulation model development: a case study of the Classic Maya collapse". In Hammond, Norman; Thompson, John L.. Social process in Maya prehistory: studies in honour of Sir Eric Thompson. Boston: Academic Press. ISBN 0-12-322050-5.

Hubbert M.K., 1956, Nuclear energy and the fossil fuels; American Petroleum Institute, Drilling and Production Practice, Proc. Spring Meeting, San Antonio, Texas. 7-25

IEA, 2006 IEA, World Energy Outlook 2006, Organisation for Economic Co-operation and Development, International Energy Agency, Paris and Washington, DC (2006).

IEA, 2007a, IEA, Coal Information 2007, Organisation for Economic Co-operation and Development, International Energy Agency, Paris and Washington, DC (2007).

IEA, 2007b, IEA, World Energy Outlook 2007, China and India, Organisation for Economic Co-operation and Development, International Energy Agency, Paris and Washington, DC (2007).

Intergovernmental Panel on Climate Change (2007d), "Climate Change 2007: Synthesis Report, Contribution of Working Groups I, II and III to the Fourth Assessment Report of the Intergovernmental Panel on Climate Change (Core Writing Team et al. (eds.))". IPCC, Geneva, Switzerland, Retrieved: 2009-05-20.

IPCC (2001b), "Figure SPM-2", In McCarthy 2001, Summary for Policymakers, Retrieved: 2011-05-18.

IPCC 2001d, "3.16", Question 3, Retrieved: 2011-08-05.

IPCC 2007d, "1. Observed changes in climate and their effects", Summary for Policymakers, CH: IPCC, Retrieved: 2011-06-17.

IPCC 2007d, "3. Projected climate change and its impacts", Summary for Policymakers. CH: IPCC.

IPCC 2007d, "3.3.4 Ocean acidification", Synthesis Report, Retrieved: 2011-06-11.

IPCC 2007d, "5.2 Key vulnerabilities, impacts and risks-long-term perspectives", Synthesis report, Retrieved: 2011-08-05, IPCC, 2001, SPM Question 3.

Jansen, E, et al., "6.3.2 What Does the Record of the Mid-Pliocene Show?", In Solomon 2007, Chapter 6: Palaeoclimate. CH: IPCC Retrieved: 2011-05-04.

Jonathan Cowie, (2007), Climate change: biological and human aspects. Cambridge University Press. ISBN 0521696194, 9780521696197.

Karl, 2009, ed., "Global Climate Change".

Klass, 1998 D.L. Klass, Biomass for Renewable Energy, Fuels, and Chemicals, Academic Press, San Diego (1998).

Klass, 2003 D.L. Klass, A critical assessment of renewable energy usage in the USA. Energy Policy, 31 (2003), pp. 353–367.

Kundzewicz, Z.W., et al., "Executive Summary", In Parry 2007, Chapter 3: Fresh Water Resources and their Management, pp. 175, Retrieved: 2011-08-14.

Le Treut, H, et al. "FAQ 1.2 What is the Relationship between Climate Change and Weather?", In Solomon 2007, Historical Overview of Climate Change. CH: IPCC.

Lior, 2008, N. Lior, Energy resources and use the present situation and possible paths to the future, Energy, 33 (2008), pp. 842–857.

Parry 2007b, ed., "Magnitudes of impact", Summary for Policymakers, CH: Intergovernmental Panel on Climate Change, pp. 17, Retrieved: 2011-05-08.

Parry 2007a, ed. "Definition of "biota". Appendix I: Glossary Retrieved: 2011-10-01.

Patterson, W.P., Dietrich, K.A., and Holmden, C., (2007), Sea Ice and sagas: stable isotope evidence for two millennia of North Atlantic seasonality on the north Icelandic shelf Arctic Natural Climate Change Workshop, Tromsø, Norway

Santley, Robert S., et al. (Summer 1986), "On the Maya Collapse". Journal of Anthropological Research 42 (2): 123–59.

Schneider, SH, et al., "19.3.2.1 Agriculture", In Parry 2007, Chapter 19: Assessing Key Vulnerabilities and the Risk from Climate Change, pp. 790, Retrieved: 2011-06-25.

Schneider, SH, et al., "19.3.1 Introduction to Table 19.1", In Parry 2007, Chapter 19: Assessing Key Vulnerabilities and the Risk from Climate Change. CH: IPCC, Retrieved: 2011-05-04.

Scott, M., et al., "7.2.2.3.1 Migration", In McCarthy 2001, Chapter 7: Human Settlements, Energy, and Industry, Retrieved: 2011-08-29.

Scott, M.J., et al., "12.3.1 Population Migration", In Watson 1996, Chapter 12: Human Settlements in a Changing Climate: Impacts and Adaptation.

Shaffer, G.; Olsen, S. M.; Pedersen, J. O. P. (2009), "Long-term ocean oxygen depletion in response to carbon dioxide emissions from fossil fuels", Nature Geoscience 2 (2): 105–109, Bibcode 2009NatGe.2.105S, doi:10.1038/ngeo420.

Shahriar Shafiee, Erkan Topal, When will fossil fuel reserves be diminished? Energy Policy, Volume 37, Issue 1, January 2009, Pages 181-189.

Singh, B.R., et al., A Study on Sustainable Energy Sources and its Conversion Systems Towards Development of an Efficient Zero Pollution Novel Air Turbine to Use as Prime-Mover to the Light Vehicle, ASME Conf. Proc., 2008, Volume 8, Paper no. IMECE2008-66803 pp. 371-378, doi: 10.1115/IMECE2008-66803.

Solomon 2007a, ed. "Direct Observations of Recent Climate Change", Summary for Policymakers, CH: Intergovernmental Panel on Climate Change.

Solomon, S, et al. 2007b "TS.3.4 Consistency Among Observations", In Solomon 2007, Technical Summary. CH: IPCC.

Solomon, S, et al. "Table TS.4", In Solomon 2007, Technical Summary, p. 52.

Stern, N (May 2008). "The Economics of Climate Change" (PDF). American Economic Review: Papers & Proceedings (UK: LSE) 98 (2): 6, doi:10.1257/aer.98.2.1, Retrieved: 2011-05-04.

Transl. with introd. by Magnus Magnusson, (1983). The Vinland sagas: the Norse discovery of America. Harmondsworth, Middlesex: Penguin Books. ISBN 9780140441543.

WHO (2009), "2.6 Environmental risks", 2 Results, Global health risks: mortality and burden of disease attributable to selected major risks. Produced by the Department of Health Statistics and Informatics in the Information, Evidence and Research Cluster of the World Health Organization (WHO). World Health Organization, 20 Avenue Appia, 1211 Geneva 27, Switzerland, WHO Press, ISBN: 978 92 4 156387 1, Retrieved: 2011-07-14.

Wilbanks, T., et al., "7.4.1 General effects: Box 7.2, Environmental migration", In Parry 2007, Chapter 7: Industry, Settlement and Society, Retrieved 2011-08-29.

# Chapter 4

# Global Warming Mitigation Using Smart Micro-Grids

Amjad Anvari Moghaddam

Additional information is available at the end of the chapter

## 1. Introduction

While focusing on some critical points such as decreasing global warming and ambient pollution, better utilization of renewable energy resources, energy management and improvement of power systems operation becomes the field of attention for many modern societies, power and energy engineers, academics, researchers and stakeholders everywhere are pondering the problems of depletion of fossil fuel resources, poor energy efficiency and environmental pollution. Besides, the conventional power grids and their assets that span large areas of the earth and form huge interconnected meshes, not only have a close relationship with social and economic activities, but also generate a substantial amount of criteria air pollutants, as evident by the continuing development of new rules under the clean air act for the electric power sector. One of the main disadvantages of such networks is their reliance on large centralized power generation units which produce particulate and gaseous emission pollutants. In other words, coal-fired power plants together with fossil fuel power stations that make a large portion of generation companies (GenCos) are the major contributors in pollutants include Greenhouse Gases (GHGs), fine particulates, oxides of nitrogen ($NO_x$), oxides of sulfur ($SO_x$) and mercury (Hg), which are thought to cause global warming. They are also contributing to carbon dioxide ($CO_2$) emissions as well as producing solid waste in the form of fly ash and bottom ash. Therefore, a new trend for modernization of the electricity distribution system and generation sector has been proposed to address these issues suitably. This plan of action mainly focuses on generating energy locally at distribution voltage level through incorporation of small-scale, low carbon, non-conventional and renewable energy sources, such as wind, solar, fuel cell, biogas, natural gas, microturbines, etc., and their integration into the utility distribution network. Generally, such energy choices are regarded as dispersed or distributed generation (DG) and the generators are termed as distributed energy resources (DERs) or microsources. On the other hand, conventional power grids which are mostly passive distribution networks with one-way electricity transportation are in the era of major modification and alteration into active distribution networks (ADNs) with DERs and bidirectional electricity transportation (Chowdhury et al., 2008a,2008b). In this regard, flexible and intelligent control systems must be incorporated in ADNs to exploit clean energy from renewable DERs. Advanced systems and key technologies should be also employed for integration of DERs. With low incorporation of renewable energy sources (RESs) the total effect on grid operations is confined, but as the penetrations of such resources increase, their mutual effects increase too (Angel & Rújula, 2009 ; Clark & Isherwood, 2004). Nevertheless, harness of RESs, even when there are good potential resources, may be problematic due to their variable and intermittent natures, thus RESs cannot necessarily be operated in a conventional manner. Instead, RESs behaviors can be predicted via expert estimators and the forecast information is exactly the kind of information that an ADN must uses to improve system efficiency (Chowdhury et al., 2008c, 2008d).

## 2. Smart Micro-Grids: A true way to mitigate global warming

It was mentioned earlier that existing transmission and distribution systems in many parts of the world use technologies and strategies that are many decades old. They make limited use of digital communication and control technologies. To update this aging infrastructure and to create a power system that meets today's growing and changing needs, developed societies try to create intelligent means which use advanced sensing, communication, and control technologies to distribute electricity more effectively, economically and securely. Additionally, there are some important side benefits for the consumers such as potential lower cost, higher service reliability, better power quality, increased energy efficiency and energy independence that are all reasons for an increased interest in distributed energy resources and focusing on what are called "Smart Micro-Grids", as the future of power systems. Although the term "Smart Grid" is frequently used today, there is no agreement on its definition. In other words, the concept of intelligence in Smart Grid design and how it will be measured is unclear. The U.S. Department of Energy (DOE) mentioned in one of its recent issues that a Smart Grid uses digital technology to improve reliability, security and efficiency of the electric system from large generation, through the delivery systems to electricity consumers and a growing number of distributed-generation and storage resources (U.S. DOE, 2008). Later, in June 2008, a meeting of industry leaders was held at the U.S. Department of Energy and seven different characteristics were declared for the Smart Grid concept:

1.  Better utilization of conventional assets, optimization and efficient operation,
2.  Accommodation of all generation and storage options in power grids,
3.  Supply of power quality as a great need of today's industry,
4.  Prediction of events and fast response to system disturbances in a self-healing manner,
5.  Robust operation against attacks and natural disasters,
6.  Active participation of consumers,
7.  Introduction of new services, products and markets.

While details vary greatly about the definition of a Smart Grid, a general definition can be made as follow: A Smart Grid is an intelligent, auto-balancing, self-healing power grid that accepts any source of fuel as its input and transforms it into a consumer's end use with minimal human intervention. It is a course of action that will result in better utilization of renewable energy resources and reduce environmental vestiges as much as possible. It has a sense of detection to understand where it is loaded beyond capacity and has the ability to reroute power to lessen overload and impede potential outages. It is a base that provides real-time communication between consumers and the utility in order to optimize energy harvesting based on environmental benefits or cost preferences. However, it should be noted that deployment of Smart Grid technologies will occur over a long period of time, adding successive layers of functionality and capability onto existing equipment and systems. Although technology is the focal point, it is only a way to achieve the goal, and the smart grid should be defined by more extensive characteristics (Anvari Moghaddam et al., 2010b, 2010c, 2011b). How the Smart Grid differs from conventional grids we know today, is illustrated in Figs. 1, 2. Conventional networks are designed to support large power units that serve faraway consumers via one-way transmission and distribution grids (Fig.1), but the future grids will necessarily be two-way real time systems, where power is generated not only by a large number of small and distributed energy resources but also by large power plants(ABB, 2009). Power flow across the network is based on a mesh grid structure rather than a hierarchical one (Fig. 2).

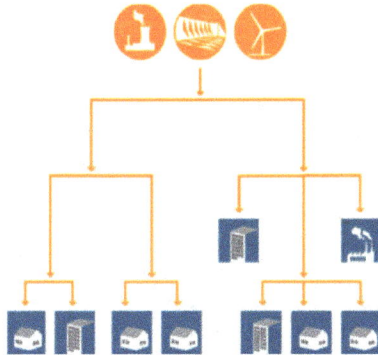

**Figure 1.** Conventional power system

Likewise, the term "Micro-Grid" in its whole vision, is an exemplar of a macro-grid in which local energy potentials are mutually connected with each other as well as with the L.V utility and make a small-scaled power grid. In such a network, DGs are exploited extensively both in forms of renewable (e.g., wind and solar) and non-conventional (micro-turbine, fuel cell, diesel generator) resources, because these emerging prime movers have lower emission and the potential to have lower cost negating traditional economies of scale (Anvari Moghaddam et al., 2011c, 2012). In addition to DGs, storage options are also used widely to offset expensive energy purchases from utility or to store energy during off-peak hours for an anticipated price spike (Divya & Østergaard, 2009;

Kaldellis & Zafirakis, 2007). In a typical Micro-Grid, DERs generally have different owners handle the autonomous operation of the grid with the help of local controllers ($\mu_c$ or MGLC) which are joined with each DER and Micro-Grid central controller ($\mu_{cc}$ or MGCC). Moreover, the central control unit (CCU), which is a part of MGCC, does the optimization process to achieve a robust and optimal plan of action for the smart operation of the Micro-Grid

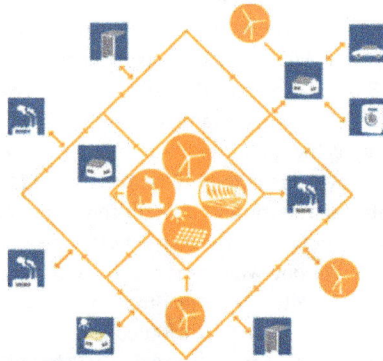

**Figure 2.** Future power system (Smart Grid)

## 2.1. Impacts of Smart Grids on energy efficiency and low carbon economy

From a utility point of view, implementation of a Smart Grid can yield several advantages over the conventional one, as shown in Fig.3. A Smart Grid can organize operations and enable utilities to tap into new paths to save energy and reduce environmental footprints to levels greater than would otherwise be attainable (Abbasi & Seifi, 2010).

As the figure shows, selected mechanisms empowered by a Smart Grid, represent pathways to energy savings and/or emission reductions, although some of these benefits overlap across the various goals. For example, indirect feedback to customers using improved billing is related to improvements both in operational efficiency and customer energy use behavior. Similarly, greater options for dynamic pricing and demand response are related to customer service enhancement as well as to demand response activation. It's also notable that some of the ways associated with the mentioned goals are slightly indirect or their energy savings potentials are difficult to express on a national standard, because they include complex market, institutional, and behavioral interactions that can vary considerably across the nations (European Commission, 2006). Considering different mechanisms have been identified earlier, this section of the work highlights the more direct ways for energy savings and reduced carbon emissions in order to get better insight into the Smart Grid environmental benefits. As shown in Fig. 3, two of the pathways can directly reduce carbon emissions while inducing energy savings: (i) higher penetration of RESs and their greater integration into the grid environment and (ii) further utilization of plug-in hybrid electric vehicles (PHEVs) and facilitation of their deployments.

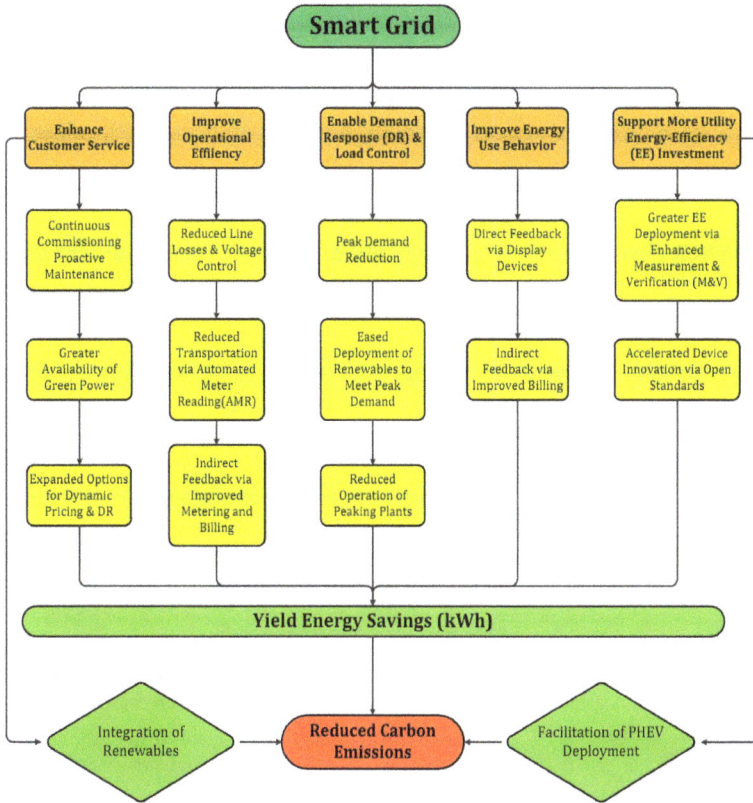

**Figure 3.** Smart Grid mechanisms for energy efficiency & emission reduction

## 2.1.1. Integration of renewable energy sources

A Smart Grid facilitates more seamless integration of RESs and other DERs including energy storage options due to its advanced control and communications capabilities. Earlier studies have indicated that energy storage can compensate for the stochastic nature and sudden deficiencies of RESs for short periods without suffering loss of load events, and without the need to start more generating plants (Anvari Moghaddam et al., 2010a, 2011a). Moreover, with higher penetration rates of renewables such as wind and solar in the overall supply mix, utilities will decrease their carbon emissions and will be better situated to meet their corresponding states' renewable portfolio standard (RPS) mandates. In a same manner, customers not only receive a greater share of green power from the utility, but also integrate their individual microsources and/or renewable options (e.g., rooftop photovoltaic systems) into the grid for participation in net metering programs. From the other point of view, all advantages of non-conventional or renewable low-carbon generation technologies and high-efficiency combined heat and power (CHP) systems can be achieved through implementation of a Smart Micro-Grid (Battaglini et al., 2009 ; Esmaili et al., 2006). In this regard, energy and power can be produced in an efficient manner through capturing waste heat by using the CHP-based DERs, while environmental pollution can be reduced extensively by generating clean power with the help of low-carbon DERs. Although DERs appear in various types and range from Micro-CHP systems based on Stirling engines, fuel cells and microturbines to renewable ones like solar photovoltaic (PV) systems, wind energy conversion systems (WECS) and small-scale hydroelectric generation, the choice of a particular DER depends strongly on the climatic conditions, regional anatomy and fuel availability (Hajizadeh & Golkar, 2007 ; Hammons, 2006). Similarly, application of biofuels and different storage technologies such as Compressed Air Energy Storage (CAES) and ultra-capacitors are investigated comprehensively with regard to a certain geographical location and some environmental constraints (Koeppel, 2008). Moreover, to lessen the greenhouse gas emissions and mitigate global warming, most of the countries including the European Union(EU) and UK, are advocating the schemes associated with further exploitation of RESs as well as better integration of DG systems as part of the Kyoto protocol. Besides, with the continuous depletion of fossil fuels as a result of growing needs in energy sector, most of the utilities are looking for non-conventional/renewable energy resources as an alternative.

### 2.1.2. Deployment of Plug-In Hybrid Electric Vehicles (PHEVs)

Implementation of a Smart Grid environment will also facilitate the market adoption and accommodate all available options for better integration and interconnection of plug-in hybrid electric vehicles (PHEVs), as the future of clean transportation systems. These battery-like vehicles can be used as substitutions for conventional non-renewable energy sources while they can be plugged into electrical outlets for recharging (EPRI, 2008a). In comparison with the current hybrid vehicles, PHEVs have different operational modes as well as flexibility in applications, thus they can further decrease the reliance on gasoline to fuel the internal combustion engines. Moreover, incorporation of PHEVs will save fuel costs, since they run on the equivalent of 75 cents per gallon or better at today's electricity prices. From an environmental perspective, the deployment of PHEVs will result in a considerable reduction in air pollutant emissions. As an illustrative example, it was announced in 2007 by EPRI and the Natural Resources Defense Council (NRDC) that PHEVs will lead to a reduction of 3.4 to 10.3 billion metric tons of GHGs by 2050, as a function of PHEV fleet penetration and the carbon-intensity of the electricity generation mix (EPRI, 2007a, 2007b, 2008b). Likewise, this EPRI study revealed that PHEVs could result in GHGs reductions of 100 to 300 million metric tons of $CO_2$ per year, based on a range of planned PHEV market share considering the reference year of 2030. From a utility viewpoint, development of Smart Grid supports further participation of PHEVs in market actions. The ability to charge PHEVs during the night with low energy price tariffs provides operational benefits through improved system load factor, environmental benefits through mitigation of GHGs emissions and economical profits though utilization of base load resources. Development of a Smart

Grid makes it possible to send signals to consumers intelligently on when to charge their vehicles or provide multi-tariff rates to encourage off-peak charging. Alternatively, PHEVs can be used for peak-shaving or power-quality applications by storing electrical energy in their onboard batteries, offering potentially powerful synergies to complement the electric power grid. With an expert coordination among smart vehicles and the Smart Grid, PHEVs may serve as a dispersed generation system itself, providing energy efficiency, stability and environmental benefits for the grid operation. Considering the attributes inherent in a typical PHEV, it is reasonable to assign some share of projected PHEV $CO_2$ reduction impact to the development of a Smart Grid. On the basis of EPRI research studies, a portion of 10% to 20% can be dedicated to PHEV $CO_2$ reduction impact in a smart grid environment, which in turn reduces the net $CO_2$ emissions from 10 to 60 million metric tons of $CO_2$ in 2030.

## 2.2. Impacts of Micro-Grids on atmospheric emissions and environmental issues

Regardless of market sensitivity, renewable energy microsources and other low-carbon generation units can effectively reduce emissions and environmental warming and this is one of the most important reasons to support Micro-Grid design and implementation. To assure eco-friendly operation, the Micro-Grid central controller (CC) should be programmed in a way to make optimal decisions for unit commitment based on the lowest net emission production, considering both displaced emission and local emission from microsources as objectives. In the presence of market sensitivity, decision-making algorithms become more complex, because market-responsive CC should include "emission minimization" as an additional criterion for dispatch decisions (Chuang & McGranaghan, 2008). This complexity can be handled suitably provided that reasonable and fair emission tariffs are introduced into the market system, i.e., the electricity supplied from the microsources would be valuated through price tariffs following the net reduction in emissions is acquired. In that case, a measure of the net emission reduction is available from the price signals itself. Emission tariffs might be also established on the basis of multi-criteria functions covering several factors such as time, season and location, so that at worst pollution times and locations, the tariffs would be most attractive. In such a situation, emission price tariffs are provided as extra input signals to the CC to dispatch microsources optimally for minimizing emissions (Cinar et al., 2010). It's also worthy of note that environmental policy initiatives and existing regulatory guidance should be given due importance for moving toward a cleaner ambient. In this regard, the US Environmental Protection Agency put limits on the amount of emissions from six air pollutants, including: nitrogen dioxide ($NO_2$), carbon monoxide (CO), sulphur dioxide ($SO_2$), lead (Pb), ozone ($O_3$) and particulates. According to recent reports on environmental pollutants it's shown that conventional power plants and fossil-fuelled vehicles are the largest producers of $NO_x$ gases. Similarly, large gas turbines and reciprocating engines result in sufficient $NO_x$ production when they are operating at high temperature. Conversely, microturbines and fuel cells emit lower amounts of $NO_x$ because of lower combustion temperatures; hence, their application as microsources would significantly reduce carbon and nitrogen compounds and total hydrocarbons (THC). Besides, many efforts are made at the resent time to develop combined environmental–

economic optimization (EED) algorithms for dispatching DGs and microsources considering cost of operation and atmospheric emissions like $NO_x$, $SO_2$, $CO_2$, etc., as weighted objective functions simultaneously (Lagorse, 2010 ; Momoh, 2009). For CHP-based microsources, usually heat optimization is at the head of concern and optimization of electricity is observed at the second step; i.e., according to customers' heat requirement, the amount of power production from CHP is determined. It's nice to mention that for large-scale CHP systems there are many technical and environmental constraints that must be met during the operation and they are mainly as follows:

1.   Demand-supply balance: at any hours of a given day, the amount of heat generation must be equal to the heat demand.
2.   Energy efficiency: the maximum electrical power generated in the process should be used to supply the electrical loads and surplus of demand must be provided thorough other microsources or purchased from the market.
3.   Emissions cap: The $NO_x$, $CO_2$ and $SO_2$ emissions must be maintained at specified limits.

According to the importance of the optimization variables, shadow prices, which are used to quantify the importance of each variable, might be developed to provide appropriate weighting factors to each of the above-mentioned constraints. Such prices might be a function of the real-time price, the demand, the time of use and the season. Shadow prices might be also defined hourly for each type of pollutants during the operation period for generating both electrical and heat energy. All the mentioned parameters would be calculated and used by the Micro-Grid's CC to arrive economically at an optimal dispatch solution using an iterative procedure. During the optimization process, both seasonal and diurnal trends and area-wise variations of emissions should be given right weighting factors for appropriate scheduling and controlling the operation of generators. For example, the amount of ozone emission is augmented in late spring and during the summer when ambient temperature is high. In a similar manner, peak ozone concentrations occur significantly downwind of emission sources, mainly due to the lengthy reaction times. Moreover, ozone concentration is increased in more crowded areas at considerable distances downwind from urban areas. Thus, during the warmer periods of times, it is wise to reduce $NO_x$ emission in or near dense populated places to decrease ozone formation. CHP-based microturbines, which play an important role in Micro-Grids, can effectively address the corresponding issue. Beyond the above-mentioned points, there are some other parameters that must be provided for controlling hazardous pollution and mitigating global warming. Such extra information includes: fair rate incentives based on specific pollutant production, displaced emissions, expected temperature, and etc.

## 2.2.1. Minimization of pollutant deposition using microsources

As discussed in previous section, Micro-Grid is considered as a small-scale, medium or low voltage (MV/LV) combined heat and power (CHP) system for supplying electrical power and energy needs of small local loads. More than one Micro-Grid may also be integrated to constitute power parks for supplying larger load demands. In this regard, the microsources inside a typical Micro-Grid serve as primary means of energy producers while they are using diverse types of low-carbon generation technologies. As an example, combined heat and power (CHP) system is a popular kind of DER useful for Micro-Grid applications. Such cogeneration system has the advantage of energy-efficient power generation via utilization of waste heat (see Fig. 4).

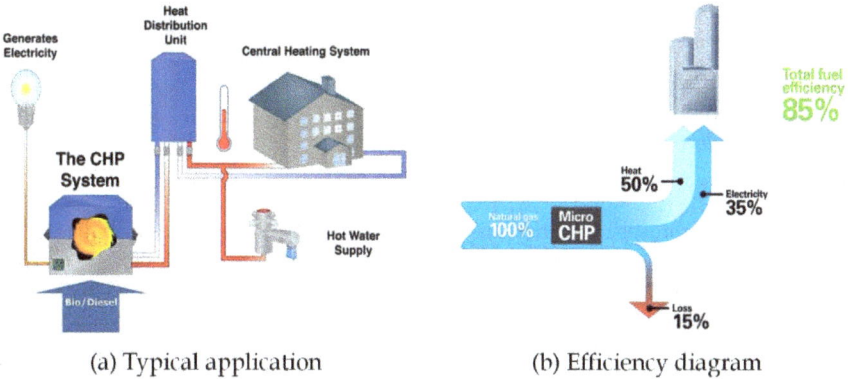

(a) Typical application               (b) Efficiency diagram

**Figure 4.** Combined Heat & Power (CHP) system

Unlike fossil-fuelled generation units, CHP-based generators capture and use the waste heat for industrial processes or other local heating purposes. Moreover, the heat reproduced at moderate temperatures (100–180 °C) can also be used in absorption chillers for cooling mechanism. In this way, a CHP system can potentially reach an efficiency of more than 80%, compared with that of about 35% for conventional power plants. The efficiency can be even more when the heat is used locally, in other words, if the produced heat is transmitted over long distances for supplying remote thermal loads, not only the overall efficiency reduces, but also the net operating cost and emission increases. On the other hand, because of lower electrical loss, CHP plants can be situated somewhere faraway from electrical loads and their produced energy can be transmitted over much longer distances, however, it should be kept in mind that such systems must always be located close to the heat loads for better efficiency and performance (Pecas Lopes et al., 2007). On the whole, it has been found that application of CHP micro-energy source yields a reduction of 35% in primary energy use in comparison with conventional power plants and heat-only boilers, 30% reduction in emission with respect to coal-fired power plants and 10% reduction in emission with respect to combined cycle gas-turbine plants. In addition to CHP-based systems, microturbines are also small and simple-cycle gas turbines that are used extensively in Micro-Grids and ranged typically from 25 to 300 kW based on the output power (Saha et al., 2008). Actually, microturbines have several advantages and inherent technologies that are briefly include: recuperation, low $NO_x$ emission technology and advanced material usage such as ceramic for the hot section parts. Microturbines have also various structural and operational features as shown in Fig. 5.

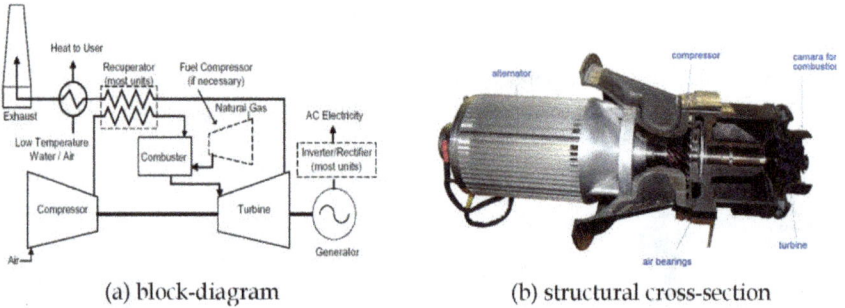

| (a) block-diagram | (b) structural cross-section |

**Figure 5.** Schematic of a microturbine

From a structural prospective, they are relatively smaller in size as compared to other DERs, have simpler installation procedure and lower level of noise and vibrations. They also have the capability of using alternative fuels, like natural gas, diesel, ethanol and landfill gas, and other biomass-derived liquids and gases. From an operational viewpoint it can be said that, microturbines are designed for 11,000 hours of operation between major overhauls with a service life of at least 45,000 hours. Total cost of such system is lower than $500 per kW which is competitive with alternatives including grid power for market applications. They can also reach the range of 25–30% in fuel-to-electricity conversion rate while the energy efficiency level can be greater than 80% if the waste heat recovery is used for CHP applications. In addition, microturbines participate actively in clean air action by producing reduced amounts of $NO_x$ emissions which are lower than 7 ppm for natural gas machines. As a matter of fact, the net emission belongs to a microturbine greatly depends on its operating temperature, power output and the control of the combustion process, therefore it can be minimized only through quick and accurate control of the combustion process which is done suitably by the microturbine's internal control system. On the other hand, the central controller of the Micro-Grid ($\mu_{cc}$) may only provide the generation set points for the microturbine considering the net emission production in relation with power level and displaced emissions for both heat and electric power output. The CC may also monitor the remaining oxygen concentration in the engine exhaust for some particular applications. In order to minimize $NO_x$ emission, microturbine manufacturers usually apply various controlling algorithms along with some combustion control methods. For instance, wet diluent injection (WDI) is a controlling approach for $NO_x$ reduction in microturbines where water or steam is injected into the combustion zone to moderate the temperature, however, this method increases $CO_2$ production, reduces efficiency and shortens equipment life.

Another type for microsources is regarded as fuel cell (FC) which converts chemical energy of a fuel into electrical energy directly (Fig. 6). Basically, it consists of two electrodes with different polarities and an electrolyte that dissociates fully or partially into ions when dissolved in a solvent, producing a solution that conducts electricity.

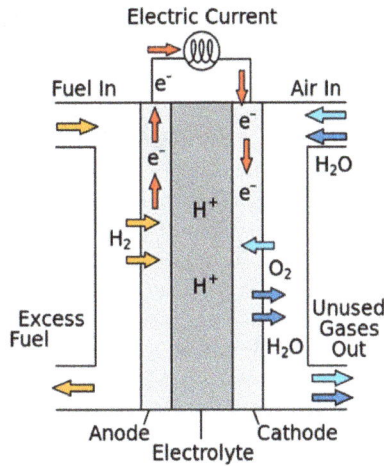

**Figure 6.** Schematic of a Hydrogen FC

From an operational prospective, the FC is very similar to a storage battery, however, the reactants and products are not stored, but are continuously fed into the cell. During operation, the hydrogen-rich fuel is fed to the anode and the oxidant which is usually air is conducted toward the cathode separately. Through an electrochemical oxidation and reduction process, electricity, as the main output, is produced at the electrodes while heat and water are produced as by-products. Compared to conventional generators, FCs have several advantages. First, they are very clean means of energy production and serves as eco-friendly sources of energy. Because of their higher efficiency and lower fuel oxidation temperature, they emit lower amounts of $CO_2$ and $NO_x$ per kilowatt of output power. Due to the absence of any rotating parts in FCs' structures, they are robust, low maintenance and almost free from noise and vibration. Moreover, they can run with different kinds of fuels like natural gas, propane, landfill gas, anaerobic digester gas, diesel, naphtha, methanol and hydrogen, therefore, this technology will not become obsolete due to versatility of consuming fuels or unavailability of energy resources. Apart from the mentioned microsources, solar photovoltaic (PV) is another kind of DER that helps the minimization of pollutants deposition by generating electricity from solar energy. PV systems have several advantages over the conventional generators as are stated in the following:

1.  Inexhaustible , clean and free nature of solar energy,
2.  Minimum environmental impact,
3.  Lower customers' electricity bills due to free availability of sunlight,
4.  Long operational lifetime of over 30 years with minimum maintenance,
5.  Noiseless operation

Owing to the above benefits, it's found that today's PV systems not only have the potential to supply a big portion of the world's energy needs in a sustainable and renewable manner, but also have the capability to reduce environmental footprints.

## 3. Technical, economical and environmental advantages of Smart Micro-Grids

It has explained earlier that effective utilization of waste heat in CHP-based microsources is one of the potential benefits of a Smart Micro-Grid. Besides, good coordination between heat generation and efficient heat utilization is a requisite task for energy optimization in Smart Micro-Grids that can be achieved by using heat generation control and thermal process control features in Smart Central Controller (SCC). Process optimization functions can also be built in the SCC to increase overall system efficiency and reliability (Molderink et al., 2009; Venayagamoorthy, 2009). On the other hand, Smart Micro-Grid has the ability to affect electricity and gas markets significantly, when its share of market participation is encouraged. In this regard, insightful market reforms must be made to allow active participation, while good financial incentives should be provided for owners to invest in Smart Micro-Grids. Once market participation is assured, Smart Micro-Grids can effectively supply quality services for distribution systems as well as ancillary services for the utilities. Similarly, with rising concern for global warming and environmental pollution, most countries are focusing on utilizing eco-friendly plants with low-carbon generators and trying to reduce their emission levels by 50% as per the Kyoto Protocol, considering the reference year of 2050. Regarding this planning horizon, Smart Micro-Grids together with cleaner microsources and RESs, strongly have the ability to reduce the overall environmental impact caused by existing infrastructures. To fulfill the clean air action, SCCs must be also programmed in a way to make smart dispatch decisions for DGs considering pollution level caused by the net emission in the locality (Figueiredo & Martins, 2010). Likewise, to give due importance and authority to Smart Micro-Grids that would help to mitigate the net greenhouse gas (GHG) and particulate emissions in the environment, rules and regulations must be made subsequently. On the whole, development of Smart Micro-Grid is very promising for the electric energy industry because of the following advantages:

- **Environmental benefits** – Smart Micro-Grids can reduce gaseous and particulate emissions and help to mitigate global warming through incorporation of low-carbon RESs together with close control of the combustion process. Moreover, local distribution of microsources and physical proximity of them with loads and consumers may help to increase the awareness of customers towards judicious energy usage.
- **Economical benefits** – Smart Micro-Grids can also end in cost savings in multiple ways: first, significant savings may be achieved from utilization of waste heat in CHP mode of operation. Second, no considerable and costly infrastructure is required for heat and power transmission as the microsources are physically situated close to the customer loads. This way of energy production gives a total energy efficiency of more than 80% as compared to a maximum of 40% for conventional power systems. Third, substantial cost savings can be obtained through integration of several microsources and construction of clean energy farms. Since each individual farm is locally operated in plug-and-play mode, the transmission and distribution (T&D) costs are drastically reduced or eliminated. Moreover, expert combination of such energy farms into a

unified Smart Micro-Grid can contribute to further cost reduction not only through eliminating the need for energy exchange with the main grid over longer transmission lines, but also by sharing the generated electricity among the local customers.

- **Technical benefits** – From a technical prospective, implementation of Smart Micro-Grids can be beneficial for both utilities and customers. The voltage profile is enhanced through better supply of reactive power for local inductive loads and the whole system as well. The congestions on transmission and distribution feeders are also reduced. Moreover, since a Smart Micro-Grid has a sense of detection to understand where it is loaded beyond capacity, thus it has the ability to reroute power to lessen overload and impede potential outages. In a similar manner, T&D losses can be cut down to about 3% by generating electricity near the load centers while the investments in the expansion of transmission and generation systems can be reduced or postponed by proper asset management. Due to decentralization of supply and better load feeding, reduction of large-scale transmission and generation outages, minimization of downtimes and enhancement of the restoration process through black start operations of microsources, power quality and reliability is enhanced consequently.

It's also noticeable that in the case of market participation, additional advantages can be achieved by Smart Micro-Grids which are mainly as follows:

- **Market power mitigation** - The grid-connected operation of Smart Micro-Grids in a market-based environment will lead to a significant reduction of market power exerted by the large dominant GenCos or through collusion of some market participants.
- **Market price reduction** - Widespread exploitation of RESs together with application of low-cost plug-and-play microsources may result in a reduction in energy price tariffs in the power market.
- **Ancillary services (AS)** - The Smart Micro-Grids may also be used to supply ancillary services such as frequency control or spinning reserve provision.
- **Long-term cost reduction** - The long-term electricity customer prices can be reduced by about 10% through appropriate economic balance between network investment and DG utilization.

## 4. Conclusion

Energy management systems and power system optimizers accompanied by integration of renewable energy resources and adoption of PHEVs which form a whole Smart Micro-Grid vision, are parts of an integrated approach to mitigate global warming and have the capability of serving as a basic tool to reach energy independence and climate change objectives. In this regard, energy efficiency mechanisms are potentially the most cost-effective, short-term options to reduce carbon emissions compared to other abatement alternatives. Moreover, energy-efficient approaches reduce GHG emissions not only through energy savings but also through the deferral of new generation with the help of technological advancements on the supply-side. In this sense, the more deferral of new generation by means of energy efficiency measures, results the more free time dedicated to improvements in bringing cleaner and more efficient generation online, thereby providing a bridge between the present and a carbon-constrained future.

On the other hand, an intelligent grid can lead to a revolution in power system operation, a revolution that will take place if new ideas and technologies along with very large penetrations of renewable energies are to be incorporated onto the grid. However, in order to efficiently operate and make good decisions, a Smart Micro-Grid must have information feeding supervisory control unit and Smart Energy Management System (SEMS). This information can be used to create better procedures and capabilities for the Smart Micro-Grid and allow more prudent investments. The optimal integration of decentralized energy storages will be also an extremely important task in the near future for the utilities. Moreover, to reach a pathway toward intelligent structures, first the barriers must be identified and then research, development and demonstrations of operation must be conducted to overcome these barriers.

## Author details

Amjad Anvari Moghaddam
*School of Electrical and Computer Engineering, College of Engineering,*
*University of Tehran, Tehran, Iran*

## 5. References

ABB. Toward a Smarter Grid. *A white paper*. ABB Inc. Cary, NC 27518, 2009.

Abbasi, A. R. ; Seifi, A. R. (2010). The Basic Concepts of Smart Grid: Initiatives, Technologies. Characteristics, Standards and Solutions. *Modelling and Simulation (IREMOS)*, Vol.3, No.1, pp. 64:70

Angel, A. ; Rújula, B. (2009). Future development of the electricity systems with distributed generation. *Energy*, Vol.34, No.3, pp. 377-383

Anvari Moghaddam, A. ; Seifi, A.R. (2010a). An advanced strategy for wind speed forecasting using expert 2-D FIR filters. *Advances in Electrical and Computer Engineering(AECE)*, Vol.10, No.4, pp. 103-110

Anvari Moghaddam, A. ; Seifi, A.R. (2010b). Improvement of Power Systems Operation Using Smart Grid Technology. *First Iranian Conference on Renewable Energies and Distributed Generation (ICREDG'10)*, March 9-11, 2010, Birjand, Iran

Anvari Moghaddam, A. ; Seifi, A.R. (2010c). Smart Grid: An Intelligent Way to Empower Energy Choices. *IEEE EnergyCon 2010*, December 18-22, 2010, Manama, Bahrain

Anvari Moghaddam, A. ; Seifi, A.R. (2011a). Study of Forecasting Renewable Energies in Smart Grids Using Linear predictive filters and Neural Networks. *IET Renewable Power Generation*, Vol.5, No.6, pp. 470 – 480

Anvari Moghaddam, A. ; Seifi, A.R. (2011b). A Comprehensive Study on Future Smart Grids: Definitions, Strategies and Recommendations. *Journal of the North Carolina Academy of Science, JNCAS*, Vol.127, No.1, pp. 28-34

Anvari Moghaddam, A. ; Seifi, A.R. ; Niknam, T. (2012). Multi-operation management of a typical Micro-Grid using Particle Swarm Optimization: A comparative study. *Renewable and Sustainable Energy Reviews*, Vol.16, No.2, pp. 1268-1281

Anvari Moghaddam, A. ; Seifi, A.R. ; Niknam, T. ; Alizadeh Pahlavani, M.R. (2011c). Multi-objective Operation Management of a Renewable Micro- Grid with Back-up Micro-Turbine/Fuel Cell/Battery Hybrid Power Source. *Energy*, Vol.36, No.11, pp. 6490-6507

Battaglini, A. ; Lilliestam, J. ; Haas, A. ; Patt, A. (2009). Development of SuperSmart Grids for a more efficient utilisation of electricity from renewable sources. *Cleaner Production*, Vol.17, No.10, pp. 911-918

Chowdhury, S. & S.P. ; Taylor, G.A. ; Song, Y.H. (2008a). Mathematical Modelling and Performance Evaluation of a Stand-Alone Polycrystalline PV Plant with MPPT Facility. *Proc. of Power Engineering Society General Meeting*, Pittsburgh, PA, July 20–24, 2008

Chowdhury, S.P. & S. ; Crossley P.A. (2008b). Islanding Protection of Distribution Systems with Distributed Generators—A Comprehensive Survey Report. *Proc. of IEEE Power Engineering Society General Meetin2008*, Pittsburgh, PA, July 20–24, 2008

Chowdhury, S.P. & S. ; Crossley, P.A. (2008c). UK Scenario of Islanded Operation of Active Distribution Networks—A Survey. *Proc. of IEEE Power Engineering Society General Meeting 2008*, Pittsburgh, PA, July 20–24, 2008

Chowdhury, S.P. & S. ; Ten, C.F. ; Crossley, P.A. (2008d). Islanding Operation of Distributed Generators in Active Distribution Networks, *Proc. of the 43rd International Universities Power Engineering Conference*, Padova, Italy, September 1–4, 2008

Chuang, A. ; McGranaghan, M. (2008). Functions of a local controller to coordinate distributed resources in a smart grid. *Power and Energy Society General Meeting - Conversion and Delivery of Electrical Energy in the 21st Century*, pp. 1 - 6

Cinar, D. ; Kayakutlu, G. ; Daim, T. (2010). Development of future energy scenarios with intelligent algorithms: Case of hydro in Turkey. *Energy*, Vol.35, No.4, pp. 1724-1729

Clark, W. ; Isherwood, W. (2004). Distributed generation: remote power systems with advanced storage technologies. *Energy Policy*, Vol. 32, No. 14, pp. 1573-1589

Divya, K.C. ; Østergaard, J. (2009). Battery energy storage technology for power systems— An overview. *Electric Power Systems Research*, Vol.79, No.4, pp. 511-520

EPRI. (2007a). Environmental Assessment of Plug-In Hybrid Electric Vehicles. In : *Volume 1: Nationwide Greenhouse Gas Emissions*, 1015325. Palo Alto, CA

EPRI. (2007b). Renewables: A Promising Coalition of Many, *EPRI Journal*, Palo Alto, CA.

EPRI. (2008a). Plug-In Hybrids on the Horizon: Building a Business Case. *EPRI Journal*. Palo Alto, CA., 2008.

EPRI. (2008b). The Power to Reduce $CO_2$ Emissions: The Full Portfolio. Palo Alto, CA, 2008. Available form http://www.iea.org/work/2008/.../2a_Tyran_EPRI%20Roadmaps.pdf

Esmaili, R. ; Das, D. ; Klapp, D.A. ; Dernici, O. ; Nichols, D.K. (2006). A Novel Power Conversion System for Distributed Energy Resources. *Proc. of IEEE PES General Meeting*, pp. 1–6

European Commission. (2006). European Smart Grid Technology Platform: Vision and Strategy. *Community research*, Luxembourg, pp.44-47

Figueiredo, J. ; Martins, J. (2010). Energy Production System Management – Renewable energy power supply integration with Building Automation System. *Energy Conversion and Management*, Vol.51, No.6, pp. 1120-1126

Hajizadeh, A. ; Golkar, M. A. (2007). Intelligent power management strategy of hybrid distributed generation system. *Electrical Power & Energy Systems*, Vol.29, No.10, pp. 783-795

Hammons, T.J. (2006). Integrating Renewable Energy Sources into European Grids. *Universities Power Engineering Conference (UPEC '06)*, Vol. 1, pp. 142 – 151

Kaldellis, J.K. ; Zafirakis, D. (2007). Optimum energy storage techniques for the improvement of renewable energy sources-based electricity generation economic efficiency. *Energy*, Vol.32, No.12, pp. 2295-2305

Koeppel, G. ; Korpås, M. (2008). Improving the network infeed accuracy of non-dispatchable generators with energy storage devices. *Electric Power Systems Research*, Vol.78, No.12, pp. 2024-2036

Lagorse, J. ; Paire, D. ; Miraoui, A. (2010). A multi-agent system for energy management of distributed power sources. *Renewable Energy*, Vol.35, No.1, pp. 174-182

Molderink, A. ; Bakker, V. ; Bosman, M.G.C. ; Hurink, J.L. ; Smit, G.J.M. (2009). Domestic energy management methodology for optimizing efficiency in Smart Grids. *PowerTech, IEEE Bucharest*, pp. 1 – 7

Momoh, J.A. (2009). Smart grid design for efficient and flexible power networks operation and control. *Power Systems Conference and Exposition (PES '09. IEEE/PES)*, pp. 1 – 8, 15-18

Pecas Lopes, J.A. ; Hatziargyriou, N. ; Mutale, J. ; Djapic, P. ; Jenkins, N. (2007). Integrating Distributed Generation into Electric Power Systems: A Review of Drivers, Challenges and Opportunities. *Electric Power System Research*, Vol.77, No.9, pp. 1189–1203

Saha, A.K. ; Chowdhury, S. & S.P. ; Crossley, P.A. (2008). Modelling and Simulation of Micro-turbine in Islanded and Grid-Connected Mode as Distributed Energy Resource. *Proc. of Power Engineering Society General Meeting*, Pittsburgh, PA, July 20–24, 2008

U.S. Department of Energy(DOE). (2008). The Smart Grid: An Introduction. Available from http://www.oe.energy.gov /1165.htm

Venayagamoorthy, G. K. (2009). Potentials and promises of computational intelligence for smart grid. *Power & Energy Society (General Meeting, 2009). PES '09. IEEE*, pp. 1 – 6

X. Energy. (2007). Xcel Energy Smart Grid. *A White Paper*, Xcel Energy Inc, Available from http://smartgridcity.xcelenergy.com/media/pdf/SmartGridWhitePaper.pdf

# Climate Change Due to Various Factors

# Chapter 5

# The Issue of Global Warming Due to the Modern Misuse of Techno-Scientific Applications

Gabrielle Decamous

Additional information is available at the end of the chapter

## 1. Introduction

Climate change has become undeniably one of the most important challenges for our generation and the ones that follow. This challenge is such that it affects the whole of our space-time spectrum: an urgent reassessment of our present-day technology is indisputably needed, while at the same time, this challenge also requires the most accurate knowledge of the shifts in temperature in the hope for a possible and still hypothetical sustainable future. Not only this, but it will also concern – and already does – the totality of the planet and of its living creatures. The problem that climate change represents is therefore very wide and has an effect on a very broad range of human activities. However, in spite of the importance of the historical and global parameters, it appears as if the cause of the rise in temperature of our planet is first and foremost technological (because of man-made carbon emissions) and that a technological solution would consequently be needed.

After all, within the history of technology and science, every once in a while, a notable breakthrough allows for the betterment of the human condition. Agriculture, electricity, steam engines, atomic science, biotechnology, etc., mark our timeline. These breakthroughs have enriched the life of many, for the sake of 'human' needs, complementing rather than overcoming the previous techniques and technologies. An example of this complementary use is the 'Green Revolution' in which mechanised agriculture played a major role in the 1960s to the 1990s, and that even saw countries such as Mexico overproducing and therefore exporting some new types of wheat. These past and contemporary techniques and technologies therefore play a major role in easing our lives. It would therefore be logical to await for the next techno-scientific generation to revolutionize our routines.

Indeed, the question left unanswered today is the following: what technological breakthrough will come next that will serve our energy needs? Nanotechnology and biotechnology holds no promise of energy quality or sustainable energy as much as hydrogen fuel is not yet ready for takeoff and probably needs more investment. Another techno-scientific revolution is therefore greatly expected. Yet, let's imagine for an instant, let's put forward the fantastic hypothesis that a solution to the production of greenhouse gases has been found, that hydrogen fuel becomes affordable and marketable, that wind and solar energy can be stored, even on the scale of our planet. The problem at the origin of global warming would probably be resolved. But would the core of the problem be really solved, the one of the misuse of technology itself?

Alongside the succession of technological innovations, we also have to acknowledge the succession of catastrophes, disasters and accidents that constitute our modern history and that are as important to underscore as the techno-scientific progresses themselves. Fukushima, Chernobyl, as the most devastating technological accidents, as much as the most inhuman misuse of science in warfare strategies, such as the gas chambers and the atomic bombings. The binary use of science and technology is evident, even today, within the continuous use of weapons around the world, the on-going oil spillages, the endless industrial pollution of all types and the uncertainties concerning the safest way to store radioactive waste for the coming thousands of years.

The binary aspect of technology is in fact recurrent, in its misuse, as much as in human's unstable control over nature and even over the technological apparatuses themselves. Consequently, even if we find a technological solution to the present state of the Earth and to the production of greenhouse gases, will the problem of the misuse of technology as such be solved? This mishandling and dual use is persistent, as we will detail in this essay, and needs an appropriate framing. Indeed, what tells us that this misuse will cease to happen once global warming no longer constitutes a threat to living beings and that another threat will not be in sight in the remote future?

Our current situation therefore needs to be tackled ontologically – that is, to think and underscore the *origin* of the problem and of this misuse at large. Through this approach, and by way of consequences, this essay will stand in the domain of the humanities. The strategy is to understand global warming beyond (and complementary to) the scientific challenge that it represents in the world today. As unexpected as it may seem, the humanities have already dedicated much thinking to similar problems and these thoughts are as crucial to consider today as they were yesterday – for we need to think about global warming from all possible angles. This essay will adopt existing theoretical, philosophical and artistic frameworks to investigate global warming. These frameworks will permit tackling the current crisis from a wide variety of angles, such as the recurrence of apocalyptical scenarios, the emergence of the so-called 'climate-sceptics', the absolute urgency to revise the Western project of modernity and the difficulty in reaching a universal agreement to resolve the problem. Scientific research progressed rapidly in the past century and a similarly emphasis on the development of thinking in the humanities is now needed, since this will help in considering possible ways out from the current situation.

## 2. Of the modern technologies and the technology of destruction: apocalyptical thinking

Even though an expertise is undeniably needed in this domain, it does not however require complex techno-scientific knowledge to understand the consequence of climate change: the misuse or excessive use of modern technology can lead to the destruction of the world – or at least to the destruction of its species and/or resources necessary for their survival, and this destruction is in sight. The problem of global warming can indeed be characterized very straightforwardly in such a form for now, by the destruction of our ecosystem. It is this simple yet frightening prediction that will constitute our starting point. This aspect is indeed very prominent, so much so that the term 'thermageddon' has even been coined, for instance by Greenpeace cofounder Robert Hunter.[1]

According to Hunter, and as he argued in *Thermageddon, Countdown to 2030,* the negative effect of the change in climate of our Earth will culminate in the year 2030, where the ice cap will have completely melted, with the result of unprecedented climatic disasters – the countdown is starting. Yet we must acknowledge here the recurrence of such apocalyptical endings. It is not the first time that humanity has been considered to be on the verge of total extinction: nuclear physicists warned us with their 'doomsday clock' as early as the 1940s after the bombings of Hiroshima and Nagasaki. We must therefore argue here, like the French philosopher Jacques Derrida did before us in the case of nuclear weaponry, the extent to which this all too human capacity of self-destruction is not new. Here is our first clue towards the framing of the current problem: the previous debates on weapons of mass destruction.

In *No Apocalypse Not Now*, Derrida argued in the 1980s that the post-war and Cold War anxiety of nuclear apocalypse and its hypothetical total annihilation was nothing new. As he stated: "One may still die after having spent one's life recognizing as lucid historian, to what extend all that was not new (…). One always die, humanity might well not escape the rule." [2] It is indeed human's finitude that is at stake in these debates over human's technological destruction of the environment, be it through nuclear warfare or emission of greenhouse gases. In the face of these repeated cataclysmic circumstances, one can even suspect the inheritance of religious believes, like German philosopher Günter Anders did. In the 1960s, Anders defined the fear for a nuclear apocalypse as a modern form of the Christian eschatology.[3] For Anders, the modern man re-created for himself the condition of the Christian apocalypse, when the believers awaited (and still are) for the last judgment, yet without the possibility of salvation that the Christian apocalypse envisioned. Global warming is consequently not the first time in which the end of our species or of our world has been discussed – and probably not the last either, given that space scientists are already planning the extinction of the sun in a few billion years from now.

This recurrence was also notably detected by Maurice Blanchot, another French philosopher, who pushed the argument even further as to state the lack of modernity of this nuclear apocalypse. In *The Apocalypse is Disappointing*, Blanchot asserted that the nuclear apocalypse, or rather the atomic terror, is in fact "a pretence", since, according to him, "what one is looking for is not a new way of thinking but a way to consolidate old predicaments. (…) it becomes clear that humanity will continue to run around old values, be it for all eternity."[4] Newness was one of modernity's leading criteria yet, as Blanchot pointed out, the modern apocalypse is merely a new way to restate ancestral fears and issues. Technological annihilation is therefore a recurrent problem, it actually *is* a technology in itself, a technology of destruction – the only difference being that nuclear apocalypse could have been (and can still be) triggered by the ruling class and the scientific-military cast, whereas the mass of Earth's citizens participates in the production of greenhouse gases, yet with no direct intention to kill anybody. Yet, to cut loose with this repeated anxiety of cataclysmic ending, we can also put forward the surprising argument, like Maurice Blanchot strikingly did before us with regard to the fear of nuclear annihilation, that even if this tipping point occurs, the span of its destruction will in fact be limited. According to Blanchot, even if it happened, the nuclear apocalypse would be "weak"[5], for it would damage our planet only and our personal universes, but it would never disrupt the rest of the course of the universe itself, that is, the whole universe.

Pushing this claim forward should not however become an excuse to avoid the imminence of the current crisis. To come back to our human scale, we also need to acknowledge, like Derrida did before us for nuclear warfare, that the two hypothetical and/or tangible events proceed from a process of *dissociation*, even if in different fashions. For Derrida, nuclear weapons intensified a dissociation of competences, when scientists and military men find themselves in the position of taking the final decision, within a situation whose uniqueness would be to have no pre-existing pattern, rendering difficult any expertise in the matter. Not only this, but in this hypothetical situation, the fate of many, of humankind and all living species, would be left in the hands of a few people, which precisely unravel here again any competency in the matter. As for climate change, the emphasis is on dissociation, yet a different one, the one dissociating the ancestral 'cause-effect' relationship itself. To give an arbitrary example, the effect of using a car in any place in the world does not directly affect the life of the driver, but the whole of the planet and, most importantly, in a differed and uncertain future. The dissociation is here dramatically obvious, even though every citizen of Earth does not equally participate in the cause of the current problem. More than a dissociation, global warming also includes a mechanism of individualism and indifference or guilt, once the awareness of the long-term effect of human pollution is acquired. Most drivers still use their car as they did ten to twenty years ago, in spite of the regular increases in expenses that this mode of locomotion implies. As Robert Hunter described, it still is a way of life, or at least for some in the West, while for others, it is merely a necessity, since the functioning of most societies is still based on fossil fuels.

Once again, however, this mechanism of dissociation, in this form, is nothing new. In the context of the use of napalm gas by the United States during the Vietnam War, the German filmmaker Harun Farocki framed this phenomenon in his fictional documentary *Inextinguishable Fire* (1969). The dissociation is suggested within the distance between factory workers and their final products, when the employees pictured do not know if they produce vacuum cleaners or some sort of specific weapons, whose effect they do not fully understand. While at the same time, the scientists are depicted as completely detached from reality, testing and discussing the effectiveness of napalm gas onto flies in a cold and clinical fashion. This dissociation, which adds to the dissociation of the 'cause-effect' relationship, was and still is the underlining theme of past and contemporary Western narratives, such as the notorious and fantastic *Faust* (1808) by Goethe, *Frankenstein* (1818)by Mary Shelley, and closer to us, the non-less fantastic opera *Dr Atomic* (2005) by John Adams. These narratives are of importance here as they underscore, by their popularity, the repeated concern towards science's doings.

In *Dr Atomic*, the physicist Robert Oppenheimer is depicted as taking an active role in convincing the American militaries to use the atomic bomb, because the bomb would have a unique effect that he qualifies in terms of iridizing luminescence and optimal dangerousness; while in Goethe's *Faust*, the erudite doctor who was in the search for the meaning of life and limitless knowledge, lets himself be taken by the irrationality of a supernatural phenomenon, to the point of opening his door to the disguised Devil, and therefore, to his perdition (Fig. 1). These narratives are indeed quite famously paradigmatic of such doubts in human betterment via modern scientific means, in today's world as much as at the time of the second Industrial Revolution – although Goethe's *Faust* was based on legends that were already popular in the sixteenth century, that is before the first Industrial Revolution, but the latter had a rather religious moralistic stance more than anything else.

**Figure 1.** Rembrandt, *Faust*, dry-point, etching and engraving with punch, 208x160cm, Rijksmuseum Amsterdam, 1650-1652.

Whereas both narratives are nonetheless very different in context and aims, the importance of these portrayals lies not solely in their unquestionable popularity, but also in their recurrence, for they reflect the reality of an on-going anxiety in the face of the malfunctioning of our interaction with nature and our technological apparatuses. This dissociation, between the scientist's research and its actual application or even the reality of the world, is therefore not new, and is also to be added to the already existing binary characteristic of the use of technology that we already mentioned in the introduction. One has not to forget that contemporary to Goethe and Shelley was Jules Verne for example, who was also famously known for all the enchanting technological devices that populated his science-fiction novels. This type of rather enthusiastic stance is still detectable today. Even out of the literary circles, scientists like the post-war quantum physicist Werner Heisenberg enthusiastically stated how modern technology would overcome the entire planet, while the contemporary French philosopher Paul Virilio even envisioned a total disembodiment of the mind thanks to advanced technology, where a 'scopic' mind would be able to travel to the other end of the universe. The dual use of the technology was therefore as present in the past as it is today.

With these mechanisms of self-destruction, of dissociation (of the relation between cause and effect) and with the binary aspect of the technological apparatus so far, we consequently have to acknowledge here the ways in which the threat of total annihilation that climate change represents today is nothing new. The aim of this essay is however not to speculate and theorize on the possible destruction of the human species' biosphere nor on the endless eschatological endings. To focus solely on the technology of destruction would be to disregard the core of the problem: the economic stakes relating to techno-scientific applications. Other characteristics of global warming consequently need to be underscored here.

## 3. Of believers, sceptics and entrepreneurs

Another unavoidable characteristic of global warming that particularly needs to be addressed here is the fact that this phenomenon is contested. There is indeed an ongoing debate between those asserting the existence of climate change, and the others, the 'climate-sceptics' or 'non-believers'. Here too, this binary opposition must be underscored as a recurrence, for it echoes once again the Cold War debates between those fearing and wanting to prevent the advent of a nuclear apocalypse, and the others, those believing that the human species will never been wiped off the surface of the Earth.

Not only this, but before its relation to economic interests, our current divide is nevertheless interesting in many ways, one of which being rhetorical: Derrida already pointed out how nuclear weapons generated their own rhetoric, of 'dissuasion' and 'deterrence', while climate change has clearly created a new set of vocabulary, distinguishing for example the 'climate-sceptics' from the others, that the sceptics strategically call 'the alarmists'. Even the term 'thermageddon', that we already noted, pertains to this rhetorical characteristic that goes far toward generating terms such as 'ecocide', to name the consequence of series of actual (and incontestable) devastating phenomenon, which in today's world unfortunately became a banality. The term was recently put into strategic use by campaigners in the United Kingdom in an attempt to have it recognized by the United Nations as a fifth "crime against peace."[6] Industrialists that are polluting, as much as "climate deniers" [7], would be liable in front of the International Criminal Court, for acts as much as for words. Such a term is paradigmatic of our modern times and reflects a reality difficult to deny – a reality that therefore goes beyond the rhetorical condition that nuclear warfare generated.

A second interesting specificity of this divide, between the climate-sceptics and the others, is that it does not engage the scientific community alone. Indeed, this particular condition becomes quickly obvious after reading a series of press articles on the topic, and goes to its paroxysm when newspapers themselves, that are supposed to be reporting current events in an objective fashion, come to accuse one another of belonging to one category or the other, therefore slipping into the binary distinction. For example, the recent article published in February 2012 by the *Guardian*, titled '*Wall Street Journal* rapped over climate change stance', is paradigmatic of this ongoing discussion – the *Guardian* accusing the *Wall Street Journal* of repeated bias towards, or rather, of holding a 'sceptic' position on climate change, via the stance of journalists and scientists.

Further on, the *Guardian*'s article however holds another representative characteristic, not simply of global warming alone, but of the misuse of technology at large: the economic stakes that advanced technology represents. To be more specific, the article was in the form of a letter, published by climate change experts, against the one published by the *Wall Street Journal* that was signed by a community of scientists expressing their scepticism. However, as the *Guardian* article pointed out quite clearly, almost half of the 'non-believer' scientists worked, more or less directly, for international corporations such as Exxon. Economic stakes and the search for benefits, on the side of corporations as much as of individuals, are obvious here. Here is another clue to help in analysing our current problem, and that leads to our initial assertion, that the origin to problems such as climate change is not essentially technological.

Entrepreneurs play a significant role in our situation, with the complicity or resistance of the scientists and/or government, and this state of affairs is, once again, not a novelty. One of the most striking and obvious examples, that dates from the 1970s and 1980s, is the insistence of the physiologist Hans Jürgen Eysenck giving credit to cigarettes, amidst the growing critics against the effects of nicotine at that time. His research was funded in the majority by the American tobacco industry. [8] To list such instances, of the interference of entrepreneurs and politicians in scientific research, would in fact be endless. Yet, historically, they can be traced back to the famous and popular extortion of Galileo's recantation by the Roman Inquisition – since the interests at stake are not always directly economic and can even be ideological. Another notorious historical example is Napoleon Bonaparte's war in Egypt. Napoleon employed more than a hundred fifty scientists and scholars during his imperialist war in Egypt that he initiated in the late 1790s to compete with the British Empire. Under the cover of realizing the ideals of the Enlightenment – that had the hope of archiving knowledge for a better understanding of our world and for possible betterment of our human condition – Napoleon covered his imperialist moves using propaganda, aided by scholars and interprets, going as far as to declare to the Egyptians that he encountered: "Nous sommes les vrais musulmans" (we are the true Muslims). [9]

The practical and ideological use of scientists, which took an infamously odious turn under the Third Reich, is therefore nothing new. This use is not always unidirectional, since some scientists themselves interfere with political matters, such as Einstein, who sent numerous letters to the American president Roosevelt in order to persuade him to start a nuclear military programme, in the eventuality that the Germans invented the atomic bomb first. Since the appearance of the profession of scientist itself there have always been embedded scientists, along with scientific resistance to ideological stakes. It must be said, however, that their activities are not the only activities to be targeted by political or economic interests since, as a parenthesis, artists are inevitably caught in similar situations. [10] Some artists were also counted among the team of scholars that Napoleon brought to Egypt.

To come back to the misuse of technology, this particular phenomenon has already been analysed in the post-war and Cold War context by the German philosopher Martin Heidegger. In an attempt at shedding light to the essence of technology, Heidegger characterized it not simply as 'a means to an end' when, to use his example, a Christian ritual requires the crafting of a chalice. In its place, and in his phenomenological way, Heidegger rather defined technology as a 'revealing', thanks to which nature would appear in particular ways to us, through technological apparatuses. Yet within this revealing lies a danger, according to Heidegger, as the essence of technology is also an "Enframing,"[11] or a *Gestell* as he called it, when men (*dasein*) thought they were harnessing nature, but find themselves challenged instead of mastering it. Here is an important claim that is still relevant today. Yet, for Heidegger, there is a major difference in between modern technology and the previous form of craft/technology, and this difference is of importance for our relation to technology, or more importantly, for our relation to nature.

It would seem, according to the philosopher, that nature has been put at our disposal when, in the case of a hydroelectric power station for example, "even the Rhine appears at our command."[12] Yet Heidegger's most striking critique of modern technology is that human interaction with nature is based on prospects of productivity, turning it into a mere stock or "standing-reserve"[13] as he calls it. As he stated: "Agriculture is now the mechanized food industry. Air is now set upon to yield nitrogen, the earth to yield ore, ore to yield uranium, for example; uranium is set upon to yield atomic energy, which can be unleashed either for destructive or for peaceful purposes."[14] This is the major difference between the pre-modern and modern technologies, according to him: a windmill does not store energy, whereas modern technology does. We must note here that Derrida, in his analysis of nuclear warfare, also underlined the race for stockpiling, that he even characterized as being capitalist, since this accumulation constitutes for him "the very movement of capitalization."[15] If Heidegger and Derrida's statements date respectively of the 1960s and 1980s, one must admit that today's technological interaction with nature has barely changed. Nature has to be generous, productive and directed toward human needs, but is put at *unreasonable* use, as Heidegger already underscored. It is well this unreasonable use that is problematic more than the need for stocking – a problem yet unsolved but of crucial importance in order to face the upcoming challenges such as global warming. The financial benefits that are at stake for the big energy groups are indeed too big to be re-centred towards a better understanding and use of the resources of the planet. This technological 'Enframing' therefore remains even today.

Ontologically, the problem of the misuse of technology is consequently not technological, as Heidegger already clearly defined the essence of modern technology as *not* being technological. As we just showed, the essence of technology is a revealing. The main issue to solve is the way it is handled. This is what is challenging for humans – and the most prominent aspect of the problem to solve today. Yet as Heidegger clearly stated with regard to this enframing, within the technological revealing, as much as danger lies, also lies the possibility of a certain form of "freedom," [16] for the fact that the misuse of technology is not humans' fate. There is, and must be, a possibility of salvation, if humans realize that they are, as he phrased it, the "one[s] spoken to" [17] by nature and not the other way around.

Consequently, we could imagine a new techno-scientific device that would be efficient enough to stop the production of greenhouse gases, we could imagine finding a new source of energy, a new fuel, but would this stop ideological and economic stakes? Ethics, in science, politics, art or any social activities, consequently constitute one key to resolving the problem. Yet to go further into finding the origin of our present situation, and in order to find a solution, we have now to acknowledge that these social disciplines, science, art, politics, are in the West components of or fuel for a larger project: the project of modernity, and this project is equally important to consider as it will help further in framing the recurrent misuse of techno-scientific devices, and show that, again, the source of the problem is not solely technological.

## 4. Modernity, technology and domination

Our relation to nature is therefore to be meticulously rethought and needs to be analysed further. Such a task cannot be avoided. Another characteristic of climate change, more subtle than the ones we just considered so far, is that it indeed is *modern* technology, or rather, modernity at large, that is at stake, which we only hinted at with Heidegger. There is indeed a more or less direct link between climate change and the advent of modernity and its spread around the world. In *We Never Have Been Modern*, French philosopher of science Bruno Latour explained the specificity of this relation, simply by showing the importance he sees, specifically, in the year 1989.

Within the moment of the fall of the Berlin wall, at the very moment when the West exulted at the victory of capitalism and liberalism over communism, came the first major conferences on global warming. As he stated: "In Paris, London and Amsterdam, this same glorious year 1989 witnesses the first conferences on the global state of the planet: for some observers they symbolize the end of capitalism and its vain hopes of unlimited conquest and total dominion over nature."[18] Climate change therefore shows the limits of such a system, the modern and capitalist system.

Yet for Latour, global warming is rather a symbol, of the malfunctioning of modernity. What is of prime importance for him is that the precepts of modernity themselves need to be revised and this is an urgent necessity. Not simply because of the consequences of the change in temperature of the planet, but also and mainly because developing countries inherit the Western modern ideals and that these ideals are partly deficient – or rather, they need to be revised and re-adjusted. First and foremost, we must note that Latour asserted his argument in the 1990s. In today's world however, modern technology is no longer the prerogative of the West alone. India has its space and nuclear programme, China its bullet train, and South Korea its nuclear programme for example. Yet the question of inheritance is of prime importance, for the reason that some of the precepts of modernity are faulty, as Latour argued, and we consequently need to underline these deficiencies.

For Latour, one of the major problems (that Heidegger also revealed before Latour), is the modern relation to nature, which is of major significance, beyond and in addition to the global economic stakes that we already underscored. This relation to nature is detectable in the West in a twofold movement: the first movement is the way in which modernity initially thought its relation to nature as completely new, because it separated the 'subject' from the 'object' – a typically modern characteristic, to which Latour gave attention. While the second movement saw this supposedly new relation to nature in terms of *domination*, as we will detail further. Here is the central problem of the technological 'Enframing' and from which global warming became representative, if not a materialization, of modernity's partly deficient precepts.

The binary separation between 'object' and 'subject' is a highly debated problem that can only be underlined here as a characteristic of Western modernity at large, and that is best epitomized by the eighteenth century/early-modern philosopher Immanuel Kant's 'Copernican turn'. In the *Critique of Pure Reason*, Kant attempted to unravel the epistemological dispute on the origin of knowledge as being either empirical or cognitive, by asserting that, even if most knowledge unquestionably has empirical roots, the mind remains the focal point and the condition *sine qua non* for knowledge to even exist – and thus becomes the necessary condition for metaphysics itself to exist. This is what Kant calls the 'Copernican turn'. For Latour, this separation – of which the 'Copernican turn' is not the starting point but the most representative moment – is what allowed Westerners to think of themselves as inherently different from 'the others', the pre-moderns, as they thought this divide absolutely new.

It is within this new separation that the second movement, the one best representing the modern relation to nature, came into action more evidently: this relation to nature was thought in terms of *domination*, and this is what is of prime importance in our context. As the German philosophers Theodor Adorno and Max Horkheimer emphasized in *Dialectic of Enlightenment*, the ideals of the Enlightenment in Europe hoped to free human beings from fear of the unknown thanks to science, and allow them to control nature and its resources yet, not in a new way, but within a religious inheritance from which human supremacy was never questioned. As the authors pointed out, the Jewish and Olympian religion (to which we can add the Christian religion), called for domination. The two authors directly quoted the book of *Genesis*, a referential text in the Jewish and Christian religion as stating: "(...)

and let them have dominion over fish of the sea, and over the fowl of the air, and over the cattle, and over all the earth, and over every creeping thing that creepeth up the earth."[19] For the authors, not only had the relation to nature barely changed, since the animals that were sacrificed in the past in place of God are still suffering the "tournament of the laboratory," with a couple of slight differences that the authors specified [20], but this also paved the way for dictatorship in Europe: the emergence of the subject and individuality better permitted, according to the author, the moulding and conformism into the mass. For the authors, this inheritance favoured the attempt at domination of man over nature, as much as of man over man. This is what is crucial to apprehend even in today's world and also what supplements Heidegger's argument.

Not only this, but the philosophical duo, Adorno and Horkheimer, even furthered their argument as far as to univocally qualify this dominative relation to nature and to humans as being "patriarchal."[21] Adorno and Horkheimer's critic relates to, and even goes beyond, the ways in which Robert Hunter described in *Thermageddon* the passion of his childhood for cars and motorcycles, when he recalled pinning posters of cars on bedroom walls and admiring the tailfins of Cadillacs, claiming that "*this* was manhood."[22] It is not exactly this masculine world that is at stake here, but something more deeply rooted than a mere addiction and dependence for car bodywork and rides. The problem is not one of gender, but of domination. For Adorno and Horkheimer, this 'patriarchal' relation to nature is in operation within these particular instances: *when the mind serves for ruling and knowledge is used as a power*. For them, this failure is the failure of the Enlightenment, when the ideals such as those of the early-modern philosopher and scientist Francis Bacon, who envisioned "the happy match between the mind of man and the nature of things," ultimately failed since, as the two authors forcefully stated: "What human beings seek to learn from nature is how to use it to *dominate it and human beings*" (the italics are mine).[23] Which in turn explains the authors' emphasis on the most paradoxical condition of the European Enlightenment: that it aimed at human's betterment, but saw the most radical and devastating dictatorships. Some of the Western precepts of modernity therefore urgently need re-assessment as Latour already pointed out.

In terms of war, even today, we have to admit that a couple of Western countries are still entangled in devastating wars, which have been displaced mainly to the Middle East. Hunter also underlined this aspect, when he stated that, beyond ecocides, there is an immediacy to look at this oil addiction politically. As he stated: "In purely political terms, oil has unleashed massacres, fortified criminal rulers, spawned atrocities, and served terrorism just about everywhere you look."[24] To which it can be added that energy firms, that are other than oil-orientated, are no less stained or necessarily greener. For example, the French nuclear corporation AREVA has long been criticized for the little regard it gives to radioactive contamination of the mines, landscape and area it exploits, particularly in the former French colony of Niger, as much as (and most importantly), for the actual workers on site.[25] Whereas paradoxically, the nuclear corporation AREVA intends to make a business out of the current crisis caused by global warming by advertising its services as 'C02-free', while still using fossil fuel powered boats to transport the processed uranium from the mines to the French continent. This shows again the exploitation of nature and humans by humans in such a concentrated form, and most importantly in a neo-colonial form, and it underscores again the repetitive problem of dominion (beyond the wars started for oil) which undoubtedly presses for a re-assessment of Western modernity. In the face of such a lack of ethics, in the case of AREVA's use of techno-scientific applications or corporate use at large, and as we already stated throughout this chapter, it becomes obvious that the cause of the current crisis is not merely technological, it is also human and cultural, and this aspect needs to be considered in the finding of a solution.

It must be said, however, that initially, the Western ideals of modernity had a reasonable justification – reasonable in the sense that it aimed at involving reason and rationality in decision making. At the origin of this modern relation to nature, the separation between 'object' and 'subject' lies, paradoxically, the particularly modern ambition for secularity. Indeed, one of the precepts of the Enlightenment, as it developed in Europe, was to prevent beliefs and superstition interfering in a given society's doings. Natural phenomena were thought to be interpreted differently, that is, empirically and assumed to be observed 'objectively' and, therefore, no longer as signs from God – which ultimately is of upmost importance. For Latour, the appearance of the scientific apparatuses, in the desire to observe natural phenomenon objectively, are in fact preceding Kant's 'Copernican turn' in the separation between the 'object' and the 'subject'. Nature and culture, object and subject therefore become separated and human beings considered as separated from nature – another typical Western turn. Nature is endowed with a history and properties of its own and this, in order to prevent the use of natural phenomenon for political, ideological and religious agendas. However, as the two philosophers Adorno and Horkheimer argued and as we have already underscored, the mythological, magical and religious original interpretation of the world persists, even in our modern times, precisely through this dominative behaviour. Here is another seminal clue to some of the imminent issues we still have to face today.

Similarly to Adorno and Horkheimer, Latour also stated that this endless quest for domination found its limit. As he stated: "The repressed returns, and with a vengeance: the multitudes that were supposed to be saved from death fall back into poverty by the hundreds of millions; nature, over which we were supposed to gain absolute mastery, dominates us in an equally global fashion, and threatens us all. It is a strange dialectic that turns the slave into man's owner and master, and that suddenly informs us that we have invented ecocides as well as large-scale famine."[26] For Latour, capitalism is also to blame, and communism did not achieve better since, by wanting to prevent the exploitation of men by men, it emphasized it instead. The nuclear corporation AREVA is here again an example of what Latour underscored, as it certainly is the capitalist rules that pushed and still pushes forward this firm to try to sustain, at all costs, its position of leader in the restricted market of nuclear reactors. Competition is not, however, the main problem. In this example, the questions that in fact remain to be posed are the following: can sensitive technology, such as nuclear technology, be used and treated in the same way as any other marketable product? In addition, can competition afford to disregard ethical consideration of nature and humans?

Not only this, but since we underscored the separation between 'object' and 'subject', we must acknowledge this modern attempt at secularity as failed once again– for even in the twenty-first century, one still thinks in terms of 'belief': on the one hand, the separation between the so-called 'climate-sceptics' and 'alarmists' forces one to claim if they 'believe' in climate change or not; while on the other hand, religious groups for instance, still question scientific empirical evidences such as Darwin's theory of evolution for instance. To come back to the climate-sceptics, this phenomenon must be analysed more carefully here, not simply for the rhetorical condition that we previously examined, the one labelling and differentiating the two categories, but also and primarily for what can be called a 'reversed religious justification'. It was precisely to prevent the justification of an act, a murder or a war in the name of God that the Western early-moderns moved toward secularity, in a joint effort that encompassed many social activities such as science, philosophy and art.[27] Yet this religious justification is reversed when the climate-sceptics use precisely the fact itself of being 'sceptics' and of not 'believing' in climate change as a justification for their economic interests (some scientists being referenced as having worked for Exxon) – interests that in some cases led to still current wars in the Middle East. The previous use of religious justification to cover economical and ideological stakes, is here re-enacted through corporate interests. Modernity's ideals of secularity clearly failed in this instance.

The present crisis, of which global warming is representative, is therefore wider than what we have tackled so far. It encompasses several cultural, political and economic stakes, about which we can do nothing but to underline them in this context. A re-assessment of the Western project of modernity is therefore crucial. However, and as the international community has been moving towards a *global* solution to refrain the change in climate, we need to pay heed to this particular *global* instance. One of the reasons for this urgency to analyse this need for a universal solution is that other ideals of the Western modernity need to be revised and it is important to consider them in our context.

## 5. Universality I: Global warming in art and science

Another of the characteristics of global warming is indeed that a *universal* solution is required, or rather, that a universal effort to reverse the current situation is demanded. In the humanities, universalism has long been critiqued, for it negates diversity and individuality. Yet in our context, the one of global warming, this demand upon us to find a universal solution is flagrant and absolutely necessary on many levels, levels that are important to highlight since they add to the list of characteristics of the Western development of modernity that need to be rethought.

In the first instance, this universal effort requires scientists from a large number of disciplines to collect data and to find a solution. The scope of this need for interdisciplinary cooperation is extremely wide and exceeds the field of climatology and environmental sciences. For example, the fields of paleo-climatology, volcanology, marine biology, meteorology, space and computer science play an important role in monitoring as well as working towards finding a solution. This need for interdisciplinary cooperation is such that the prefix 'climate' is sometimes added to some already pre-existing scientific disciplines.

After dedicating much writing to demonstrate the extent to which the situation of climate change is nothing new, we must accentuate here that the scope of this interdisciplinary need is what is perhaps the most specific characteristic of the present situation, because the post-modern interdisciplinary cooperation is extensively required, more than it ever has been since the advent of modernity and even post-modernity.

The urgency of our current state is such that a new scientific field was created, with its 'climate experts', and this with no direct or major technological apparatus to define it. The international cooperation of scientists in this matter dates from the nineteenth century, but never did it encompass such a wide range of disciplines, that have only recently been grouped under the name 'climate science'.[28] As we already emphasized, techno-scientific disciplines succeeded one another, creating new fields according to new resources or apparatuses, as biotechnology would, for example, be characterized and defined by bio-technological apparatuses and a focus on genetic material. In the modern era, we have seen a succession of nuclear technology, space technology, biotechnology and even nanotechnology. Yet climate change has created a field of experts that cuts *across* the wide range of pre-existing disciplines. This is perhaps the most important and specific characteristic of climate change. The reason for this is simple: the effect of greenhouse gases affects the *totality* of the planet and of our ecosystem, whereas the scientific research, as it has been conducted in the modern West so far, has been based predominantly on the *specialisation* and *autonomy* of the fields. Specialization and autonomy are another of the precepts of modernity that we have previously emphasized, such as secularity and the separation between 'object' and 'subject' or 'nature' and 'culture'.

The specialization of the fields is a problem that Heidegger already spotted in is magnum opus *Being and Time* – so prominent in the humanities – even though the main argument of his piece is not directly directed at solving techno-scientific problems. Yet for him, the ontical sciences, or the empirical ones, make the mistake of trying to understand natural phenomenon from the restrictiveness of the delimitation of their own discipline. While at the same time, he also clearly stated the absolute necessity of the specialization of the fields in another essay titled *Science and Reflexion*, when he stated that "specialization is also not merely an unavoidable evil. It is a necessary consequence, and indeed the positive consequence, of the coming to presence of modern science."[29] The current paradox is now clearly evident and calls for interdisciplinary cooperation as perhaps never before, while requiring at the same time an advanced specialization of the scientists in the direct tradition of modernity's autonomy of the fields – which can sometimes become an obstacle as each discipline has its own specific language.

Yet we must acknowledge that this tension is however not always an obstacle and can become a force. The necessity of interdisciplinary cooperation has become so important that even the arts are being included into the effort and here in another from than a mere cultural barometer of interest or anxiety towards science that we previously analysed with Goethe's *Faust* for instance. For the last couple of decades, the two disciplines, art and science, have been deploring the ways in which they are separated and so distant from one another – another result of the modern predicates of autonomy of the fields. Yet in this instance, one has to admit that the challenge that climate change represents today favours, if not forces, a proximity and cooperation. One of the reasons for this is that artists are no less Earth's citizens than are scientists, activist and politicians, and their statements and actions are therefore as important to cover as those from scientists, philosophers and experts.

At first glance one might think these artistic activities done in response to climate change are of peripheral importance. Some art protagonists themselves look at this type of art with a doubtfuleye, but disregarding the importance of artistic strategies would be a mistake. The efficiency is more subtle than one may think, for they work at a social level. A perfect example of this, among the myriad of artworks produced in relation to climate change that are equally as inventive as one another, is the work of the collective HeHe (Helen Evans and Heiko Hansen). In 2008, they organized a city-wide event in Helsinki, Finland, during which a green laser was flashed, over night, onto the cloud produced by the chimney of the Salmisaari power plant – a coal power plant. The laser was interactive in the sense that it would, in real-time, reduce or expand according to the consumption in electricity of the city. The piece was titled *Nuage Vert* (Green Cloud, Fig. 2) – the collective having an interest for human-made clouds at large such as cigarette smoke. [30] In order to realize the project, a joint effort between the artists and some experts was necessary, which is paradigmatic again of the current need for an interdisciplinary approach that post-modernity had already called for. The artistic event was made possible thanks to the teamwork of experts in various areas such as air quality monitoring, electrical engineering, laser technology and computer science.

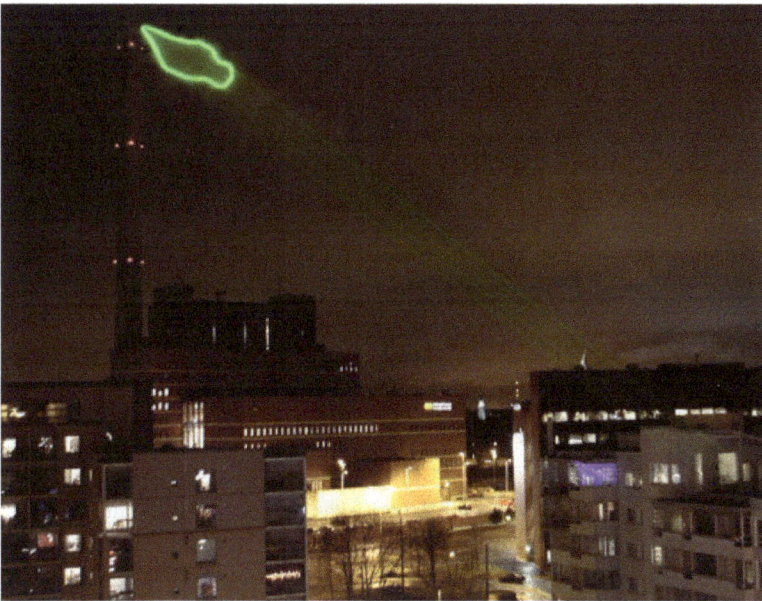

**Figure 2.** HeHe, *Nuage Vert* (Green Cloud), Helsinki, Finland, 2008.

The specificity of the project is however manifold: on the one hand, not only did the artists give visibility to at least one type of greenhouse gas – a phenomenon that still remains an abstract idea for many since most power plants are usually remotely located and that the greenhouse gases are mainly invisible; while on the other hand, they also worked within the city itself, with an intensive mediation of the project on a local level. The artists asked the inhabitants to reduce their use of electrical appliances in order to change the length of the green cloud during the week of the event. HeHe not only displayed posters throughout the city, but they also toured primary schools to explain the project to children, knowing that they would, in return, ask their parents to participate in the project and, by way of consequences, that they would 'educate' their parents on their level of consumption – education being perhaps one of the most forceful technologies used by the collective in this project.

Another specificity of the project of importance here is the link that was finally re-established between humans and their local and direct environment. We have already showed the modern dissociation between 'object' and 'subject', the separation between 'nature' and 'culture', and most importantly, the dissociation of the relationship cause-effect. Yet with *Nuage Vert*, more than a direct relation to 'nature', a cause-effect relationship is re-established through the inhabitants' direct link with the remote production of greenhouse gases that were made visible – a link that has been ruptured generations ago and definitely needs to be re-established. The tension between the specialisation of the fields and the cross-disciplinary discussion is here clearly a strength..

It is on a similar plane of action that the fascinating artistic approach of the American artist and biologist Brandon Ballengée is also located. For decades Ballengée has been locally collecting deformed frogs, reptiles and amphibians in different parts of the world in order to determine the cause of their deformities. This is the 'Malamp Project' that he initiated in the late 1990s, among other equally important projects of the artist. [31] As with the collective HeHe, interdisciplinary work is of central importance. Ballengée collaborated with numerous other scientists, such as Professor Stan Session, an amphibian specialist at Hartwick College, in New York, yet he also opens up his field trips to the public, in what he calls his 'eco-actions'. Here the link human beings and nature is re-built in a different fashion than the modern one, through the awareness of one's own local ecosystem (and therefore consistent with the modern ideals of enlightenment through knowledge). Ballengée's artworks take different forms, between the 'eco-actions' (Fig. 3) and the exhibition of physical artworks: the artist developed a process to preserve the specimens, that colours their bones and cartilages, therefore highlighting the deformities (Fig. 4). In the gallery space, large prints of the scanned specimens, as well as the preserved specimens themselves, are exhibited, therefore constituting unavoidable empirical evidence of these birth defects.

These birth defects have many causes, from parasites to predators and, unmistakably, from chemical pollution – a fact that has recently been under the spotlight of the media given the suddenly high numbers of eyeless prawns, fish with lesions and crabs without claws that have been fished up from the Gulf of Mexico and for which the 2010 BP oil spillage is believed to be the cause on the 2010 BP oil spillage.[32] Ballengée's strategy is therefore located at the tension between the modern ideals and mistakes, as much as at the tension between the global and the local and a re-connected cause-effect relationship.

**Figure 3.** *Amphibian Eco-Action*, Yorkshire Sculpture Park, Wakefield, England, Photograph 2008 by Jonty Wild.

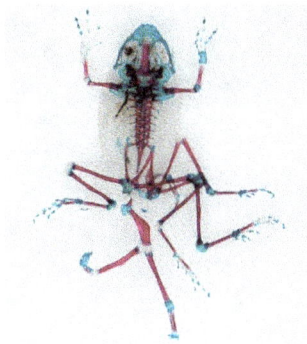

**Figure 4.** Brandon Ballengée, *DFA 23, Khàrôn*, Unique scanner photographic print of cleared and stained multi-limbed Pacific Treefrog, 2001/2007.

It must be said that this type of art is not the privileged strategy of individual artists on their own and sometimes encompass the cooperation of curators, museums and governmental or even independent founding bodies. The exhibition *Lovely Weather* for example was an artists-in-residency project in Donegal, Ireland, that was co-curated by Annick Bureaud and John Cunningham, with the help of Letterkenny Regional Cultural Centre, the Public Art Project of the Donegal County Council and the art/science journal Leonaro/Olats. [33] The primary aim of the exhibition however, and the specificity of its strategy, was to have a local approach to the problem of global warming. One of the artists in residence for instance, the Canadian artist Seema Goel, initiated a spinning and knitting workshop whose final items, some socks and hats, used patterns from translated data coming from the Irish Malin Head meteorology station and other meteorological data (Fig. 5). Yet the workshop has more ambition than simply materializing opaque data for individuals. This project was initiated after the artist came across a labelled 'Donegal sweater' that therefore appeared to be produced locally, but was in fact made in Thailand for the clothing corporation American Eagle. The intention of the project was primarily to use sheep's wool collected from local farms (which has almost no value for the farmers) and to spin and knit with it, therefore producing some typically and finally authentic Inishowen/Donegal wool. With this artwork, the problems of the exploitation of humans by humans, of the disappearance of local industries in the age of globalization and of the use of fossil fuel means of transportation in the exportation of products such as those produced by American Eagle, is here solved in one spin of a bobbin: the workshop started a wool cooperative which is still active today.

**Figure 5.** Seema Goel, "Carbon Footprint", Lovely Weather project, exhibition view, photo Annick Bureaud.

To come back to the need for interdisciplinary work, it must be said that if this post-modern need for teamwork is of crucial importance and a characteristic of the consequences of climate change, this interdisciplinary effort cannot, once again, be considered entirely new. In the past, and in the pre-modern West, art and science were linked by the fact that they were directed towards cult and the scriptures. Religion was the unifying force. In Western art (or 'craft' since art was considered as such at that time), Renaissance painters or even the Greek sculptors and architects devoted their work for communication with the Gods and for the illustration of religious or divine characters. The form and the content of the work of art was principally directed toward a cult and was an object of cult, as with the famous Renaissance frescoes and Greek temples. The same applied in 'science' – or in science as 'knowledge' –since science as a professional discipline only emerged in the late 1830s. During the Middle Ages for example, Paul of Burgos, a prominent Spanish religious character and exegete stated and published that the Earth had two centres of gravity, one for the soil and one for water.This was done in order to conform to the narrative of the creation of the universe by God, as described in the book of *Genesis*, when God immersed the Earth with water and ordered on the third day for dry land to appear.[34] The development of new techniques, craft, and the knowledge of the cosmos and political decision where unified through religious beliefs.

Trans-disciplinary cooperation is therefore not new and it is not the first time to be put to work since the advent of the modern autonomy and specialization of the fields. Other examples, and closer to us, are the totalitarian regimes, such as Nazism in Germany and Communism in Russia, that saw the enforcement of cooperation between art and science, as with other social activities, in the service of the regime. Yet their aims were ideological and their scope national. The rarity of the current situation is such that there is no political-ideological input, nor particular national stakes – or at least not at first glance. The scope is transnational and trans-disciplinary (even though we have to note here that the bulk of what some call 'climate art' is mainly created by Western artists). However, climate change and the threat of it represents a call for a wider universal effort and, paradoxically, a demand for cooperation and simultaneously specialization, therefore putting modernity's autonomy of the fields into question. We previously showed how the origin of the current crisis is not technological and we have now to acknowledge that the finding of a solution does not necessarily include technological devices. Not only this, but in the ontological approach we have conducted so far, human beings were considered as a whole, as if there was only one human category, one unconsciously thinking of its relation to nature in terms of domination. Yet are all humans the same? With the same background, history or culture? Do they all share the same role in the triggering the current crisis? Acknowledging this will help us to give perspective to the call for a *universal* solution for global warming and underline better the role played by the Western project of modernity in the present situation.

## 6. Universalism II: International politics, from the competition of the empires to the Cancun agreement

At another glance and after underlining the paradox between universalism (or the need for interdisciplinary work) and the specialization of the fields, climate change implies not only a joint scientific effort, but also a joint effort on behalf of all Earth's citizens, no matter their origins, nationality or activities. This is what is utterly new in this situation, climate change is a global phenomenon that concerns each of us. Yet what we must clearly ask is the following question: is ultimate universalism even achievable?

Once again, we can put forward the claim that ideally 'we', the citizens of the Earth, 'we' the scientists, the corporate energy agencies, the politicians of the world, the non-experts - in a word 'we' the human beings at large, can, should and will combine our forces and will work together. After all, that is precisely what Albert Einstein hoped for when he called for unity in his *Manifesto* against weapons of mass destruction, written with Bertrand Russell in 1955. As they stated: "We appeal as human beings, to human beings: Remember your humanity, forget the rest."[35] For Einstein and Russell, no matter what the issue is, "whether Communist or anti-Communist, whether Asian or European or American, whether White or Black, then these issues must not be decided by war."[36] We must be optimists and imagine that this plea will be met today, given the necessity of a transnational unification to solve the problem. Yet we must admit that, in the case of nuclear weapons, this universal solution was certainly not reached, even today. The recent wrestles between Western countries and Iran on Iran's nuclear programme or the possible use of depleted uranium in the 1991 Iraqi war are evidence enough. In the case of nuclear weapons, a universal agreement still appears hardly achievable. Consequently, can a universal agreement be achieved in order to battle against the changes of climate of the Earth, given the imminence of the threat?

The divides that prevent a unique solution are in fact numerous and can only be mapped out here in a restricted form, before looking at a couple of solutions. One of the main causes of dissension is of course economic, caused by the many competing national interests. The American refusal to sign the Kyoto protocol, followed by the Russian hesitation, is paradigmatic of this difficulty in reaching a universal agreement (and therefore a solution) because of national interest. The George W. Bush administration declared following the protocol that it would gravely endanger the US economy, while Russia had a similar claim before finding more support from the European Union.

A second important cause, preventing universal agreement, which we have already underscored, is the divide between the climate-sceptics and the others. The interests of the large industrial and energetic corporations – or the 'Carbon Club' as Hunter calls them – are still too important for them to feel concerned by ethics in any form. Preventing ecocides does not guarantee economic benefits – of this we already know the problem. Yet this divide does not concern the Carbon Club alone, nor even the divide among the scientists at large, but is spread among the world population to the point that a very controversial publicity film was launched – then withdrawn – as part of a British and global campaign called 10:10. The campaign aimed at reducing carbon emission by ten percent in 2010, following the Copenhagen accords, and in order to give a sense of responsibility to individuals, companies and businesses globally.[37] Yet the four minute film, titled *No Pressure* and directed by Richard Curtis depicts a series of graphic explosions of persons, including children, in their daily life, in a work environment, at school or on a football pitch. The individuals are being killed on screen because they showed signs of scepticism or are lacking enthusiasm with regard to the 10:10 campaign.[38] Although withdrawn from the campaign even before broadcast, the video spread on the Internet and caused a public outcry. This particular instance is of importance, first because climate change is still contested even today and tends to fade away in public debates. As the organization stated, to explain their video: "With climate change becoming increasingly threatening, and decreasingly talked about in the media, we wanted to find a way to bring this critical issue back into the headlines whilst making people laugh."[39] Secondly, this aspect is of importance because it shows that it still takes a lot of marketing strategies, provocation, financing and imagination to sustain concern from the public with regard to the problem. As Franny Armstrong, the founder of the 10:10 campaign stated: "We've 'killed' five people to make *No Pressure* – a mere blip compared to the 300,000 real people who now die each year from climate change."[40] Sensitizing individuals appears to be as difficult as forcing energy corporations comply. Most of all, we have to realize that it takes time, since the first conferences on climate change in the 1980s, for the idea of the absolute necessity of cultural change to make its way into people's mind and for this idea to remain.

Another peripheral yet still significant cause preventing general agreement that was quite apparent during the moments of the Kyoto protocol, and beautifully pictured by Hunter in *Thermageddon*, is based on political regimes and cultural differences. The most obvious example is again the George W. Bush administration's justification for its reluctance in taking part to the global effort, that this reluctance had to do with the fact that it would affect the 'American life style itself' and its car-addiction habit. Another cultural tension also described by Hunter was the dismay generated by a discussion between the members of an Arab and a Canadian delegation. According to Hunter, in order to reply to the statement of the Arab representatives, who argued for the importance of keeping low prices for petrol, so "the poor can drive to work,"[41] the Canadian delegate suggested a redistribution of wealth instead. Not only was the difference between the two understandings of the problem blatant, but the Canadian delegate was a female and this caused her to be swiftly dismissed, as Hunter recalled, and to which he added "but what could they do in a foreign democracy where women have the vote?"[42] How can one find a general consensus in spite of the cultural, ideological, economic and political differences, which should be a force enriching the debate but which in this condition rather appears as an obstacle? Here again, as we already pointed out the finding of a solution does not involve purely technological concerns.

Adding to this aspect is the divide between the East and the West, or between the developed and industrialized countries. Apprehending this divide is of prime importance, for it relates to the urgency of revising the Western project of modernity itself and also adds to the cultural histories that consciously or unconsciously precondition our relation to nature and human beings. This problem was again portrayed by Hunter, for instance in the lecturing of an American representative to delegates of developing countries such as China, Brazil and India, on the effort the latter countries would have to produce to reduce carbon emission – to which somebody in the audience finally said in order to point out the paradox: "A nation of sports utility vehicles lecturing nations of bicycles?"[43] This paradox is representative of a wider problem whose causes are, again, deeply rooted, politically and historically. The divide between the East and the West is a consequence of the Western project of modernity as this project was linked to the building of the European Empires.

On the one hand, it would be absolutely unethical to avoid mentioning here that the emission of greenhouse gases is an unpredicted yet direct outcome of the modernization of the West through the two Industrial Revolutions, which was initiated in Europe and in the West at large. It is this problem that now affects the *totality* of the planet and countries – hence Bruno Latour's instance on the importance to revise Western modernity. In *We Never Have Been Modern*, Latour also alerted us to the importance of considering this divide. He asserted that modernization was brutal toward the pre-moderns and that post-modernization now represents another challenge. As he stated: "Imperialist violence at least offered a future, but sudden weakness on the part of the conquerors is far worse for, always cut off from the past, it now also breaks with the future. Having been slapped in the face with modern reality, poor populations now have to submit to post-modern hyperreality."[44] Because the Western techno-sciences are being exported, a revision of their use and applications is doubtlessly necessary.

However, on the other hand, we must also consider the fact that the idea of modernity itself was precisely constructed through this divide between the East and the West. The great Universal Exhibitions that were staged in the late nineteenth century constitute the best example. In these exhibitions, Western scientific devices and art objects were opposed to colonized cultures presented as backward and archaic. The 1889 exhibition, that was held in Paris, even gathered from the colonies some inhabitants and their habitats, that were exhibited in human zoos and contrasted to reconstituted medieval houses (Fig. 6). The narrative suggested via these juxtapositions was the superiority of the Empire, illustrated by the demonstration of technological superiority. Here is a particularly flagrant example of the patriarchal use or understanding of technology and humans as critiqued by Adrono and Horkheimer. The narrative in turn was used especially by the French Empire to justify its imperialist wars that were nonetheless initiated in the midst of a competition of strength and power between the few Empires of Europe.

This implemented fracture, between the East and the West, is also accentuated by the fact that the industrialization of the developing countries is now asked to be curbed because of the problem of gas emission – a crisis which originated in Western and illustrates again the importance to consider parameters that are no technological in the finding of a resolution.

Moreover, in this necessary attempt at finding a common consensus, how can international organizations, such as the United Nations and UNESCO succeed? By trying to universalize a method of calculation of anthropogenic emissions, by quantifying limitations or reduction commitment according to countries as the Kyoto Protocol tried to do, in such an attempt at finding a common equation to filter all discrepancies and level them down to equality in the task of reducing greenhouse gas emission, how can one rationalize and put into numbers the richness and variety of paths that each country follows (whose borders are not always stable) as much as the differences in power that still divide the world?

**Figure 6.** Medieval and Renaissance Houses (left), and Algerian/Kabyle House and Family (right), Paris, Universal Exhibition of 1889, Library of Congress, Washington. See also Alfred M. Picard, Exposition universelle internationale de 1889 à Paris: Rapport general, ed. 10 vols. Imprimerie nationale, Paris, 1891-1892.

Universalism and diversity are here again clearly at stake and a recurrent tension. The problem of climate change is therefore taken here into another context, beyond the techno-scientific or even philosophical problem. The greatest challenge for today's world is therefore located in this particular divide, *when the intricate geopolitical interests that were at play in the advent of modernity in the West are still at stake today*. The example of the exploitation in a uranium mine by the French company AREVA already described, the lack of ethics in the exploitation of mines and workers of the former colony of Niger, clearly shows the extent to which a more profound change in human activities at large is needed.

In the midst of these difficulties, we must however consider the late appearance on the horizon of international and socio-political breakthroughs. The Barack Obama administration, for instance, re-oriented the US policy towards more contribution with the

Copenhagen Accord of 2009, then with the Cancun Agreement (reached at a United Nations conference on climate change in Mexico in 2010). An international fund, called Green Climate Fund, has even been established. This fund would help developing countries financially for climate-oriented programmes, projects and policies, which constitutes a fantastic step forward even though its financing is not yet defined. However, no matter how great these political decisions are, they are also dependent on the integrity, good will and the seriousness of the political leaders as individuals, and are therefore bound to them – showing here again that the solution to the problem is not merely technological.

## 7. Conclusion

As we just argued, even if we develop a revolutionary technological device, the current crisis might well be fixed, but will this also prevent the misuse of technology? In other words: will the technological apocalypse be permanently stopped or simply delayed? At present, they are some solutions in sight, along with international cooperation. Yet if the search for sustainable sources of energy is developing, how sustainable are the efforts on a socio-political and economic level given that the world is still affected by international conflicts and exploitation of humans by humans? Furthermore, within the continuous effort that is incontestably and unconditionally needed on the part of the scientific research, how will the numerous future technological innovations adequately serve humans with respect to the diversity of their place of birth, history and culture and this, beyond the Eastern/Western divide? In order to face the persistent misuse of scientific applications that started to be dramatically recurrent since the 19th century, the parameters to consider in addition to the techno-scientific ones are therefore numerous. Without revising the interaction of humans towards their environment, without realising that humans are 'the one spoken too', as Heidegger stated, and not the ones controlling nature, can the centuries-long technological misuse (or 'Enframing' to use Heidegger's vocabulary) be stopped? Without a re-adjustment of the *patriarchal* understanding of the world – and most importantly: of humans – that Adorno and Horkheimer underlined, and the inheritance of the religious call for domination over nature, will we solve the problem? In the search for a solution, the separation between developed and industrialised countries, the impact of colonization and of post-modernity, the impact of the lobbyists and of the patriarchal/dominative behaviour in any place of the world must be addressed.

To be more precise, within these ideal yet necessary revisions, of the interaction of humans with their environment and with humans, it in fact is the *modern* split between nature and culture, as much as the modern secularity, that are at stake. As we already pointed out, it indeed is the western project of modernity at large that needs to be revised as Bruno Latour already argued in the 1990s. In *We Never Have Been Modern*, Latour urged for a revision of some of the predicates of modernity, in particular the separation between 'object' and 'subject'. Not only this, but for him, it is the modern idea of progress itself that needs to be reconsidered, since the modern mistakes, accidents and disasters should equally be considered instead of an idea of technological and social betterment, where the future is envisioned as necessarily better than the present. Is it not indeed this precise mythical understanding of modernity that has led an entire generation to leave a legacy of pollution of thousands-year radioactive waste in the hope that a solution will be found by the future generations, then imagined to be ultra-techonologised? To paraphrase the notorious nineteenth century French poet Charles Baudelaire: if no one cannot dispute the fact that our lives are better today than they would have been a hundred years ago, how can one bet on their betterment in the future? A century later, the same question still imposes itself, which finally urges to ask: will we learn from our mistakes?

## Author details

Gabrielle Decamous
*Faculty of Language and Cultures, Kyushu University, Fukuoka, Japan*

## 8. References

[1] Hunter R. Thermageddon, Countdown to 2030. New York: Arcade Publishing Inc; 2002.

[2] Derrida J. No Apocalypse, Not Now, (full speed ahead, seven missiles, seven missiles). Diacritics. Summer 1984; p. 23.

[3] See Gunter A. La Menace atomique. Considérations radicales sur l'âge atomique (The Atomic Threat. Radical Considerations). First published in 1960.

[4] Blanchot M. The Apocalypse is Disappointing. In Friendship. Op. Cit., p. 104

[5] Op., Cit., p. 107.

[6] See Jowit J. British campaigner urges UN to accept 'ecocide' as international crime. The Guardian Friday 9 April, 2010.
http://www.guardian.co.uk/environment/2010/apr/09/ecocide-crime-genocide-un-environmental-damage (accessed 14 April 2012). For the ecocide campaign, see http://www.thisisecocide.com/.

[7] Idem.

[8] See Buchanan R. D. Playing with Fire. USA: Oxford University Press; 2010.

[9] Said E. Orientalism, London: Penguin; 1977; p.83.

[10] On the use of artists by corporations, along the activist/artistic responses to nuclear warfare see my article Nuclear Activites and Modern Catastrophes: Art Faces the Radi-oactive Waves. Leonardo Journal MIT Press 2011; 44 (2) 124-132.

[11] Heidegger M. The Question Concerning Technology. In The Question Concerning Technology and Other Essays. New York: Harper Torchbooks; p. 23.

[12] Op. Cit, p.16.

[13] Op. Cit, p.17.

[14] Op. Cit, p. 15.

[15] Derrida. J. No Apocalypse, Not Now, (full speed ahead, seven missiles, seven missiles)'. Op. Cit., p. 23.

[16] Heidegger M. The Question Concerning Technology. Op. Cit., p. 25.

[17] Op. Cit., p. 27.

[18] Latour B. We Never Have Been Modern. Cambridge: Harvard University Press; 1993; p. 8-9.

[19] Genesis I. Quoted by Adorno, T. W. and Horkheimer M. In Dialectic of the Enlighten-ment, Philosophical Fragments. Stanford: Stanford University Press; 2002. First pub-lished in 1944; p. 5.

[20] According the author, there is a major difference, according to the author, between the ritual sacrifice and laboratory experiments, in the way that the experiment needs the specificity of the one being experimented, being an atom or a rabbit, while sacrifice was more flexible. Op. Cit., p. 7.

[21] Op. Cit., p. 2.

[22] Hunter R. Thermageddon: Countdown to 2030. Op. Cit., p. 26.

[23] Adorno T. W. and Horkheimer M. Dialectic of the Enlightenment. Op. Cit., p. 2.

[24] Hunter R. Thermageddon: Countdown to 2030. Op. Cit., 32.

[25] Greenpeace conducted a research onsite and concluded to unacceptable conditions for locals and workers, see
http://www.greenpeace.org/international/en/news/features/AREVAS-dirty-little-secrets060510/ (accessed October 2011). From personal sources, I have been explained that, even if AREVA constructed a hospital for its workers, it only is open to current workers, not retired ones. The retirement age on site is however very young, which for-bids any treatment of eventual cancer due to the radiations. At the time of my inquiries (summer 2009), I have been told that, as an example, a former mineworker had died of cancer at age 30, leaving a family behind.

[26] Latour B. We Never Have Been Modern. Op. Cit., p. 8-9.

[27] On the joint cultural effort between science, philosophy and art toward a secular society in Europe, as much as on the importance to underscored this failed attempt in a world where non-secular countries such as Iran are targeted, see my in my article: Bridging the Gap: Art, Science, Philosophy–Modernity in Question. The International Journal of the Humanities 2012; 9 (8) 185-194.

[28] On the history of climate change and of scientific professions relating to global warming, see Weart S. R. The Discovery of Global Warming, March 2011.
http://www.aip.org/history/climate/internat.htm (accessed March 2012).

[29] Heidegger M. Science and Reflection. Lecture given in August 1954 in preparation of The Question Concerning Technology for the series of conferences The Arts in the

Technological Age. In The Question Concerning Technology and Other Essays. New York: Harper & Row; 1977; p. 169.

[30] For more information on the project see http://www.pixelache.ac/nuage-blog/. The project won the Environmental Art Award, the Ars Electronica Golden Nica and the 01SJ Green Prix for Environmental Art in 2008. In 2010, the collective recreated the event in France, in Ivry-sur-Seine. A video of the event is available at

http://vimeo.com/17350218 (accessed April 22nd).

[31] For more information, see Brandon B. Malamp, The Occurrence of Deformities in Amphibians, Nicola T. and Miranda P. Editors. The Arts Catalyst and the Yorkshire Sculpture Park; 2010. See also:

http://greenmuseum.org/content/work_index/img_id-371__prev_size-0__artist_id-19__work_id-86.html (accessed 22 April 2012) and the fantastic article by Sevier L. Deformed toad artist hopes to win public sympathy for amphibians . Ecologist. 30 March 2010.

http://www.theecologist.org/how_to_make_a_difference/wildlife/451908/deformed_toad_artist_hopes_to_win_public_sympathy_for_amphibians.html (accessed 22 April 2012).

[32] See Dahr J. Gulf seafood deformities alarm scientists. Al Jazeera 20 April 2012. http://www.aljazeera.com/indepth/features/2012/04/201241682318260912.html, (accessed 22 April 2012).

[33] For more information, see Bureaud A. Lovely Weather: Reflecting on the Letterkenny Donegal Art & Climate Residencies and Exhibition'. Leonardo Journal MIT Press 2011; 45(2) 182-183.

[34] See Grant E. Star, and Orbs, The Medieval Cosmos, 1200-1687. Cambridge: Cambridge University Press; 1996.

[35] Russell B. and Einstein A. The Russel-Einstein Manifesto. London. 9 July 1955. http://www.pugwash.org/about/manifesto.htm (accessed December 2010).

[36] Idem.

[37] See http://www.1010global.org/.

[38] See Curtis R. No Pressure. 2010.
http://www.youtube.com/watch?v=sE3g0i2rz4w&feature=player_embedded (accessed 23 April 2012). Surprisingly enough, Curtis is also credited for films such as Love Actually an Four Weddings and a Funeral.

[39] Organisers quoted by Singh A. In Richard Curtis and an explosion of publicity. The Telegraph. 2 October 2010.

http://www.telegraph.co.uk/news/celebritynews/8038113/Richard-Curtis-and-an-explosion-of-publicity.html (accessed 22 April 2012).

[40] Idem.

[41] Hunter R. Thermageddon: Countdown to 2030. Op. Cit., p. 114.

[42] Idem.

[43] Op. Cit., p. 120.
[44] Latour B. We Never Have Been Moder. Op., Cit., p. 180.

# Chapter 6

# Environmental Benefit of Using Bagasse in Paper Production – A Case Study of LCA in Iran

Sotoodehnia Poopak and Amiri Roodan Reza

Additional information is available at the end of the chapter

## 1. Introduction

World paper and paperboard demand is expected to grow by about 2.1% till year 2020 and the growth will be fastest in Eastern Europe, Asia (except Japan) and Latin America [1]. There are two kinds of paper production: (a) using wood (virgin) as raw materials and (b) using non-virgin material like kanaf and bagasse [1]. There are several studies that applied LCA in pulp and paper products (Merrild et al.,2008; Murphy & Power.,2007; Schmidt et al.,2007; Holmgren and Hening,2005; Dias,2007;Wiegard,2001; Fu et al.,2005 and Dias et al.,2002). In their research they discovered that energy and water consumption, Greenhouse Gases (GHG) and methane emissions, chlorine and raw materials used for non-virgin papers is less than virgin material. This study focused on LCA of non-virgin material (baggasse) in paper factory in Iran. The Pars Paper Factory is a government owned factory and is located in Southwest Iran and is 500 m from Hafttapeh Sugarcane factory. It was established in 1963 with a production capacity of 35,000 metric tonne per year. Nowadays, the production of this factory has reached 40,000 metric tonne per year. Hafttapeh Sugarcane factory was supplying bagasse to the paper factory. Water for this process is provided from the Dez River which is also near the factory. Source of energy for this factory is hydroelectricity and mazut. Mazut is a brownish-black petroleum fraction consisting largely of distillation residues from asphaltic-type crude oils, with a relative density of about 0.95.

## 2. Problem statement

Paper is made from plant fibers called cellulose, which are found in wood. Cellulose must be converted into pulp before being used to manufacture paper. To begin the papermaking process, recovered fiber is shredded and mixed with water to make pulp. The pulp is washed, refined and cleaned, then turned to slush in a beater. During the papermaking process emission can released into air and water and caused the air and water pollution. In additional, nowadays, by rapid economic development and population growth the demand for paper also increased in the world. More demand on paper needs more harvesting of woody materials. Unsustainable harvesting of wood can caused deforestation, climate change, etc. However, producing one metric tonne of paper from non virgin materials such as bagasse, kanaf and bamboo can save 17 trees,3.3 cubic meter ($m^3$) of landfill space, 360 L of water, 100 L of gasoline, 60 pounds of air pollutants, 10401 kilowatt of electricity [2,3]. On the other hand greenhouse gases (GHGs) such as carbon dioxide ($CO_2$), methane ($CH_4$) and nitrous oxide ($N_2O$) are critical components of the earth's atmosphere. Without these gases, the earth would be in deep freeze. These gases act like a blanket, trapping heat around the earth and keep temperatures necessary for human life. However, anthropogenic activities such as fossil fuel burning, land clearing and deforestation can cause 'thickened the greenhouse blanket' which means can effected on climate changes. This paper was aimed to identify all impacts of paper making process in Iran, using LCA as a tool.

## 3. Pars Paper Factory, Iran

Pars Paper Factory is a government owned factory and is located in southwest of Iran. It is 1500 km from Tehran the capital city of Iran and 100 kilometers north of Ahvaz, 45 kilometers south of Andimeshk and 20 kilometers from Shush. This factory was near the Hafttapeh Sugarcane Factory which supplied the raw material for produce paper.

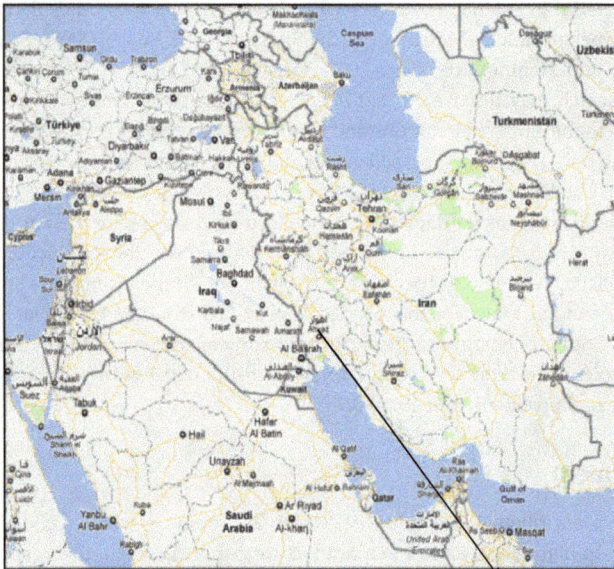

**Figure 1.** Location of Pars Paper Factory

Pars Paper Factory was established in 1963 with 35,000 tonne paper production capacity per year. Nowadays, the production of this factory has reached 40,000 tonnes per year.

**Figure 2.** Pars Paper Factory

As shown in Figure 3, paper production process divides into three parts.

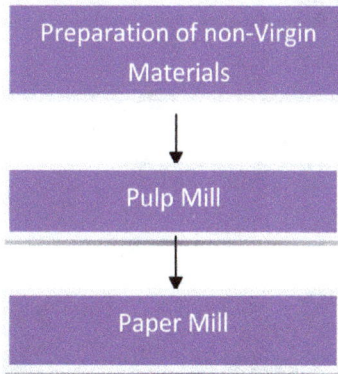

**Figure 3.** Paper Production Process

In preparation of non-virgin materials, bagasse was provided from sugarcane factory.

## 3.1. Preparation of Bagasse

Bagasse is the fibrous residue remaining after sugarcane is crashed to extract its juice and is currently used as a renewable resource in the manufacture of pulp and paper products and building materials. Using agricultural crops rather than wood has the added advantages of reducing deforestation. Due to the case with which Bagesse can be chemically pulped, Bagasse requires less bleaching chemicals than wood pulp to achieve a bright, white sheet of paper. Because of this reason there is less impacts of materials that used in the bleaching section such as Chlorine on the environment. The fibers are about 1.7mm long and are well suited for tissue, corrugating medium, news print and writing paper. Bagasse contains 65-68% fibers, 25-30% pith, 2% sugar and 1-2% minerals. This factory brings the raw materials from sugarcane factory which is 500 meter from the paper factory .The raw materials send to paper factory through the pipe or conveyer belts. The energy that use for transported the Bagasse to the paper factory is water or air.

**Figure 4.** Transported the Bagasse through the conveyer belts from the sugarcane factory to the paper factory

**Figure 5.** Transported the Bagasse through pipe from the sugarcane factory to the paper factory

This is the section that separate fiber from the pith. This process called Depithing. The paper production is continuously for year because of that the factory store the Bagasse, therefore; the harvest of the sugarcane was on just 6 months and factory storage the Bagasse as it shown in figure 6. The Bagasse has potential to fire so, it become wet to prevent the fire.

**Figure 6.** Bagasse Storage place in paper factory

The Bagasse before send to the Pulp mill part mixed with water and send to the

Bagasse Dewatering section to separate the pith and fibers.

## 3.2. Pulp mill

In this part there are five sections as follow: Cooking, Pulp washing, Pulp Screening & Cleaning, Pulp Thickening and Bleaching.

### 3.2.1. Cooking

The fibers reduce the water by cooking in this section and the energy that use is steam. Approximately 10-15 minutes need to cook the fibers. This factory had 5 boilers however currently it just use 3 of them the reason is the rest is out of service.

### 3.2.2. Pulp washing

After cooking process the remainder called Pulp and it is in black color. The black pulp called Black Liquor, washed for three times to change the color.

### 3.2.3. Pulp screening and cleaning

This was the third section on the pulp mill process .In this stage remove all the sand and useless fibers. It was shown in Figure 9 as below.

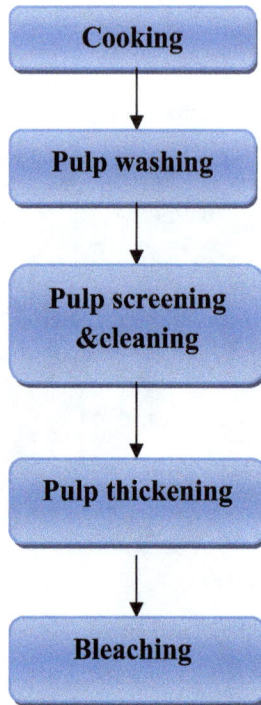

**Figure 7.** Pulp Mill Process

**Figure 8.** Cooking process on pulp mill in paper factory

**Figure 9.** Pulp Screening & Cleaning in paper factory

### 3.2.4. Pulp thickening

In this section, the pulp reduced the water around 12% in this stage.

### 3.2.5. Bleaching

The final section by using the Cl (Chlorine) gas and NAOH the Black Liquor changed color to the white color and the process done in three times.

## 3.3 Paper mill

This is the final part in the Pars Paper Factory. Final production is paper. Paper milling is the last process in producing paper. The pulp will go through several processes to finally become paper. The paper which is white in color is cut to A4 size. At this stage, the moisture in the paper is reduced to 55-60%.

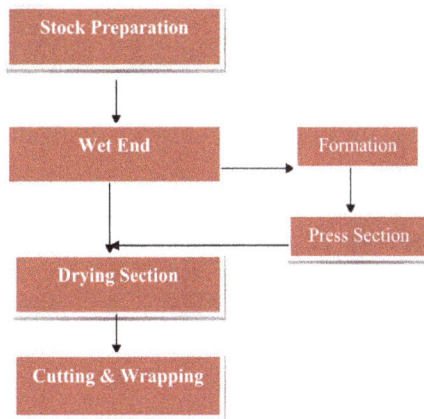

**Figure 10.** Paper Mill Process

### 3.3.1. Stock preparation

Pulp is insufficient for making paper so, in this section adds materials such as Kraft to improve the pulp. In this stage additional material such as Kraft that use for this stage bought from Malaysia or Thailand.

### 3.3.2. Wet end

This section is very sensitive in Paper mill so, it is divided to the 2 subsection as follow:

### 3.3.3. Formation

### 3.3.4. Press section

After materials are added to the pulp it will go to the Flow Box and be spread on the mantle.

During this process the pulp gives up some water and it is then sent to the Drying Section.

**Figure 11.** Wet End (Formation) Section

### 3.3.5. Drying section

Wet paper after Press Section with 55-60% moisture passing from some cylinder .These cylinder heat with steam so, the paper will loss all the moisture. In this stage for improve the qualities of printing and surfacing of paper, it pass from the Callevdering (using iron) section and cover with starch. Inputs for this stage are corn starch and resin.

### 3.3.6. Cutting and wrapping

This is the final stage in the Paper mill .The cutting of the paper to size depends on the requirements of customer.

**Figure 12.** Wet End (Press Section)

**Figure 13.** Drying Section

**Figure 14.** Callevdering Section

## 4. Life Cycle Assessment (LCA)

LCA is the assessment of the environmental impacts of a given product or process throughout its lifespan [4]. Life cycle of a product include four main stages; production stage, manufacturing stage, use stage and end-of-life stage. The environmental evaluation using the LCA approach is done by applying four steps; defining the goal and scope of the study, establishing a Life Cycle Inventory (LCI), Life Cycle Impact Assessment (LCIA) and finally interpretation of environmental burden associated with the product [5].

### 4.1. Goal and scope definition

The goals of the LCA study are to:

i.   Evaluate the environmental performance of paper manufacturing process, and
ii.  Identify inputs that have environmental potential from the paper manufacturing.

### 4.2. Scopes of the LCA study

• **System boundary**

In this study, the A4 size paper commonly used for writing, printing and copying a document was chosen as an assessing subject in the life cycle assessment. The life cycle of an A4 paper which starts from the raw material extraction stage, production stage, use stage and end-up at the disposal stage. However, the system boundary of the study only focused on the paper production stage (dotted line in Figure 15). In general, the three stages of paper production are: preparation of non-virgin materials, pulp mill and paper mill process [6].

• **Functional unit**

The Functional Unit was set as the production of one metric tonne of paper for one year.

• **Assumption**

In the LCA study, the following assumptions were made:

There were no wastes or emissions to air and water nor by-products during paper production process and the transportation from each stage is not taken into account because of lack data.

### 4.3. Life Cycle Impact Assessment (LCIA) method

The methodology used to develop this research includes observation, data collection, Site visit and interviews. Several approaches were used for data collection for this study, which are as follow:

Site visit and observation: this was done by visiting the Pars Paper Factory in Iran.

Interviews: some interviews conducted for this research.

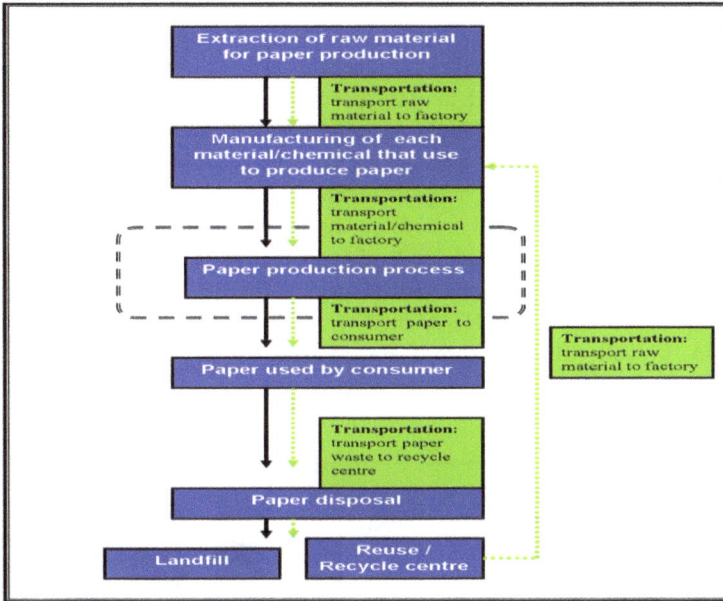

**Figure 15.** Life Cycle of Paper and System Boundary

Impact assessment is an important step in measuring the environmental impacts in LCA. SimaPro comes with a large number of standard impact assessment methods. In this study, CMLBaseline2000 method was used for Life Cycle Impact Assessment (LCIA). The CMLBaseline2000 provides ten types of impact categories with its unit as shown in Table 1 The emissions inventory data are in terms of the mass released into the environment—such as 1 kg—per functional unit it also means the impact of a unit mass (1 kg) of an emission to the environment [7].

| No | Impact category | Unit |
|---|---|---|
| 1 | Abiotic depletion | Kg Sb eq |
| 2 | Acidification | kg $SO_2$ eq |
| 3 | Eutrophication | Kg $PO_4$eq |
| 4 | Global warming | kg $CO_2$ eq |
| 5 | Ozone layer depletion | kg CFC-11 eq |
| 6 | Human toxicity | kg 1,4-DB eq |
| 7 | Fresh water aquatic ecotoxicity | kg 1,4-DB eq |
| 8 | Marine aquatic ecotoxicity | kg 1,4-DB eq |
| 9 | Terrestrial ecotoxicity | kg 1,4-DB eq |
| 10 | Photochemical oxidation | kg $C_2H_4$ eq |

Sb: Antimony    CFC: Trichlorofluoromethane    DB: Dichlorobenzene

**Table 1.** Impact Categories and Units

## 5. Results and discussion

CMLBaseline2000 was used to analyze the potential environmental impact using Simapro 7.0 database. The graph is scaled to 100% per impact category, in order to allow the description of widely dispersed values per impact category in one diagram (Figure 16). Colour difference of the graph is representing the different types of input. The negative value of the impact means benefit to the environmental. The impact value for each impact was contributed from inputs that were used during the paper making process. In this factory, there were 12 types of inputs involved in the process and had been analyzed. They were; Bagasse (farmed tree 1), Kraft (farmed tree 2), Electricity, Heavy fuel oil (Mazut), Water, Sodium hydroxide (NaOH), Aluminum sulphate (Al$_2$(SO$_4$)$_3$), Optical Brightness Agent (OBA), Chlorine (Cl). Clay, Corn Starch and Resin.

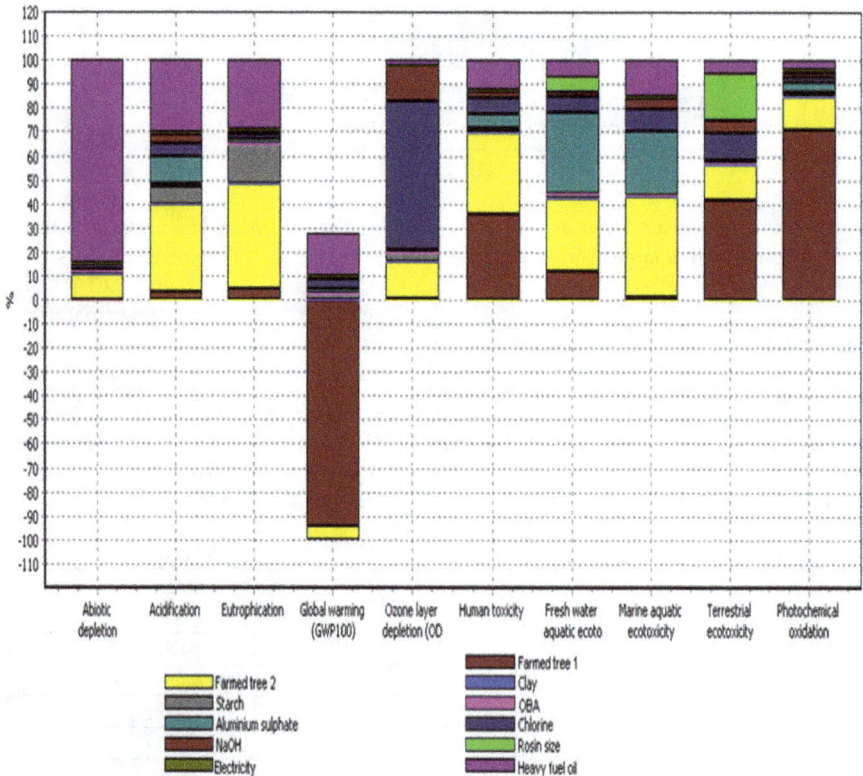

**Figure 16.** Impacts of paper production process from all inputs

## 5.1. Abiotic depletion

Inputs that Cause Abiotic Depletion Impact shown in Figure 17. From the total, mazut (fuel oil) contribute the highest impact value of 85% followed by kraft with 11%, of the total impact. The resin, bagasse, OBA (Optical Brightness Agent), NaOH, corn starch and Aluminum sulphate make up smaller impacts in a range of 0.1-2.0%. Clay and electricity contribute very little impact which are $2.90 \times 10^{-3}$ kg Sb eq and $2.91 \times 10^{-5}$ kg Sb eq, respectively and this is the reason why these two inputs give almost no impact value. It was identified that for abiotic depletion, mazut is the main input that contributes the highest impact value while electricity was the lowest. In the paper production process, mazut is used as an energy source for heating and steam-raising for furnaces, kilns and boilers. Mazut is a brownish-black petroleum fraction consisting largely of distillation residues from asphaltic-type crude oils, with a relative density of about 0.95. This means coal contains the highest amount of carbon per unit of energy, so it emits more greenhouse gases than the other fossil fuels [8]. The consumption of hydroelectric power will reduce environmental degradation because of renewable sources [9]. Electricity derived from fossil fuels can increase global greenhouse gases (GHG) while hydroelectricity or nuclear electricity may not increase the GHG emissions [8, 10].

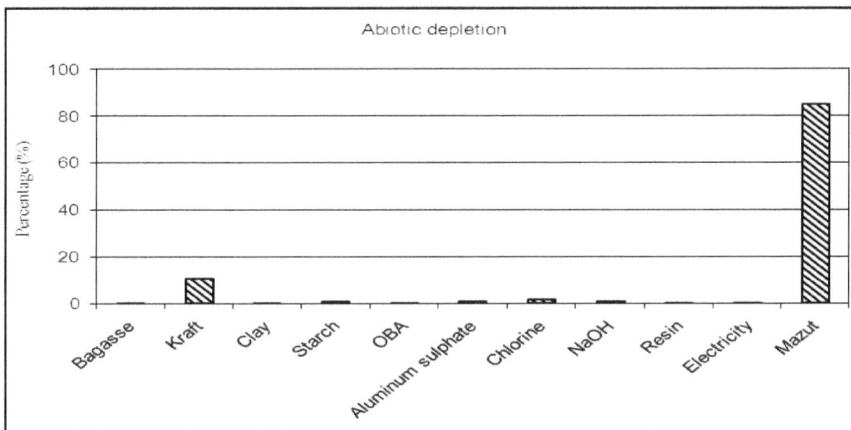

**Figure 17.** Inputs Cause Abiotic Depletion Impacts

## 5.2. Acidification

Acidic gases such as sulfur dioxide, nitrogen oxides (released during the burning of fossil fuels) contribute to the acidification of the soil and fresh water ecosystem. The category indicator for acidification was measured in kilograms of sulfur dioxide equivalent (Kg $SO_2$ eq). Weigard, (2001) indicated that $N_2O$ is produced naturally through human activities such as the burning of fossil fuels, deforestation, land-use changes and some industrial processes (Figure 18).

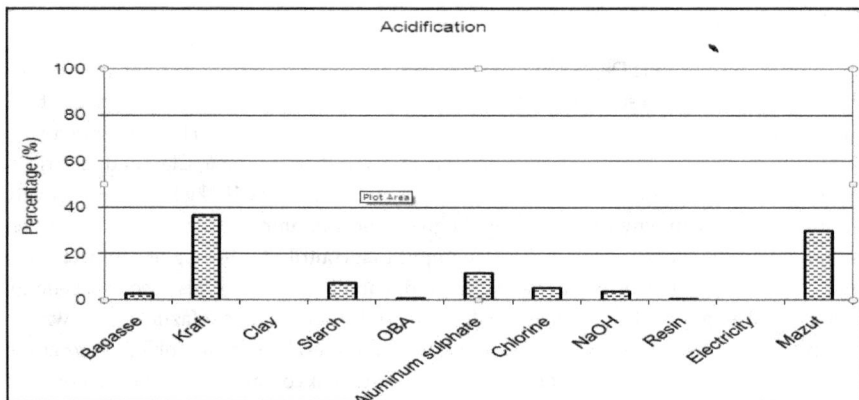

**Figure 18.** Inputs for Acidification Impact

## 5.3. Eutrophication

As it shown in Figure 19, Kraft gives the highest impact (44%) in eutrophication followed by mazut (29%), starch (16%), bagasse (5%), Aluminum sulphate (2%), chlorine (2%), resin (1%), NaOH (1%), OBA (1%), clay (0.3%) and electricity (0.0001%). The enrichment of soil and water by nutrients is measured by the EP (Eutrophication) impact category. An increased EP could lead to algal blooms in lakes with reduction in sunlight penetration and other adverse consequences, and similar undesirable effects on soil. Release of nitrates and phosphates continuously to fresh water and marine water can cause increased nutrient buildup. During the combustion of fossil fuels and fuel production high $NO_x$ is produced [11, 12]. This can result in accumulation of nitrates, phosphates and dissolved oxygen content [13].

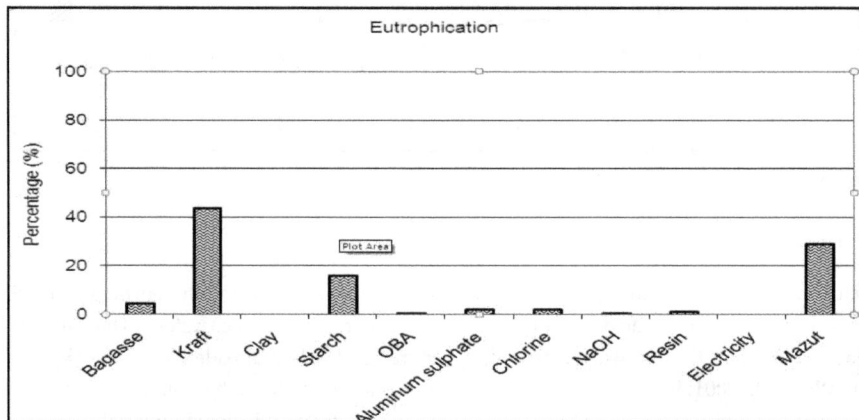

**Figure 19.** Inputs for Eutriphication Impact

## 5.4. Global warming

Impact to global warming was measured in kg $CO_2$ equivalent. Figure 20 shows the impact value to global warming from each input. Bagasse gave a negative impact value which was -130% while kraft gave value by -9%. Other inputs such as clay, starch, OBA, aluminum sulphate, chlorine, resin, electricity and NaOH gave positive impacts which are 0.02%, 3%, 0.5%, 3%, 5%, 2%, 0.1%, 0.002% and 25%, respectively, of the total impact. For global warming impact, mazut contributed the highest impact value (25%) and bagasse the lowest (-130%). Weigard, (2001) illustrated that burning fossil fuels can release 6.2 (GtC) into the atmosphere each year. His study was done using LCA for quantification of greenhouse gases at Visy industries. Based on his research, Weigard, (2001) believes that changing land-use (eg. deforestation) results in increased emissions of carbon. Likewise, the results of this study show that using bagasse as a renewable raw material for paper production could lead to less deforestation because bagasse has a negative impact on the environment. The negative value of the impact indicates benefits to the environmental

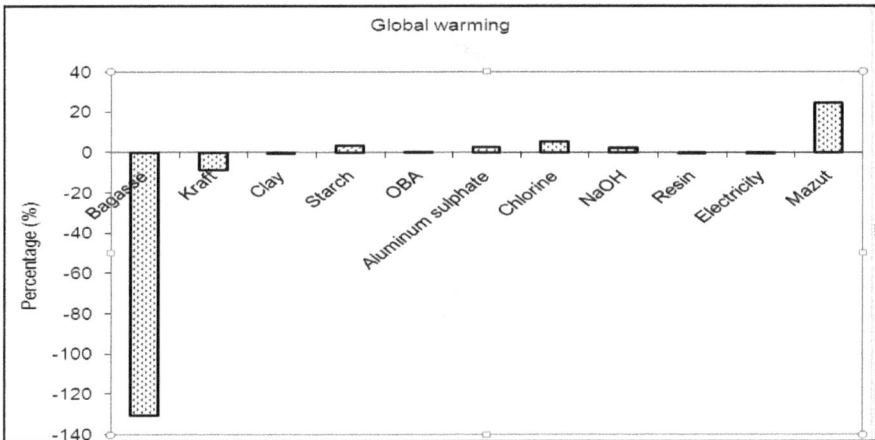

**Figure 20.** Inputs for Global Warming Impact

## 5.5. Ozone layer depletion

Ozone layer depletion was measured as CFC-11 equivalent. Inputs that contributed to this impact are shown in Figure 21. Chlorine contributed the major impact value with 62% and electricity the lowest at (7.8x $10^{-05}$ %). Kraft was the second major contributor (16%) and NaOH was the third (14%). Others made up a small range of impacts which was less than 5% each; starch (4%), mazut (2%), aluminum sulphate (1%), OBA (0.4%), bagasse (0.4%), resin (0.2%) and clay (0.01%).

Before the 1980s and early 1990s, free chlorine was used to bleach paper; however, nowadays, the use of free chlorine has ceased and chlorine–dioxide or other means of bleaching such as ozone which have taken over [14].Chlorofluorocarbons (eg. CFC-11 and

CFC-12) were first manufactured in the 1930's but were not present in the atmosphere in any appreciable quantity before 1950. Up until the 1990's, they were widely used as propellants, refrigerants and foaming agents. They act as a GHG in the troposphere but also damage the ozone layer in the stratosphere. The study shows that man-made chemicals can cause ozone layer depletion [8].

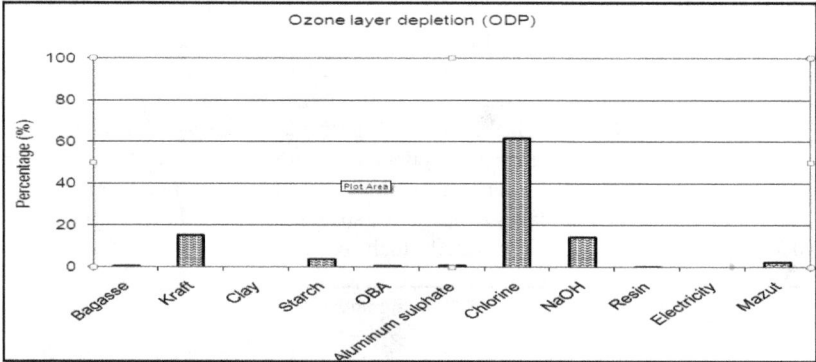

**Figure 21.** Inputs for Ozone Layer Depletion

## 5.6. Toxicity

The toxicity impact was measured as 1, 4-DichloroBenzene equivalents/ kg emission (Kg 1,4-DB eq). In the CML2BaseLine2000 method for LCIA, toxicity to human environment, fresh water, marine and terrestrial ecosystem were considered. Figure 22 explains the toxicity impacts of the various materials/elements. From the total impact, kraft contributed the highest impact of about 42%. Aluminum sulphate was in second place with 26% followed by mazut (15%), chlorine (10%), NaOH (4%), bagasse (1%), starch (1%), resin (1%), OBA (0.2%), clay (0.02%) and electricity (0.0005%).

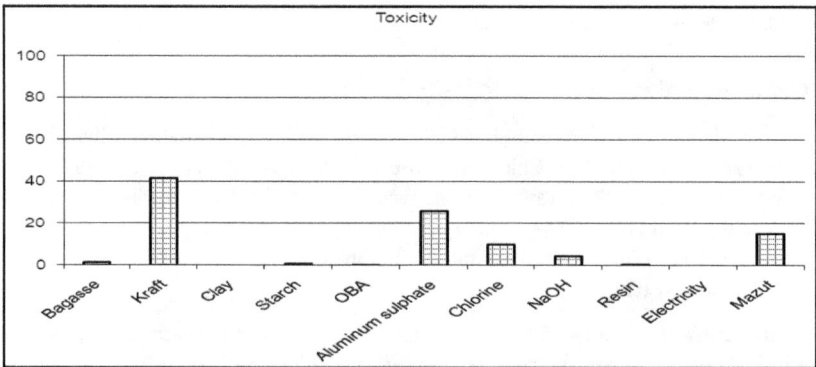

**Figure 22.** Inputs for Human, Freshwater, Marine and terrestrial ecotoxicity Impacts

## 5.7. PhotoChemical oxidation

Figure 23 shows the impact value of each input for photochemical oxidation. Bagasse gave the highest impact value in photochemical oxidation with 71%. Kraft contributed 14%, aluminum sulphate and mazut 4% each which chlorine and resin contributed 2% each, starch and NaOH 1% each. OBA, clay and electricity were at the lower and of the range at 0.2%, 0.008% and 0.001% respectively.

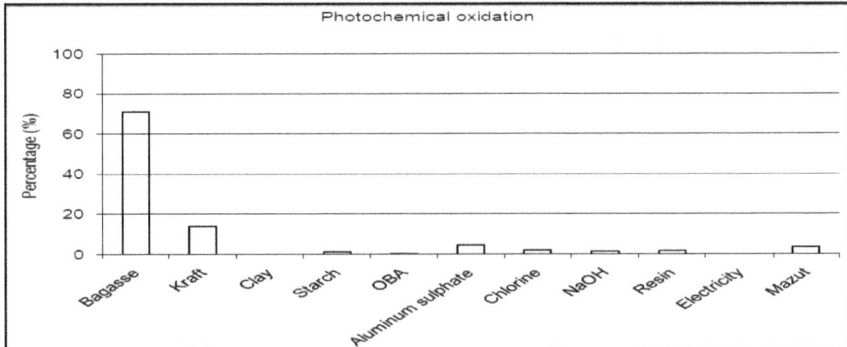

**Figure 23.** Inputs for Photochemecal Oxidation Impacts

# 6. Conclusion

Based on the above study, following conclusions are drawn:

1.  Nuclear energy, hydroelectricity or even using pith as a source of friendly energy should be used in place of Mazut for paper making.
2.  More environmentally friendly materials for bleaching should be used against chlorine.
3.  Paper recycling is to be utilized in place of Kraft to reduce the amount of impact of Kraft on.

## Author details

Sotoodehnia Poopak*
*Institute of Biological Science, Faculty of Science , University of Malaya, Kuala Lumpur, Malaysia*

Amiri Roodan Reza
*Department of Knowledge Management, Faculty of Creative Multimedia, Cyberjaya, Malaysia*

# 7. References

[1]  Honnold, V (2009).Developments in The Sourcing of Raw Materials for the Production of Paper. United States International Trade Commission. J of Int Commerce and Economics.

* Corresponding Author

[2] Malaysian Newsprint Industries (2007). Paper Recycling Report. http://www.newsprint.com.my.

[3] WasteCap, 2008. WasteCap of Massachusetts.68 Hopkinton Road/ Westboro/MA 01581, March 2008.
http://www.WasteCap.Org/WasteCap/Commodities/Paper/Paper.htm.

[4] Curran M A (2006). Life Cycle Assessment Principal and Practice , National Risk Management Research Laboratory, U.S Environmental Protection Agency, Cincinnati, Ohio 45268.1-15.

[5] Murphy R(2004).Green Composites. Imperial College, London, United Kingdom.23-26.

[6] Dias A C, Arroja L, Capela I(2007). Life Cycle Assessment of Printing and Writing Paper Produced in Portugal. Int J of LCA, 12(7).521-528.

[7] Pennington D W, Potting J, Finnveden G, Lindeijer E, Jolliet O, Rydberg T, Rebitzer(2004). Life Cycle Assessment Part 2: Current Impact Assessment Practice. J of Environment Int, 30,721-739.

[8] Wiegard Jean (2001). Qualification of GreenHouse Gases at Visy Industries Using Life Cycle Assessment. M Tech Thesis, Swinburne University of Technology, Australia.

[9] Fress N, Hansen M S,Ottosen L M, Toenning K, Wenzel H(2005). Update of the Knowledge Basis on the Environmental Aspects of Paper and Cardboard , Environmental Project No 1057, Danish Environmental Protection Agency , Copenhagen, Denmark.

[10] Elliason, B (2000). Energy in the 21st Century , The Role of GHG Control Technologies, Paper Presented at the 5th International Conference on Greenhouse Gas Control Technologies (GHGT-5)/August , Cairns.

[11] Eriksson E, Gilespie A R, Gustavsson L, Langvall O, Olsson M, Sathre R (2007). Integrated Carbon Analysis of Forest Management Practices and Wood Substitution. J of Forest Resource, 36.671-681.

[12] Ally Jamie, Pryor T (2007). Life Cycle Assessment of Diesel, Natural Gas and Hydrogen Fuel Cell Bus Transportation System. J. of Power Sources, 170. 401-411.

[13] Gordon G (2003). Interior Lighting for Designers. New Jersey: John Wiley and Sons Inc,197-198.

[14] Villanueva A, Wenzel H (2007). Paper Waste-Recycling, Incineration or Landfilling? A Review of Existing Life Cycle Assessments. J of  Waste Management ,27,29-46.

# Chapter 7

# Solar Dynamo Transitions as Drivers of Sudden Climate Changes

Silvia Duhau and Ernesto A. Martínez

Additional information is available at the end of the chapter

## 1. Introduction

There is a consensus about the origin of the increase of global surface temperature of the 20th century is the fast process of industrialization, that is producing an exponential increase in CO2 and other greenhouse gases in the boundary layer of the Earth atmosphere. However at 1924 a transition to a new configuration of the solar dynamo system occurred [1] that seated this system in the XX century Grand Maximum at which the highest values of solar activity of the last 400 years occurred. Therefore, the sharp increase of global temperature has been not only synchronic with the fast process of industrialization but also with a sudden increase of solar activity.

At the Schwabe polar cycle #24 that started at year 2000 , maximized at year 2008 and would end at mid of 2013-14, a new solar dynamo transition is occurring that is leading to lower values of solar activity [2,3], and as a consequence the flux of solar energy on the Earth atmosphere is decreasing fast. Therefore by observing the future evolution of climate variables we will be able to evaluate the relevance of solar activity variability on climate changes.

A thorough determination of the contribution of solar activity to climate change is hindered by the fact that the only source of solar origin that is included in the climate models is total solar irradiance, TSI, for example see [4-8] This source of solar energy increased in the average in only about 0.13% along the last 400 years, which might explain at most a 30% [8] of the temperature increases along that period. However, besides TSI there are other sources of solar energy that might modify climate by mechanisms that have been proposed and studied by the last thirty years, for a review see [9,10]. The aim of the present work is to assess the impact of the 2008 dynamo transition on Earth's surface temperature. This will assist us to be well prepared in studying the unique experiment that nature is currently bringing to us, i.e. the solar dynamo transition and its consequences for climate that will be now documented worldwide and monitored with modern technical means. This may contribute to improve climate models by indicating the variables that are ignored in those models and that future observations finally prove to be relevant to climate changes.

The prediction of the impact of the 2008 solar dynamo transition on climate relies in the prediction of solar activity after the transition. Based in the regularities found in the time series of sunspot maxima along the last millennium Schove [11] predicted a value of sunspot maximum #24 well below to those prevailing since 1924 and that fall near the low values that this cycle is having at only a year of its date of occurrence [12,13]

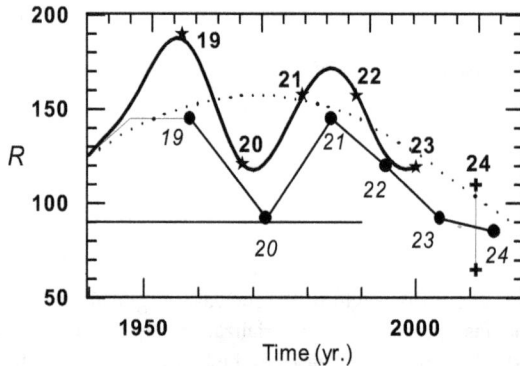

**Figure 1.** The predictions of Schove [11] for sunspot maxima #19 to #24 (filled circles) and the observed values (stars) together with an estimation of sunspot maxima 24 for the descending transition (crosses). The thick line is the envelope of sunspot cycle maxima. The constant level is the sunspot maxima average as determined prior 1923 and the dashed line is the secular variation in sunspot maxima time series for a descending transition (from [14]).

Unless in [9,10, 15] the only variable that is usually taken into account in the field of solar-terrestrial physics for quantifying solar activity variability is sunspot number, of which direct observations does exist for the last 400 years. This variable gives a measure of the strength of the toroidal component of the solar magnetic field [16]. As solar activity ultimately depends on solar magnetic field strength [17-21] all the variables related to solar activity must bear some relationship with sunspot number. However, it was found [1, 14, 15, 22] that this relationship is non linear and so we need to resort to other proxies than sunspot number that give a better estimation of the true variables. These variables and its proxies are introduced in section 2.

In our view, the success of Schove's [11] prediction of sunspot maximum #24 by 50 years in advance, is based in the fact that being the solar dynamo a bounded system it undergoes natural modes of oscillation. However, the dynamics of this system is described by a set of non-linear differential equations, therefore its natural oscillations are non-stationary which impede us applying the Fourier base function to describe them. A mathematical methodology suitable for the description of natural oscillations in solar dynamo system has been developed by us [2, 3, 15]. This method and the way on which it is applied to predict sunspot maximum #24 is briefly summarized in 3, and applied in section 4 to determining solar dynamo natural oscillations in the variables defined in 2. The same method is applied in section 5 to look for the signatures of solar dynamo transition in surface temperature and from these results and the analysis of the latitudinal variation on temperature, that is presented in section 6, in section 7 the impact of the 2008 transition on the evolution of surface temperature along the XXI century is evaluated. Finally the conclusions are presented in section 8.

## 2. The solar dynamo transition in solar activity

### 2.1. The solar variables and its proxies.

Solar activity has several manifestation, of these, those that are relevant to climate change are:

a.  Total solar irradiation, TSI. This is the only source of solar origin that is considered in climate modeling , for example, see [5-9]. To study it here we resort to the reconstruction from Wang, Lean and Shelley [23-24].

b.  Solar Flares produce the hard part of the electromagnetic spectrum (gamma and X rays and ultraviolet radiation) and also solar proton events. We resort to quantify it the Flare index . Data for this index [25] exist for the interval 1976.5-2008.5

c.  Coronal Mass ejections , CME's, that are usually going together with flares and which main effect is the acceleration in the heliosphere-magnetosphere system of height energetic particles [26] leading , when interacting with the magnetosphere, to high energetic particles events. To quantify the geo-effectiveness of CME`s we will resort to the Sudden Storm Commencement, SSC, index as defined by Duhau [27] that is based in the amplitude and rise time of geomagnetic storm sudden commencement, defined by Mayaud [28], and computed in the interval 1868-1998 from the ISGI data [29] .

d.  Variability of the solar coronal magnetic field, the solar 'open flux' that modulates cosmic ray particles flux. The solar open flux is well described by the geomagnetic index aa [30] defined by Mayaud [31]. The data in the interval 1844-1985 is from [32] and from 1986 onward there are two version of this index , one is the standard from ISGI [29] and the other is the data from [33].

Solar activity ultimately depends on solar magnetic field variations, that has two components: the toroidal and the poloidal one [17-21]. Since now on we will call 'strength' of the polaidal and toroidal field to the amplitude of the Schwabe cycle on the respective solar magnetic field component.

The sunspot number gives a measure of the total toroidal field at the spots [16]. Therefore sunspot number at solar maximum, Rmax, is a measure of the strength of this component. While at solar minimum the polar field is mainly dipolar and the open flux is coming mostly from this field , as a result geomagnetic index aa at solar minimum, aamin, gives a measure of the dipolar field of the Sun [2, 15, 34-37]

Rmax is determined here from the yearly values of Group Sunspot Number [38] and of sunspot number International time series value as provided by NOAA [39] for the intervals 1610-1704 and 1705-2011 respectively. Aamin is determined from the two data sets of geomagnetic index aa as detailed in paragraph (d). Note that while aa index is always a proxy for the solar open flux, only at solar minimum is also a proxy for the strength of the dipolar component of the global magnetic field.

## 2.2. The solar dynamo transition as seen in the solar variables.

Rmax, and aamin fluctuated around constant levels of 93, 4 spots and 10.3 nT respectively [1]. We have called these couple of constants 'transition point ', that is a point in the 'phase diagram' of Rmax vs. aamin (as proxies for the toroidal and polar magnetic field strengths) (Figure 2). The transition point has the property that when the path of successive points in this diagram is close enough to it, a transition occurs in the solar dynamo system that leads to a sudden change in the strength and length of its natural oscillations.

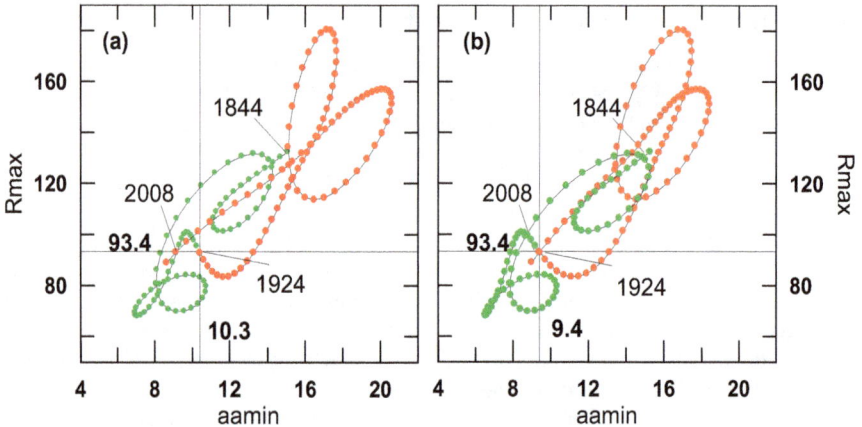

**Figure 2.** The diagram for the period 1844-2011 per year for (a): the Standard data from ISGI and (b) the Lockwood homogenized time series data (from [37]).

There are three types of solar dynamo episodes: Grand Maximum, Grand Minimum and Regular oscillations. Which of then occur after a transition depends on how close to the transition point is passing the path when coming back to that point. The paths determined by the two different data sets of Figure 2, passed exactly (within the experimental error) by the transition point at 1924 and the Grand Maximum Episode of the XX century (red points) started. The same happened at the 2008 transition for the data at the right, which indicates that a Gran Minimum is coming after this transition. But for the data at the left the path at 2008 is far enough from the transition point to lead to predict that a Regular Oscillations type episode, alike to that occurring prior 1924 (green points), would instead follows after the transition. From the properties of the oscillations in proxy time series of Rmax from Schove [11] and of Usoskin [40] for the last 1700 years, we [37] have estimated that, among the two possible type of episode that may occurs after a Grand Maximum, the forthcoming episode would be of a Regular type, alike to the one previous to 1924 (green points in Figure 5) and that this episode would endure for the rest of the present millennium.

A full sequence of the three types of episodes occurred since 1610 [1]. The solar dynamo transitions produces a sudden change in the average value of the successive maxima of the Schwabe, ~11 year, cycle (Figure 3) , that occurs in synchronicity in all the related variables.

This indicates that a substantial change in solar dynamo configuration occurs after each transition that is affecting all the layers that compose the solar dynamo system.

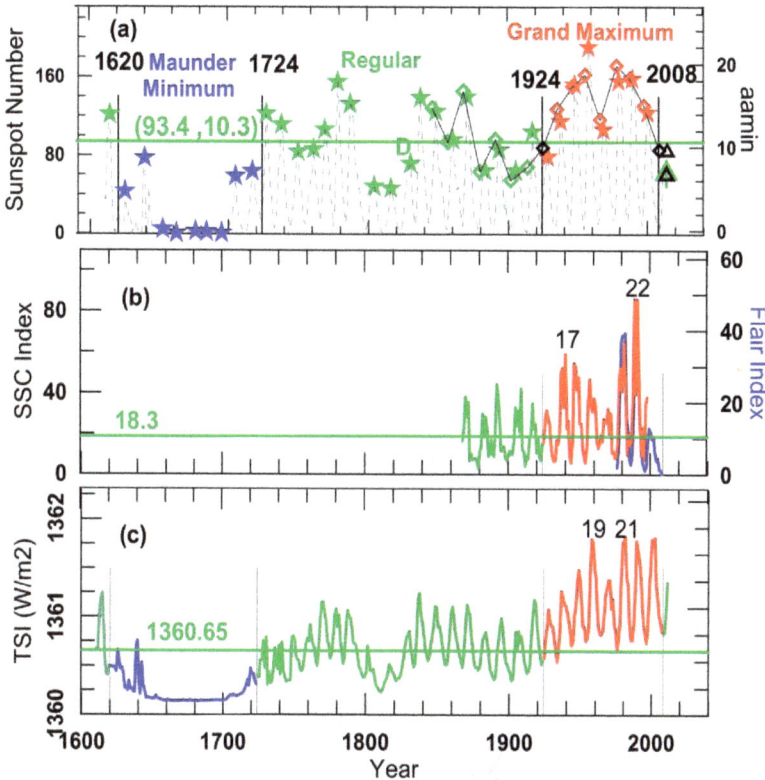

**Figure 3.** Solar dynamo transitions in (a) sunspot number maxima (stars) and geomagnetic index at minima (diamonds). The black diamonds indicates the polar cycles at which the transitions occur and the horizontal green line the transition point (93.4 spot, 10.4 nT) level [1]. The letter D indicates the short type Dalton Minimum. The green and black triangles are the predictions from [11] (upper black),[2] (green) and [3] (lower black) , respectively. (b): SSC Index and Flare Index (blue line). (c): total solar irradiance, TSI. In (b) and (c) the horizontal green lines are at the average value along the Regular Oscillations episode and the black numbers are the conventional numbering of the strongest maxima occurring after 1924.

It may be observed that at and above the secular time scale all the variables have a very similar behavior, and so the corresponding oscillations appear having the same length and nearly the same phase. The same do not happen with oscillations with shorter length, as much as for example, the two relative maxima that occurs in all the variables after 1924 transition are not synchronic, but are at solar cycle maxima #17 and #22, in SSC and Flare Indexes (Figure 3b) , and at solar cycle maxima #19 and #21 in TSI (Figure 3c). This is studied further in the section 4 and 5 by means of the mathematical methodology that is discussed in the next section.

The effect of CME's and flares on the heliosphere-magnetosphere system is generically called 'solar storm'. It is know that both phenomena are often going together and in fact Flare index, along the short interval on which is known, follows closely to SSC index that is a proxy for the geo effectiveness of CME's. Therefore since now we will consider SSC as an acceptable proxy for the geo-effectiveness of solar storms.

## 3. A base function of compact support for representing solar dynamo natural oscillations

The Fourier spectrum of sunspot number time series, see [41] and references therein, have prominent peaks. The most conspicuous among them, like the Gleissberg, the Süess and the Hallstatt one, are usually called 'cycles'. However, we do not know a priori the real nature of the phenomena that leads to these peaks and its variability. As much as to call cycle to any new peak that appear in the spectra of solar variables has been called by Hoyt and Schatten [42] 'cycle mania'.

We have introduced a methodology that allows splitting solar activity related time series in natural oscillations, based in a study of the variability on time of the peaks in the spectra and a suitable base function to represent the time series in real time, as is summarized below.

### 3.1. Natural modes of oscillation in the wavelet spectrum.

It is apparent in the time series of Figure 2, natural modes of oscillations in a non-liner non-stationary system, like the solar dynamo system, have a transitory behavior which make it impossible representing them by the discrete Fourier base function, that presupposes that the waves are linear and stationary. As the solar dynamo system is bounded we still may apply a discrete transform method, but due to the transient nature of its natural oscillations the applied base function must be of compact support [43], as are the wavelet base function. The shape of the selected wavelet must be alike to that of the signal that is being represented [44]. The wavelet spectrum of the Rmax time series (see Figure 4) has peaks with a Gaussian envelope that indicates the presence of oscillations with fairly well defined periodicities. On the other hand, the variables as a function of time (Figure 3) exhibit oscillation with rapid time changes in their amplitude. Therefore we have selected the Morlet wavelet base function that is a harmonic function with a Gaussian envelope.

The wavelet spectrum of sunspot number time series and its proxies have peaks at periods that goes from seconds, to millenniums. As the time series analyzed here is Rmax, oscillations with lengths below the Hale cycle length (~22 years) are not contained in the times series. Besides the time changes in amplitude of all the peaks on the spectrum, at and below the lower Gleissberg band, the Fourier period of the dominant peaks changes strongly on time too, as an example of this we have computed the spectra (Figure 4) for two different time interval.

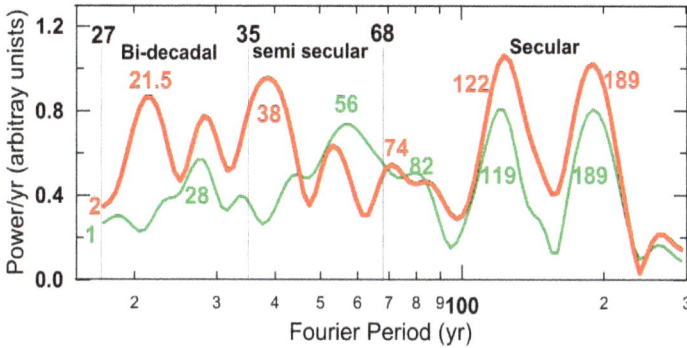

**Figure 4.** The Morlet wavelet spectrum for two time spans: 1665- 1865 (1) , 1865-2000 (2). The vertical bars delimitates the Fourier periods of the wavelet components that are included in each of the three period bands. The colored numbers are the values of the Fourier periods of the nearest peak. Below the 90 year period , only the most prominent peak within a given period band is indicated, for each time span .

For each of the spectra on Figure 4, the Fourier period of the dominant peak in the bi-decadal band is the first quasi-harmonic of the Fourier period of the dominant peck in semi-secular band. As the Fourier period of the dominant peaks change from a time interval to the other, the length of the corresponding natural oscillation changes too. This is further analyzed next.

## 4. Time changes of the natural modes of oscillations.

We will study here only the solar variables that are relevant to climate change and that has a proxy time series larger than a century , these are TSI, aamin, that is proxy for the open flux strength and SSC, that is a proxy for the frequency and intensity of solar storms. As a result of the analysis summarized in 3 we have split the time series in three oscillations (Figure 5). These oscillations are found by adding all the wavelets components which Fourier periods are in the respective bands as defined in Figure 4. In the case of the secular oscillation the linear trend is added and the transition level is subtracted. A preliminary interpretation of the phenomena underlying the three oscillations is given in [3].

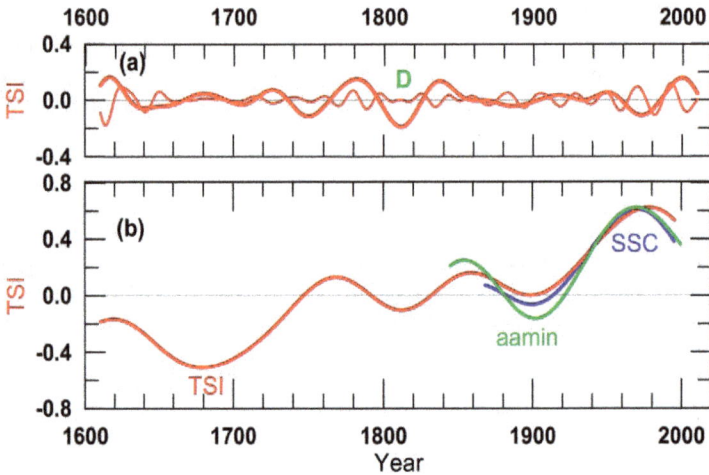

**Figure 5.** The three solar dynamo natural oscillations (a): The bi decadal (thin line) and the semi secular (thick line) in TSI. (b) the secular oscillations in TSI , aamin and SSC, were SSC and aamin are scaled to fit TSI at the secular oscillation that produced the XX century Grand Maximum.

The short Dalton Minimum that occurred around 1820 is the result of the synchronization of a relative minimum in a strong semi-secular (Figure 5a) occurring in synchromicty with a relative minimum in a weak secular (Figure 5b) oscillations. Note that while the semi secular oscillation around de Dalton minimum has a length of about 60 years, it has a length of only 40 years along the Grand Maximum (Figure5a). Therefore a increases of the amplitude of the semi-secular oscillation is synchronic with a increases of its length. This behavior dramatically occurs in the secular oscillation (Figure 5b).

On base of the relationship between the length of the oscillations and its intensity we have predicted the date of occurrence of solar maxima #24 , to be at 2013.5 [2]. And in base of the time changes of the three oscillations we have predicted the value of sunspot maximum #24. After a descending transition, like the 2008, either a Grand Minimum or a Regular Oscillations episode (blue and green starts in Figure 6) may occur. These two possible cases are barely distinguishable between them This is due to the fact that sunspot maximum #24 would occur at 2013.5 [2,12] and the secular oscillation passed by cero only at 2008 and so this oscillation will have only six years and a haft to develop. Only by sunspot maximum #25 the two cases will differentiate unambiguously [37]

The sequence #23-#24 is found to be alike to the -#13 –#12 and also to the #11-#12 ones (see numbered maxima in Figure 6). These three sequences are similar but no equal, since a given sequence never repeat identically due to the variability in relative phase and amplitude of the three natural oscillations on solar dynamo system.

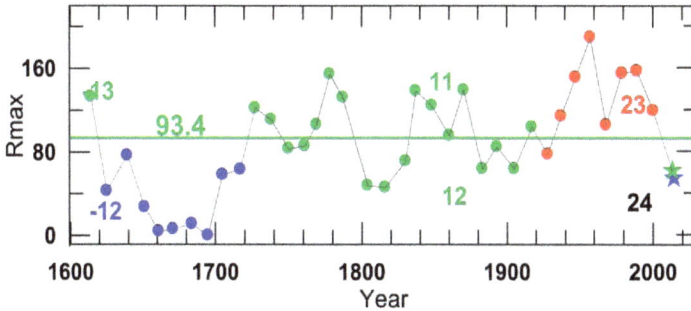

**Figure 6.** Rmax time series (the data are the same that in Figure 3a) and the predicted values for sunspot cycle #24 , for the Regular episode [2] and for the Grand Minimum [3] cases (blue and green stars, respectively) . The two sequences of sunspot maxima that are alike to the #23-#24 sequence are indicated in the figure by its conventional numbering. To assign them prior the Dalton Minimum we have assumed that the sunspot cycle that occurred prior the #5 was lost in the sunspot number time series, as suggested in [45]

## 5. Solar dynamo transition as seen in the Earth surface temperature

To find the signature in Earth surface temperature of the sudden changes on solar dynamo natural oscillations we represent the time series of north hemisphere temperature for the last 400 years by the same base function that we have used when representing the natural oscillations in solar variables in section 4. By adding wavelet components with periodicities above the 17 years we include (Figure 7) the three natural modes of modes of oscillations in solar activity and at the same time we filter the natural oscillations in the climate system.

The temperature increases since 1610 has not been steady but occurred in four steps (blue, green, red and violet horizontal lines in figure 7), three of them occurred in synchronicity with the solar dynamo transition, and the fourth, that is seen in the surface data (black line) but not in the satellite data (blue line), started at 1970. The fact that each of the first three steps occurred in synchronicity with the date of occurrence of the solar dynamo transition suggest that the increase of temperature is linked to some solar variables that have had sudden increases after each transition. This appears to be the case of solar storms (see its proxy, SSC, in Figure 3b). Moreover, the last step follows the sudden increases of solar storms that started at 1950 and reached a value that quadruplicates those prior 1924 at sunspot maximum #22, peaking at 1989. This is studied further in the following by analyzing the relationship between the secular and the semi secular oscillation in temperature with that in the solar variables.

The secular variation is alike in all the solar variables (see fig 5b). As a result, if solar activity were the main source of climate changes the secular oscillation in temperature must be alike to the secular oscillation in any of the variables related to solar activity. Only one of the four relevant variables, TSI, have a time series long enough to test this similitude. The result is in Figure 8. There is a good agreement between the secular oscillation in TSI and temperature, more if we take into account that the secular oscillation in the other involved variables (Figure 5b) has a similar but not an equal time variation.

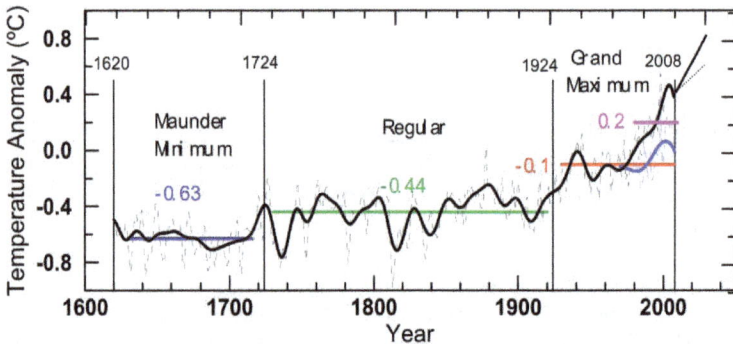

**Figure 7.** Solar dynamo transitions in global surface temperature. The dashed black line is the yearly averages of paleodata [46] in the interval 1610-1849 to which ground based data [47,48] from 1850 onward are pasted (dashed black line) The dashed blue line is satellite UAH MSU lower troposphere temperature data [49] and the thick line are the smoothed values (see text). The vertical lines indicate the dates of occurrence of the four historically documented solar dynamo transition (cf. black diamonds in Figure 3a). The projections of IPCC [50] for the forthcoming twenty years are shown for two cases: one on which the emissions of greenhouse gases would continue at the same rate as today (full) and the other on which it would remain in the actual level (dashed black line) .

The secular oscillation in surface temperature (Figure 8) started increasing above the secular oscillation in TSI at 1860 and reached a value that is 0.3ºC above the one expected from the secular oscillation in TSI. We cannot jump immediately to the conclusion that this departure is entirely due to the industrial revolution, because it was not gradual, but occurred mostly in the interval 1860-1900 and after 1980, that is after the last two relative maxima in the secular oscillation in solar activity that occurred at 1957 and 1977 (see Figure 5b).

A recovery from the little ice age (LIA) is occurring since 1800-1850 [51, 52]. And, after the mid-1970s, ice mass loss has accelerated [52-54]. This is consistent with the suggestion [9-10] that the rapid increases of surface temperature along the XX century Grand Maximum was due to the increases in the frequency and intensity of solar storms which geo-effectiveness (see figure 3b), increased suddenly after the 1924 transition to quadruplicate its value prior 1924 at sunspot cycle 22 maxima, peaking at 1990.

The semi -secular oscillation in temperature and TSI (Figure 9) are alike. It is the strongest in SSC. After 1924 the semi secular oscillation in all the variables has two relative maxima, one prior and the other after 1970, like it happened with temperature (see also Figure 7).

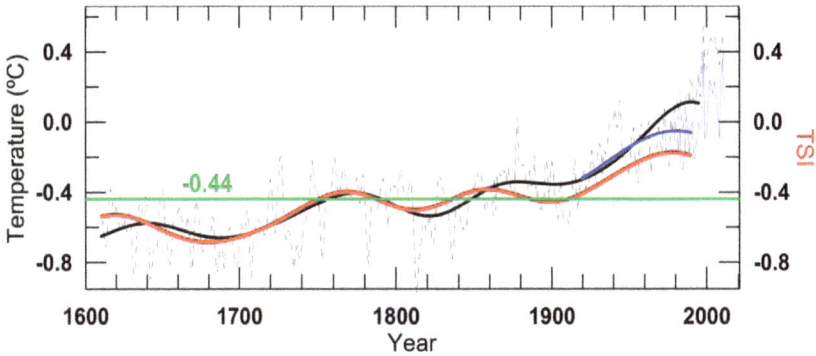

**Figure 8.** The secular oscillation in the temperature data of Figure 7. and in TSI of Figure 3c , this last scaled to fit temperature prior 1800.

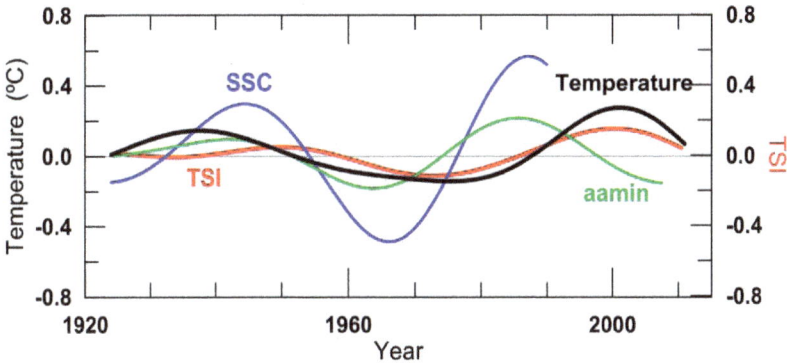

**Figure 9.** The semi secular oscillation in surface temperature and in the three relevant solar variables as indicated in the figure. TSI, SSC and aamin are scaled by the same factor that in Figure 8.

## 6. An analyses of the latitudinal variation on temperature

There is a hiatus in average global temperature increase is recent years. This may be the first indication of the impact on climate of current decreases of solar activity. However, at latitudes below 24 º (see Figure10d) the fast increases that started at all latitudes at 1970 , is still going on.

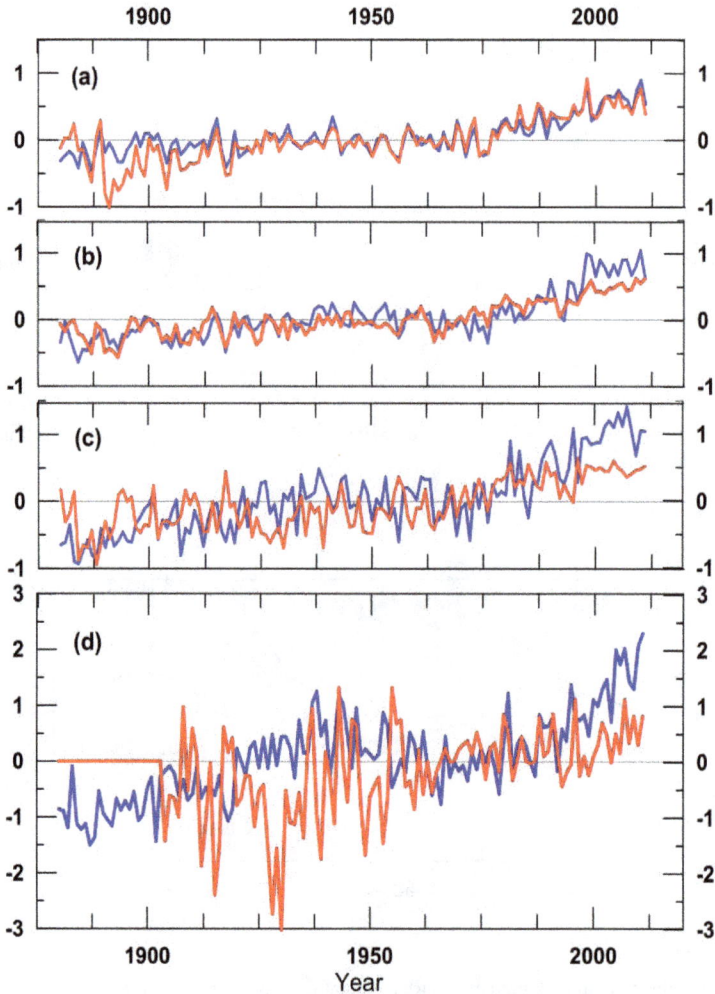

**Figure 10.** The average temperatures for the latitude bands (a) 0-24°, (b) 24 to 44,(c) 44 to 64 and (d) 64 to 90, for the south (red) and the north (blue) , hemisphere, respectively . The data is from [55]

Ice melting has accelerated since 1980 [52-54] and some models indicates that the ice feedback albedo mechanism due to this acceleration is appreciable at the poles [56] and also at height latitudes [57]. This feedback mechanism is still not well represented in IPCC climate models that underestimate Arctic see ice thinning by a factor of 4 and fail to capture the recent see ice kinematic acceleration [53]. From this we may conclude that, either ice-albedo feedback is underestimated in the IPCC models [54], or some source of heating mainly operating in the polar cap and at high latitudes is missing in those models. The synchronicity between ice kinetic acceleration with the occurrence of the strongest solar storms (Figure 3b and Figure 9 blue line), that after 1970, in only 20 years duplicates its strength , is consistent with this hypothesis, because these solar events, when interacting with the Earth magnetosphere operates mainly in the polar cap and at the subauroral region [58-62]. The mechanism involved are reviewed and some evidences that a 70% of global temperature increases after the 1924 transition was due to this strong increases on solar storms are presented in [9, 10].

If the above source of heating were significant, they may explain the step like appearance of the temperature time series (Figure 7) if we assume that not only after the 1924 transition, that is the only one along which a proxy, SSC, for the geoeffectiveness of solar storms is known, but also after any solar dynamo ascending transition, solar storms substantially increases in less that 25 years, while it takes to TSI (compare blue and red lines in Figure 9) by more than 40 years to reach its first relative maximum. At that moment, following its semi-secular oscillation the strength of solar storms has already decreased and therefore the principal source of heating at high latitudes is fading out. However, the ice is still melted and the secular oscillation in TSI is still at its maximum, so the temperature is maintained at the same levels than before due to enhancement of the TSI heating by positive ice-albedo feedback.

Solar dynamo is well over its cycle #24 and the average value of TSI (Figure 3c) is still well above the average prior 1924. The decrease of TSI (red line in figure 9) is being delayed with respect to the decreases of the other solar variables. As was delayed its increases when emerging the Sun form the Maunder Minimum at 1724 (ccompare Figure 3c with Figure 3b). By the contrary, solar storms (see figure 2b), has already started decreasing by 20 years ago and will fall to a half of its 1990 values (sunspot maxima 21, see Figure 3b) by 2013.5, if the above mechanism sustain surface temperature would start decreasing in the near future. In the next section we present an estimation of this decrease.

## 7. An estimation of the future evolution of surface temperature

There are sources of climate change other than anthropogenic gases and solar activity that are:

1. Natural oscillations of the climate system and volcanism. These sources vary in the bi-decadal time scale and below [5]. To filter then we have taken into account oscillations in the semi-secular time scale and beyond.
2. Time changes in the strength of the dipolar component of the geomagnetic field and its tilt angle, that is the position of the geomagnetic poles, that modulates the geoeffectiveness of solar storms in the atmosphere [58-62]. Appreciable changes in geomagnetic dipolar field strength and its tilt angle occurs in time scales at and above

the millennial one [63-65]. As a result the contribution to climate change of geomagnetic field along the current century may be disregarded.

3.  It has been suggested [66, 67] that periods of acceleration of the Earth's rotation rate correspond to years of increasing intensity of the zonal circulation and to global-surface warming, and periods of deceleration correspond to years of decreasing zonal-circulation intensity and to a global decrease in surface temperatures. A measure of the acceleration of the Earth's rotation rate is provided by the time derivative of the excess of the length of the day, LOD. This variable follows solar activity with a delay of 94 years [68] and so along the time interval analyzed in [66, 67] that is 1850 to 1960, the semi secular oscillation in LOD was the strongest of the last 400 years, because it was the one around the Dalton Minimum episode. Only along this exceptionally strong semi secular oscillation LOD has appreciably contributed to temperature [10] , and so we may disregard its contribution to the secular and the semi secular oscillation along the XXI century.

We conclude that the main sources of global warming at the XXI century will be the industrial increase of greenhouse gases and solar activity. In view of the inability of climate models in reproducing present ice melting acceleration and the mechanism that we have presented here by which solar activity increases might explain this phenomena, there are some possibility that the strong increases in solar activity after the 1924 transition was the principal driver of the climate warming of the XX century. However, as a careful evaluation of this mechanism is still lacking we can not rule out the possibility that, by that contrary most of the atmospheric heating of the XX century was of anthropogenic origin.

In the case that solar activity was the source of the decreases of global temperature along the XX century, global temperature will start decreasing not later than about 2020 when solar dynamo will be fully settled in its new Grand Episode. It has taken by 150 years for ice melting reaching current levels. After the 1924 transition the increases on solar storms from the values sustained along the Regular oscillation episodes occurred in two steps, one culminating at 1947, and the other at 1989, but solar storms has already decreased to values comparables to those along the Regular episode, that is at 25% of its value around 1989 in only one step, and the decreases in TSI to its values prior 1924 will culminates by ~2024 , as a result we expect that it will take unless a century to ice cover to be at the same level than prior 1850 and so, as solar activity will continue with the moderate levels corresponding to a Regular Oscillation episode the temperature would be oscillating around the same level than prior 1923, that is -0.44 °C, by the end of the present century.

In the other hand, in the case that most of the warming since 1970 were of anthropogenic origin the contribution of solar activity to temperature change would be of -0.34 °C . As for the next two decades [50] a increases of about 0.2°C per decade is projected for a range of SRES emission scenarios. Even if the concentrations of all greenhouse gases and aerosols were kept constant at year 2000 levels, a further warming of about 0.1°C per decade would be expected. Therefore the heating due to greenhouse gases along the forthcoming 20 years will be between 0.2°C to 0.4°C and so the effect of solar activity would lead to the temperature stay constant or decreasing at the most in 0.14 °C till about 2030 to start increasing again after that, unless greenhouse emission were severely limited in the future.

## 8. Conclusions

The sudden increases of solar activity that occurred after the 1724 and 1924 solar dynamo transitions, has been accompanied by a sudden increases of average surface temperature of 0,2ºC , and 0.34º after 1974 and 1924, respectively. Therefore, of the total increases of the average temperature level, that was of ~ 0.8ºC along the last 400 years, less than 0.3 º may be of non solar origin, in agreement with previous results [15]

A solar dynamo transition to a new Grand Episode of lower solar activity is occurring, that would be settled at sunspot cycle #24 [1, 14, 15]. It would be alike [37] to the 1724-1924 Regular Oscillations episode. In fact sunspot cycle maximum #24 , that would occur at 2013.5, is being the weakest of the last 100 hundred years [13], being alike to sunspot cycle maximum #12 occurring at 1883.

Wile greenhouse gases emission continued increasing at present, there is a hiatus in temperature increases since 10 years ago. This may be the first indication of the impact of the current solar dynamo transition on climate cooling. However, at latitudes above the 64° (North and South) temperature it is still increasing fast as much as there is a acceleration of the ice-melting since 1980 [52-54] and some models indicates that the ice feedback albedo mechanism due to this acceleration is appreciable at the poles [56] and also at height latitudes [57]. This may indicates that this feedback mechanism is still not well represented in IPCC climate models, since they underestimate Arctic sea ice thinning by a factor of 4 and fail to capture the recent sea ice kinematic acceleration [54]. Or it may indicate that a heating source that mainly operates at the polar cap and high latitudes is still missing in those models, and, as suggested in [9, 10] this source may be solar storms, which after the mid-1970s have duplicated their average intensity and frequency as compared with those occurring at the XIX century.

Natural sources of climate changes has been reviewed here, and from present knowledge of them it was concluded that in the long term (time scales above the semi-secular) the main sources of climate change along the past century were greenhouse gases and solar activity, and that the same would happen along the current, XXI century. The episode of Regular Oscillations in solar activity that is starting by now would endure for the rest of the present millennium [37]. If this prediction and the principal source of polar and high latitudes atmosphere heating were solar storms, the sudden decreases of the geoeffectiveness of solar storms to a 50 % of its values prevailing along the XX century Gran Maximum will lead to global temperature to decreases from the present average level of ~0.2°C to the same level that along the 1724-1924 Regular Oscillation episode, that is -0.44°C. The time that it will take to the climate system to react to current decreases of solar activity sensitively depends on the ice-albedo feedback mechanism that is still not well known. An estimation of this time is possible by observing that is has taken ~150 for the recovering from the Little Ice Age [51, 52]. The fact that the principal solar source of atmosphere heating at the poles and high latitudes has already decreased to its values prevailing prior 1924, allows us estimating that the expected decreases in 0.64°C would occurs in about 100 years. Solar storms has decreased yet to values alike to that prevailing at the XIX century, but TSI will decreases substantially only by sunspot cycle maximum #25 (that will occur at ~2024), a appreciable decreases of temperature would be observed only by 2030.

In the case that the main source of heating of the last 100 years were greenhouse gases, the solar activity decreases would contribute with a cooling of only ~0.3°C, and taking into account the projections of temperature from climate models [50], the sudden decreases of solar activity that is going on would mitigate the impact of greenhouse gases on global warming only by the forthcoming 20 years.

## Author details

Silvia Duhau
*Departamento de Física, Facultad de Ingeniería, Universidad de Buenos Aires and Consejo Nacional de Investigaciones Científicas y Técnicas, Argentina*

Ernesto A. Martínez
*Dirección General de Cultura y Escuelas, Buenos Aires, Argentina*

## 9. References

[1] Duhau, S., De Jager, C. The solar dynamo and its phase transitions during the last millennium. Solar Phys. 2008; 250, 1.

[2] De Jager, C., Duhau, S. Forecasting the parameters of sunspot cycle 24 and beyond. J. Atm. Solar Terr. Phys. 2007; 71, 239.

[3] Duhau, S., De Jager C. The forthcoming Grand Minimum of solar activity. J. of Cosmology 2010; 8, 1983.

[4] Kiehl, J. T., Hack, J. J., Bonan, G. B. , Boville, B. A. Williamson, D. L., Rasch, P. J. The National Center for Atmospheric Research Community Climate Model: CCM3*. J. Climate 1998; 11, 1131.

[5] Solomon, S., Manning, D.J., Qin, M , Chen, Z., Marquis, M., Averyt, K.B., Tignor M. and Miller, H.L. (eds.). IPCC. Contribution of Working Group I to the Fourth Assessment Report of the Intergovernmental Panel on Climate Change, Chapter 8 Climate Models and Their Evaluation. Cambridge University Press, Cambridge, United Kingdom and New York, NY, USA 2007; http://www.ipcc.ch/pdf/assessment-report/ar4/wg1/ar4-wg1-chapter8.pdf (accessed 07 May 2012).

[6] Dwyer, J., Norris J. R. and Ruckstuhl, C. Do climate models reproduce observed solar dimming and brightening over China and Japan? . J. of Geophys; Res. 2010; 115, D00K08, 8 PP., 2010 , doi:10.1029/2009JD012945.

[7] Kopp, G. and Lean, J. L. A new, lower value of total solar irradiance: Evidence and climate significance. Geophys. Res. Lett. 2011; 38, L01706, 7 PP., doi: 10.1029/2010GL045777.

[8] Lean, J., and Rind, D. Evaluating Sun-climate relationships since the Little Ice Age . J. of Atmosph. and Solar-Terr. , Phys. 1999; 61, 25.

[9] Duhau, S. Long Term Variations in Solar Magnetic Field, Geomagnetic Field and Climate, Procceding of 9th Asian-Pacific Regional IAU (APRIM 2005) edited by Sutantyo, W., Premadi, P. W., Mahasena, P., Hidayat, T. and Mineshige, S. 2005; 18.

[10] Duhau, S. Solar activity, Earth's rotation rate and climate variations in the secular and semi-secular time scales. Phys. and Chemistry of the Earth. 2006; 31, 99.

[11] Schove, D. J. The sunspot cycle, 649 B.C. to A.D. 2000. J. of Geophys. Res. 1955; 60, 127.

[12] Hathaway, D. A Standard Law for the Equatorward Drift of the Sunspot Zones. Solar Phys. 2011; 273, 221, DOI 10.1007/s11207-011-9837.

[13] Marshal Space Flight center. http://solarscience.msfc.nasa.gov/predict.shtml (accessed 02 July 2012)

[14] Duhau S. An early prediction of sunspot maximum 24, Solar Phys. 2012; 213, 203, DOI: 10.1023/A:1023260916825.

[15] De Jager, C. and Duhau, S. The variable solar dynamo and the forecast of solar activity; effects on terrestrial surface temperature; in J. M. Cossia (ed), Proceedings of the global warming in the 21th century. NOVA science publishers, Hauppauge, NY, 2010; 77.

[16] Nagovitsyn Y. To the description of long-term variations in the solar magnetic flux: The sunspot area index. Astron. Lett. 2005; 31, 557. Translated from Pis'ma v Astronomicheski˘ı Zhurnal, 31, No. 8, 622.

[17] Fisher, C.H., Fan, Y., Longcope, B.W., Linton, H.G., Pevtsov, A.A. The Solar Dynamo and Emerging Flux - (Invited Review), Solar Phys. 2000; 192, 119

[18] Tobias, S.M. The solar dynamo . Phil. Trans. R. Soc. Lond, A 2002; 360, 2741; DOI1.1078/rsta.2000.1090.

[19] Ossendrijver, MUnderstanding the solar dynamo, Astron. Astrophys. Rev. ., 2003; 11, 287.

[20] Dikpati M., de Toma, G. and Gilman, P. A. Predicting the strength of solar cycle 24 using a flux-transport dynamo-based tool, Geophis. Res. Lett. 2006; 33, L05102, doi:10.1029/2005GL025221.

[21] Charbonneau, P. Dynamo models of the solar cycle, Living Rev. Solar Phys. 2012; 7. tirl:http://solar physics.livingreviews.org/Articles/lrsp-2010-3/. (accessed 20 January 2012).

[22] Duhau S. and Chen, C. The sudden increase of solar and geomagnetic activity after 1923 as a manifestation of a non-linear solar dynamo, Geophys. Res. Lett. 2002; 29, 10.1029/2001GL013953.

[23] Wang, Y. M, Lean, J. L. and Sheeley, N.R. . Jr. Modeling the Sun's Magnetic Field and Irradiances since1713. The Astrophysical Journal 2005; 625, 522.

[24] The laboratory for Atmospheric and Space Physics. http://lasp.colorado.edu/sorce/tsi_data/TSI_TIM_Reconstruction.txt (accessed 01 , May 2012).

[25] The Geophysical Data Center (ftp://ftp.ngdc.noaa.gov/STP/SOLAR_DATA/SOLAR_FLARES/FLARES_INDEX/Yearly/ (accessed 01 May 2012).

[26] Lario D. and Simnett. G. M. Solar Energetic particle Variations, in Solar Variavility and its Efects on Climate, J. M. Papaand P. F. Fox Eds. AGU, Geophysical Monograh 2003; 141, 195.

[27] Duhau, S. Global Earth surface temperature changes induced by mean Sun dynamo magnetic field variations In: Solar variability as an input to the Earth's environment. International Solar Cycle Studies (ISCS) Symposium, 23 - 28 June 2003, Tatranská Lomnica, Slovak Republic. Ed.: A. Wilson. ESA SP-535, Noordwijk: ESA Publications Division, ISBN 92-9092-845-X, 2003; 317.

[28] Mayaud, P.N. Analysis of storm sudden commencements for the years 1868-1967. J. Geophys. Res. 1975; 80 (A1): doi:10.1029/0JGREA0000800000.

[29] Service International des Indices Geomagnetiques , ISGI publication Office http://isgi.latmos.ipsl.fr/lesdonne.htm (accessed 02 January 2012).

[30] Lockwood, M., Stamper, R. and Wild, M. N. A Doubling of the Sun's Coronal Magnetic Field during the Last 100 Years. Nature. 1999; 399, 437.

[31] Mayaud, P. N. The aa indices: A 100-year series characterizing the magnetic activity, J. Geophys. Res. 1972; 72, 6870..

[32] Nevanlinna, H., and Kataja, E.: An extension of the geomagnetic index series aa for two solar cycles (1844-1868). Geophys. Res. Lett. 1993; 20; 2703.

[33] Lockwood, M., D.. Hancock, W., B , Henwood, R. Ulich, R. , Linthe, H. J., Clarke, E. and Clilver, A. M. The long-term drift in geomagnetic activity: calibration of the aa index using data from a variety of magnetometer station , 2006;
(http://www.eiscat.rl.ac.uk/Members/mike/publications/pdfs/sub/241_Lockwood_aa_co rrect_S1a.pdf (accessed 02 January 2012)

[34] Legrand, J. P. and Simon, P. A. A two component solar cycle, Solar Phys. 1991; 121, 187.

[35] Russell, C. T., On the possibility of delivering interplanetary and solar parameters from geomagnetic records, 1975. Solar Phys., 42, 259.

[36] Layden, A. C., Fox, P. A., Howard, J. M., Dsarajedini, K. H. and Sofia, S. Dynamo based scheme for forecasting the magnitude of solar activity cycle. Solar Phys. 1991; 132, 140.

[37] De Jager C. and Duhau, S. Sudden transitions and grand variations in the solar dynamo, past and future, J. Space Weather Space Clim. 2012; 2 , A07, DOI: 10.1051/swsc/2012008.

[38] Hoyt, D. V., and Schatten, K. Group sunspot numbers: A new solar activity reconstruction. Solar Phys. 1998; 179, 189.

[39] Naval Research laboratory.
ftp://ftp.ngdc.noaa.gov/STP/ SOLAR_/SUNSPOT_NUMBERS/AMERICAN/X (accessed 01 May 2012).

[40] Usoskin I. G. , Solanki, S., Schüssler, M., Mursula. K., Alanko, K. Millennium-scale sunspot number reconstruction; evidence for an unusually active sun since the 1940s. Phys. Rev. Lett. 2003; 91, NO 21, 211101-1.

[41] De Jager, C. Solar forcing of climate. 1. Solar variability. Space Sci. Rev. 2005; 120, 197.

[42] Hoyt, D. V. and Schatten, K. The Role of the Sun in Climate Change, Oxford University Press.1997.

[43] Farge, M. Annu. Rev. Fluid Mech. 1992. Wavelet transforms and their applications to turbulence 1992, 24, 395.

[44] Torrence, C., Compo, G. P. A Practical Guide to Wavelet Analysis. Bull. Amer. Meteor. Soc. 1998; 79, 61.

[45] Usoskin I., Mursula G. K. and Kovaltsov, G. A. Lost sunspot cycle in the beginning of Dalton minimum: New evidence and consequences Geophys. Res. Lett. 2002; 29, 2183, doi:10.1029/2002GL015640, 2002

[46] Moberg, M., Sonechkin, D. M., Holmgren, K.. Datsenko N. M. and Karlén, W. Highly variable Northern Hemisphere temperatures reconstructed from low- and high-resolution proxy data Nature 2005; 433, 613.

[47] Jones, P. D., Lister, D. H., Osborn, T. J., Harpham, C. Salmon, M., Morice, C. P. Hemispheric and large-scale land-surface air temperature variations: An extensive revision and an update to 2010. J. of Geophys. Res. 2012; 117, D05127, 29 PPdoi:10.1029/2011JD017139.

[48] University of West Anglia, Climate Research http://www.cru.uea.ac.uk/cru/data/temperature/hadcrut3nh.txt (accesed 02 January 2012).

[49] NOAA http://vortex.nsstc.uah.edu/data/msu/t2lt/uahncdc.lt (accessed 02 January 2012).

[50] Solomon, S., Manning, D.J., Qin, M , Chen, Z., Marquis, M., Averyt, K.B., Tignor M. and Miller, H.L. (eds.). IPCC. Contribution of Working Group I to the Fourth Assessment Report of the Intergovernmental Panel on Climate Change, Projections of futre change in climate, Cambridge, United Kingdom and New York, NY, USA. 2007; http://www.ipcc.ch/publications_and_data/ar4/wg1/en/spmsspm-projections-of.html (accessed 15 July 2012).

[51] Akasofu, S. On the recovery from the Little Ice Age, Natural Sciences 2011; 2, 1211.doi:10.4236/ns.2010.211149.

[52] Masiokas , M. H., Luckman , B. H., Villalba , R., Ripalta , A., Rabassa, J. , Little Ice Age fluctuations of Glaciar Río Manso in the north Patagonian Andes of Argentina, Quaternary Res. 2010; 73, 96.

[53] Stroeve, J., Holland, M. M., Meier, W. Scambos, T. and Serreze, M. Arctic sea ice decline: Faster than forecast, Geophs. Res. Lett. 20016; 34, L09501, 5 PP., doi:10.1029/2007GL029703.

[54] Rampal P., Weiss, J., Dubois, C and , Campin, J. M. IPCC Climate models do no capture ice arctic sea ice drift acceleration : consequences in term of projected sea ice thinning and declining , J. Geophys. Res. 116, C00D07, 17PP.

[55] NASA GSFC http://data.giss.nasa.gov/gistemp (assecced 02 May 2012).

[56] Perovich, D., Light, B., Eicken, H., Jones, K. F., Runciman K. and Nghiem, S. V. Increasing solar heating of the Arctic Ocean and adjacent seas, 1979–2005: Attribution and role in the ice-albedo feedback. Geophsy. Res. Lett. 2007; 34, L19505, 5 PP.doi:10.1029/2007GL031480.

[57] Austin, J. and Coleman, S. M: Lake Superior summer water temperatures are increasing more rapidly than regional air temperatures: A positive ice-albedo feedback. Geophys. Res. Lett. 2007; 34, L06604, doi:10.1029/2006GL029021.

[58] Bucha, V. Conclusions. In: Magnetic Field and the Processes in the Earth's Interior, Ed.-in- Chief V. Bucha, Co-edited by G. Petrova, S. Burlatskaya, I Cupal, V. P. Golovkov, H. Kautzleben and W. Webers., Prague Academia, 1983.

[59] Bucha, V. and Bucha, V. Jr. Geomagnetic forcing of changes in climate and in the atmospheric circulation, J. of Atmosph. and Solar Terr. Phys. 1998; 60, 146.

[60] Jackman, Ch. H. and MacPeters, R. D. The effect of solar proton events on ozone and other constutents. In: Solar Variability and its Effects on Climate , J. Pap and P. Fox, eds. Geophysical Monograph, 141, American Geophys. Union, 2005; 305.

[61] Pudovkin M. I. and Morozova, A. L. Time variation of atmospheric pressure and circulation associated with temperature changes during Solar Proton Events, J. of Atmos. and Solar Terr. Phys. 1998; 60, 1729.

[62] Morozova A. L. , Pudovkin M. I. and Thejll, P. Variations of atmospheric pressure during solar proton vents and Forbush decreases for different latitudinal and synoptic zones. Int. J. of Geomagnetism and Aeronomy 2002; 3, 181.

[63] Wilson R. L. Dipole Offset–The Time-Average Palaeomagnetic Field Over the Past 25 Million Years, Geophys. J. of the Royal Astron. Soc. 1971; 22, 491.

[64] Yang, S., Odah H., and Shaw, J. Variations in the geomagnetic dipole moment over the last 12,000 years Geophys. J. Int. 2000; 140, 158.

[65] Korte, M and Constable, C. G. Continuous geomagnetic field models for the past 7 millennia: 2. CALS7K Geochemistry Geophysics. Geosystems 2005, 6, Q02H16, 18 PP., 2005doi:10.1029/2004GC000801.

[66] Lambeck, K. and Cazenave, A. Long Term Variations in the Length of Day and Climatic Change Geophys. J. R. Astr. Soc. 1976; 46, 555.

[67] Hunt, N. B. G. The effects of past variations of the Earth's rotation rate on climate Nature 1979, 281, 188; doi:10.1038/281188.

[68] Duhau S. and Martinez, E. On the origin of the fluctuations in the length of day and in the geomagnetic field on a decadal time scale Geophys. Res. Lett. 1995; 22, 3283. doi:10.1029/95GL03285.

# Chapter 8

# Impact of Regional Climate Change on Freshwater Resources and Operation of the Vanderkloof Dam System in South Africa

Oluwatosin Olofintoye, Josiah Adeyemo and Fred Otieno

Additional information is available at the end of the chapter

## 1. Introduction

Concerns about climate change have prompted calls for action at every level of government and across many sectors of economy and society. It is therefore pertinent to establish a suite of coordinated activities that will examine the serious and sweeping issues associated with global climate change, including the science and technological challenges involved, and provide advice on actions and strategies nations can take to respond to it [1]. Therefore, a proper and good understanding of what climate is and disruptions that variation in climate (climate change) may cause, as we consider its impact on social and economic stability is of paramount importance. Global warming is no longer a speculation. The threat is real and has far reaching consequences. It is absolutely necessary therefore, to sensitize peoples of all nations about the imminent danger posed by global warming and depletion of fresh water resources.

According to [2], water scarcity has emerged as a global issue in recent times and South Africa, currently categorized as water stressed country is forecasted to experience physical water scarcity by the year 2025 with an annual freshwater availability of less than 1000 m³ per capita. The main cause of the scarcity is growing extensive water demand and availability of limited water resources. The situation is further being aggravated by population growth, economic development, urbanization and in more recent times by anthropogenic climate change ([3]; [4]).

Global climate change is a sensitive subject which affects the environment, ecology and quality of life on the earth. Global variations in climate have brought about extreme events like flood and drought which have had drastic impacts on river basin development structures. Such structures include dams on which nations in Africa have depended for most of its renewable energy generation [5].

[6] defined global warming as an average increase in the earth's temperature which in turn causes changes in climate and reported that global warming enhances the water cycle by intensifying the cycle of water. It is speculated that as a result of global warming, more cloud will form and there will be more rain and snow especially in areas close to water whereas in areas particularly away from water sources, excessive evaporation would dry out soil and vegetation, resulting in fewer clouds and less precipitation. Thus, the area will probably get more droughts, rivers and lakes will become shallower and the amount of groundwater decreases.

Global warming or climate variability is expected to alter the timing and magnitude in runoff and soil moisture. As a result, it has important implications for the existing hydrological balance and water resources as well as for future water resources planning and management. Quantitative estimation of the hydrological effects of climate change is therefore essential for understanding and solving potential water resource problems that may occur in the future ([7]; [8]). In the past, decisions relating to the management of extreme climatic conditions, especially as they affect developments within river basins, have either been experimental or experiential [9]. Such subjective approaches have not been able to provide quantitative measure for predicting future climate change impact. This study aimed at using statistical and mathematical modelling approach to provide quantitative measures of past, present and future climate change impacts in the Vanderkloof river basin, South Africa. It focused on the specific impact of global warming as related to the operation of the Vanderkloof dam in South Africa. The specific objectives of this study include studying the impact of global warming on the Vanderkloof River catchments in order to determine its effect on the hydrology of the basin especially on the municipal water supply, hydropower and irrigation systems in the basin and suggesting ways of maintaining and improving on the existing outputs from the existing system in spite of the climate change phenomenon. The outcome of the study will help determine the impact of climate change on the Vanderkloof River basin. It will also be useful in suggesting water management options and preparing operational guides for the dam system so as to optimize the use of available water resources. This will help in making recommendations to policy makers and the authorities of the dam system to enhance the future operation of the dam.

## 2. Methodology

### 2.1. Data and statistics

Long term time series data of 14 Hydro-meteorological variables of the Vandekloof watershed was analysed using mathematical and statistical methods with the aim of developing quantitative models that can be used to forecast future climate change scenarios in the basin and evaluate the performance of the dam system. The variables used include the average annual values of minimum temperature, maximum temperature, average temperature, wind speed, watershed precipitation, dam surface precipitation, dam surface evaporation, reservoir inflow, reservoir outflow, reservoir elevation, reservoir storage, turbine release, irrigation water release and municipal water supply. Range of minimum temperature, maximum temperature, average temperature, wind speed and watershed precipitation spans from 1977 to 2011 (35 years), dam surface precipitation, dam surface evaporation, reservoir inflow, reservoir outflow, reservoir elevation, reservoir storage and turbine release ranged from 1977 to 2008 (32 years), while the span of data for irrigation water release and municipal water supply is from 1990 to 2008 (19 years). Sample statistics such as mean, standard deviation, variance, skewness coefficient and kurtosis were computed to provide an insight into the characteristic (e.g centre and dispersion) of the respective population parameters. Normality of the data was examined using skewness coefficient and kurtosis.

## 2.2. Estimation of autocorrelation coefficient and pre-whitening of data series

Autocorrelation or serial correlation may cause an increase in the expected number of false-positive trends. If autocorrelation exists in the time-series data, an approach to remove this trend needs to be adopted. The approach used to detect lag $k$ autocorrelation is based on the equation ([10]; [11]):

$$r_k = \sum_{t=1}^{N-k} \frac{\sum_{t=1}^{N-k}(x_t - \bar{x})(x_{t+k} - \bar{x})}{\sum_{t=1}^{N}(x_t - \bar{x})^2} \tag{1}$$

where $x_t$ is the time-series data value at time $t$ and $N$ is the number of samples for constant sampling interval. The range of $r$k is $0 \leq r$k $\leq 1$ with a value of 0 meaning that the time-series is independent, and a value of 1 meaning that autocorrelation exists [10]. Autocorrelation is tested for at the 95% significance level. The first-order autocorrelation coefficient $r_1$ is especially important because for physical systems, dependence on past values is likely to be the strongest for the most recent past. For the one-sided test, the World Meteorological Organization recommends that the 95% significance level for $r_k$ be computed by [11]:

$$r_k(95\%) = \frac{-1 + 1.645\sqrt{N - k - 1}}{N - k} \tag{2}$$

where $N$ is the sample size and $k$ is the lag. If the computed value of autocorrelation is greater than the critical value at a significance level of 95%, then the existence of autocorrelation in the time series data is not by chance and a method for removing this may be adopted. The most common approach for removing the impact of serial correlation in time-series data is the pre-whitening method as follows [10]:

$$xp_t = x_{t+1} - r_k x_t \tag{3}$$

where $xp_t$ is the pre-whitened series for time interval $t$, $x_t$ is the original variable $x$ for time interval $t$, and $r_k$ is the estimated serial correlation coefficient at lag k. In this study, equations 1 to 3 were adopted to detect the presences of serial correlation in the time series of the hydro-meteorological variable.

## 2.3. Statistical trend analysis

Trend analysis is a statistical method widely implemented to analyze hydrological time series of temperature, streamflow, precipitation and other climatic variables [10]; [12]; [8]; [13]. The statistical trend analysis in this study was carried out in three phases. First the nonparametric Mann-Kendall test was applied to detect the presence of monotonic increasing or decreasing trend in the time series of the hydro-meteorological random variables. Second, the slope of a linear trend in the data series was estimated using two methods of linear regression analysis. Last, the Pearson Product Moment Correlation Coefficient of the variables and time were computed to determine the strength of the linear relationship between the variables and time.

## 2.3.1. Trend detection

The time series of some hydrological random variables often exhibit significant trends over time. Trend detection in hydro-meteorological time series is of practical importance in analyzing the impacts of global warming and climate change in the various ecosystems of the earth. Statistical procedures may be adopted for the detection of the gradual trends in hydrological series over time. The purpose of trend testing is to determine if the values of a random variable is generally increasing (or decreasing) over some period of time in statistical terms [14]. The Mann-Kendall test is a non-parametric test which is commonly adopted to detect monotonic trends in hydrologic data analysis. This test does not require the assumption of normality of the random variable and only indicates the direction but not the magnitude of significant trends. Trends detectable by the Mann-Kendall method are not necessarily linear. Moreover, the Mann-Kendall test is less affected by the presence of outliers because its test statistic $S$, is based on the sign of differences and not directly on the values of the random variable. Due to its advantages, this test has been applied in a series of recent climate studies ([15]; [16]; [13]; [10]; [12]). The null hypothesis $H_o$ in the Mann-Kendall test is that there is no trend and data are independent and randomly ordered. This is tested against an alternative hypothesis $H_1$, that there exists a trend in the time series. The Mann-Kendall test statistic $S$ is estimated using the equation ([12]; [14]):

$$S = \sum_{k=1}^{n-1} \sum_{j=k+1}^{n} \text{sgn}(x_j - x_k), \tag{4}$$

where $x_j$ and $x_k$ are the annual values in years $j$ and $k$, $j > k$, respectively, and

$$\text{sgn}(x_j - x_k) = \begin{cases} 1 & \text{if } x_j - x_k > 0 \\ 0 & \text{if } x_j - x_k = 0 \\ -1 & \text{if } x_j - x_k < 0 \end{cases} \tag{5}$$

A high positive value of the $S$ statistic indicates an increasing trend, while a low negative value indicates a decreasing trend in the time series of the random variable. The evaluation of the probability associated with $S$ and the sample size, $n$, is however necessary to determine the statistical significance of the trend [17]. The variance of $S$ is computed as

$$VAR(S) = \frac{1}{18} \left[ n(n-1)(2n+5) - \sum_{p=1}^{q} t_p(t_p - 1)(2t_p + 5) \right] \tag{6}$$

where $q$ is the number of tied groups and $t_p$ is the number of data values in the $p^{th}$ group. For a sample size of $n > 10$, the sampling distribution of $S$ is known to follow a standard normal distribution $Z$. The computed values of $S$ and $VAR(S)$ are used to compute a $Z$ test statistic as follows [14]:

$$Z = \begin{cases} \dfrac{S-1}{\sqrt{VAR(S)}} & if \ S > 0 \\ 0 & if \ S = 0 \\ \dfrac{S+1}{\sqrt{VAR(S)}} & if \ S < 0 \end{cases} \tag{7}$$

The statistical significance of the $Z$ values is tested for at the 95% and 99% levels of significance. The critical values of $Z$ at 95% and 99% significance levels are $Z_{0.025}=1.96$ and $Z_{0.001}=2.58$ respectively. The trend is said to be decreasing if $Z$ is negative and the absolute value of $Z$, computed using equation (7), is greater than the critical value, while it is increasing if $Z$ is positive and greater than the critical value. If the absolute value of $Z$ is less than the critical value, there is no trend and the alternative hypothesis that there is a trend is rejected [17]. The significance of a trend simply implies that the occurrence of the trend is not by a process of chance in the selection of the random sample, it has a definite cause. If a trend is significant at the 99% level of significance, then it is said to be highly significant [18].

## 2.3.2. Development of regression models

Regression analysis involves the use of mathematical and statistical techniques for modeling and analyzing several variables with the aim of developing quantitative relationships between a dependent variable and one or more independent variables. Specifically, regression analysis provides insight into how the typical value of a dependent variable changes when one of the independent variables is varied while the others are held constant. Regression analysis is widely used for prediction and forecasting. This study employs the use of two methods of linear regression to develop quantitative statistical models for climate change analysis in the Vanderkloof River Basin, South Africa. The first method is a parametric method while the second is a nonparametric method of linear regression analysis. The parametric method employs the method of least-squares deviations while the non-parametric method involved the use of the Thiel-Sen estimator of slope to develop the respective linear model equations. Parametric methods are suitable when the population can be assumed to conform to a particular probability distribution, whereas non-parametric methods are distribution free methods. Non-parametric methods are often used due to theirs advantages: simplicity, capability of handling non-normal and missing data distributions, and robustness to the effects of outliers and gross data errors ([10]; [19].

### 2.3.2.1. Method of least squares

The least-square method of linear regression requires the assumptions of normality of residuals, constant variance, and true linearity of relationship [12]. Normality implies that the population from which the sample was drawn is normally distributed. Many statistical procedures rely on population normality. The null hypothesis for a normality test states that the population is normal. The alternative hypothesis states that the population is not normal. The regression equation is obtained as [20]:

$$Y = aX + b \qquad (8)$$

where, $X$ =time (year), $a$ =slope coefficient, $b$ = leastsquare estimates of the intercept.
a and b are evaluated using the following equations [20]:

$$a = \frac{n \sum(xy) - \sum x \sum y}{n \sum(x^2) - (\sum x)^2} \qquad (9)$$

$$b = \frac{\sum y - b \sum x}{n} \qquad (10)$$

### 2.3.2.2. Thiel-Sen method of linear regression

The Sen's slope estimator also known as the Kendall robust line-fit method is a non-parametric method of robust linear regression that chooses the median slope among all lines through pairs of two-dimensional sample points. This method offers many advantages and competes well against simple least squares even for normally distributed data. It can be computed efficiently and is insensitive to the presence of outliers; it can be significantly more accurate than simple linear regression for skewed and heteroskedastic data. Missing values are allowed and the data need not conform to any particular distribution. Moreover, the Sen's method is not greatly affected by single data errors ([21]; [19].

The Sen's method can be used in analysis where the trend can be assumed to be linear i.e.

$$f(t) = Q_t + B \qquad (11)$$

where Q is the slope, B is a constant called the intercept and $t$ is time. To evaluate the slope estimate Q in equation (11) the slopes of all data value pairs is computed using the equation

$$Q_i = \frac{x_j - x_k}{j - k} \qquad (12)$$

where $j > k$ . If there are $n$ values $x_j$ in the time series there will be as many as $N = n(n-1)/2$ slope estimates $Q_i$. The Sen's estimator of slope is the median of these $N$ values of $Q_i$. To obtain an estimate of B in equation (11) the $n$ values of differences $x_i - Qt_i$ are calculated. The median of these values gives an estimate of the intercept, B [21].

## 2.4. Computation of correlation coefficients

Correlation coefficient measures the strength of the linear relationship between a dependent and an independent variable ([18]; [22]). In using the Pearson Product Moment Correlation Coefficient, The sample correlation $(r)$, is obtained using equation (13), [20]:

$$r = \frac{n\ \Sigma(xy)-\Sigma x\Sigma y}{\sqrt{[n\ \Sigma(x^2)-(\Sigma x)^2][n\ \Sigma(y^2)-(\Sigma y)^2]}} \tag{13}$$

$r$ ranges from -1 to 1. A value of $r$ close to 0 indicates that there is no association between the variables. R-square ($R^2$), or the square of the correlation coefficient, is a fraction between 0.0 and 1.0. A $R^2$ value of 0 means that there is no correlation between the variables and no linear relationship exist between them. On the other hand, when $R^2$ approaches 1.0, the correlation becomes strong and with a value of 1.0, all points lie on a straight line ([18]; [20]).

In this study, the Microsoft Excel software was used for the computation of relevant statistics and plotting of figures. A program was also written in visual basic for applications to facilitate the computation of the Man-Kendall statistics $S$, Sen's slope $Q$, and intercept $B$.

## 3. Results and discussion

A summary of the computed values of various parameters resulting from the statistical analysis performed in the study is presented in this section. Table 1 presents a summary of statistics for the hydro-meteorological variables analysed in this study. A summary of the autocorrelation analysis is presented in Table 2. The correlation coefficients between the climatic variables and time are presented in Table 3. Table 4 summarizes the result of the Mann-Kendal analysis while the developed Sen model equations and regression model equations are presented in Table 5. Figures 1 to 14 depicts plots showing the time trend of the variables.

| Variable | Statistics | | | | |
|---|---|---|---|---|---|
| | Mean | Variance | Std. Deviation | Skew | Kurtosis |
| Minimum Temperature (°C) | 11.68 | 0.30 | 0.55 | 0.816 | 0.588 |
| Maximum Temperature (°C) | 23.16 | 0.93 | 0.96 | -1.072 | 0.477 |
| Average Temperature (°C) | 16.78 | 0.13 | 0.35 | -0.378 | -0.240 |
| Wind Speed (km/h) | 17.99 | 1.75 | 1.32 | 0.327 | -0.860 |
| Reservoir Elevation (m) | 54.16 | 23.57 | 4.86 | -1.446 | 2.187 |
| Watershed Precipitation (mm) | 533.19 | 11685.65 | 108.10 | -0.073 | -0.607 |
| Dam surface Precipitation (ML) | 3972.78 | 3089745.86 | 1757.77 | 0.617 | -0.190 |
| Dam surface Evaporation (ML) | 18696.50 | 8125912.00 | 2850.60 | 0.078 | 1.680 |
| Reservoir Inflow (ML) | 428598.34 | $6.60 \times 10^{10}$ | 256883.91 | 1.579 | 3.281 |
| Reservoir Outflow (ML) | 455289.63 | $1.03 \times 10^{11}$ | 321278.35 | 2.066 | 4.623 |
| Reservoir Storage (ML) | 2430909.53 | $2.35 \times 10^{11}$ | 484349.18 | -1.082 | 0.859 |
| Turbine Release (ML) | 342055.06 | $2.61 \times 10^{10}$ | 161538.93 | 0.895 | -0.048 |
| Irrigation Water Release (ML) | 28280.55 | 89411891 | 9455.79 | 2.173 | 7.313 |
| Municipal Water Supply (ML) | 68.09 | 1075.60 | 32.80 | 2.627 | 6.550 |

**Table 1.** Statistical summary of hydro-meteorological variables at Vanderkloof dam

| Variable | Lag 1 Autocorrelation | |
|---|---|---|
| | Computed | Critical Value at 95% significance level |
| Minimum Temperature (°C) | 0.622 | 0.255 |
| Maximum Temperature (°C) | 0.847 | 0.255 |
| Average Temperature (°C) | 0.616 | 0.255 |
| Wind Speed (km/h) | 0.505 | 0.255 |
| Reservoir Elevation (m) | 0.445 | 0.262 |
| Watershed Precipitation (mm) | 0.156 | 0.287 |
| Dam surface Precipitation (ML) | -0.043 | 0.262 |
| Dam surface Evaporation (ML) | 0.431 | 0.262 |
| Reservoir Inflow (ML) | 0.219 | 0.262 |
| Reservoir Outflow (ML) | 0.237 | 0.262 |
| Reservoir Storage (ML) | 0.442 | 0.262 |
| Turbine Release (ML) | 0.272 | 0.258 |
| Irrigation Water Release (ML) | 0.423 | 0.262 |
| Municipal Water Supply (ML) | 0.734 | 0.262 |

**Table 2.** Summary of autocorrelation analysis of hydro-meteorological variables at Vanderkloof dam

| Variable | Correlation Coefficient | $R^2$ |
|---|---|---|
| Minimum Temperature (°C) | -0.073 | 0.0053 |
| Maximum Temperature (°C) | 0.790 | 0.6241 |
| Average Temperature (°C) | 0.164 | 0.0269 |
| Wind Speed (km/h) | 0.341 | 0.1163 |
| Reservoir Elevation (m) | 0.137 | 0.0188 |
| Watershed Precipitation (mm) | -0.190 | 0.0361 |
| Dam surface Precipitation (ML) | 0.253 | 0.0640 |
| Dam surface Evaporation (ML) | -0.265 | 0.0702 |
| Reservoir Inflow (ML) | -0.029 | 0.0008 |
| Reservoir Outflow (ML) | -0.197 | 0.0388 |
| Reservoir Storage (ML) | 0.154 | 0.0237 |
| Turbine Release (ML) | -0.026 | 0.0007 |
| Irrigation Water Release (ML) | 0.687 | 0.4720 |
| Municipal Water Supply (ML) | 0.606 | 0.3672 |

**Table 3.** Correlation coefficients between meteorological variables and time at Vanderkloof dam

| Variable | S | Variance (S) | Z | Trend Significance | |
|---|---|---|---|---|---|
| | | | | 95% | 99% |
| Minimum Temperature (°C) | 55 | 4115.667 | 0.842 | No | No |
| Maximum Temperature (°C) | 334 | 4145.333 | 5.172 | Yes | Yes |
| Average Temperature (°C) | 67 | 4112.333 | 1.029 | No | No |
| Wind Speed (km/h) | 113 | 4151.000 | 1.738 | No | No |
| Reservoir Elevation (m) | 49 | 3461.667 | 0.816 | No | No |
| Watershed Precipitation (mm) | -30 | 1833.333 | -0.677 | No | No |
| Dam surface Precipitation (ML) | 67 | 3461.667 | 1.122 | No | No |
| Dam surface Evaporation (ML) | -93 | 3461.667 | -1.564 | No | No |
| Reservoir Inflow (ML) | -19 | 3461.667 | -0.306 | No | No |
| Reservoir Outflow (ML) | 17 | 3461.667 | 0.272 | No | No |
| Reservoir Storage (ML) | 53 | 3461.667 | 0.884 | No | No |
| Turbine Release (ML) | -28 | 3802.667 | -0.438 | No | No |
| Irrigation Water Release (ML) | 111 | 817.000 | 3.848 | Yes | Yes |
| Municipal Water Supply (ML) | 83 | 817.000 | 2.869 | Yes | Yes |

**Table 4.** Summary of Mann-Kendall analysis at Vanderkloof dam

| Variable | Sen Model Equation | Regression Model Equation |
|---|---|---|
| Minimum Temperature (°C) | y = 0.0085x - 5.3524 | y = -0.004x + 19.93 |
| Maximum Temperature (°C) | y = 0.0600x - 96.32 | y = 0.078x - 133.7 |
| Average Temperature (°C) | y = 0.0064x + 4.1428 | y = 0.006x + 4.852 |
| Wind Speed (km/h) | y = 0.051x - 83.7370 | y = 0.046x - 74.79 |
| Reservoir Elevation (m) | y = 0.0667x - 77.2877 | y = 0.072x - 91.15 |
| Watershed Precipitation (mm) | y = 5486.6458 - 2.4863x | y = -2.788x + 6079 |
| Dam surface Precipitation (ML) | y = 48.8098x - 93325.5258 | y = 48.84x - 93325 |
| Dam surface Evaporation (ML) | y = 189266.7823 - 85.6231x | y = -83.13x + 18430 |
| Reservoir Inflow (ML) | y = 2343832.67 - 971.18x | y = -815.2x + 2E+06 |
| Reservoir Outflow (ML) | y = 1377.9458x - 2403626.43 | y = 6966.x - 1E+07 |
| Reservoir Storage (ML) | y = 7609x - 12607593.25 | y = 8202.x - 1E+07 |
| Turbine Release (ML) | y = 3689198.6061 - 1709.9297x | y = 243.0x - 13841 |
| Irrigation Water Release (ML) | y = 831.7600x - 1634940.88 | y = 1155.x - 2E+06 |
| Municipal Water Supply (ML) | y = 1.633x - 3207.08 | y = 3.530x - 6990 |

x= time, and y represents the hydro-meteorological variable

**Table 5.** Developed Thiel-Sen and Regression Model Equations

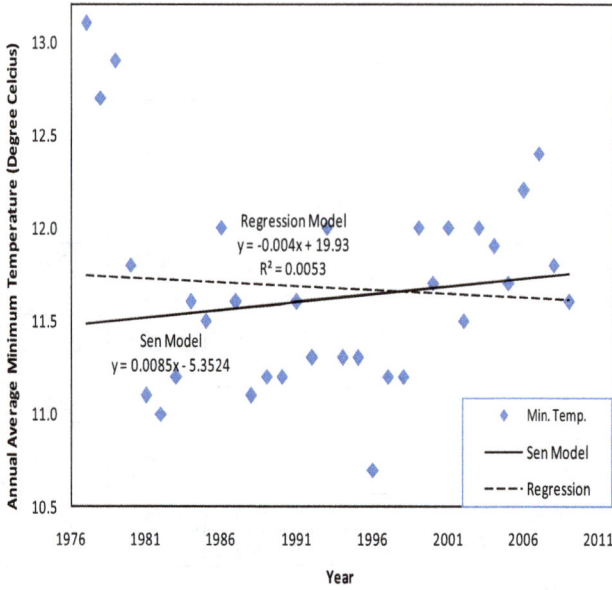

**Figure 1.** Vanderkloof Annual Average Minimum Temperature Trend (1977-2011)

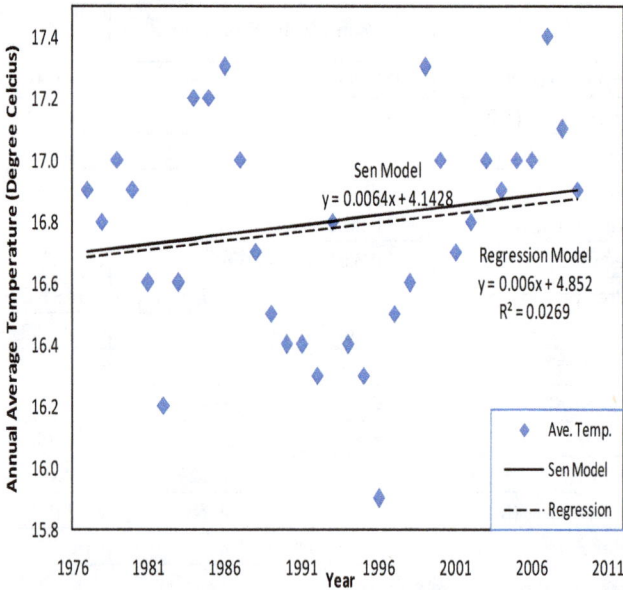

**Figure 2.** Vanderkloof Annual Average Temperature Trend (1977-2011)

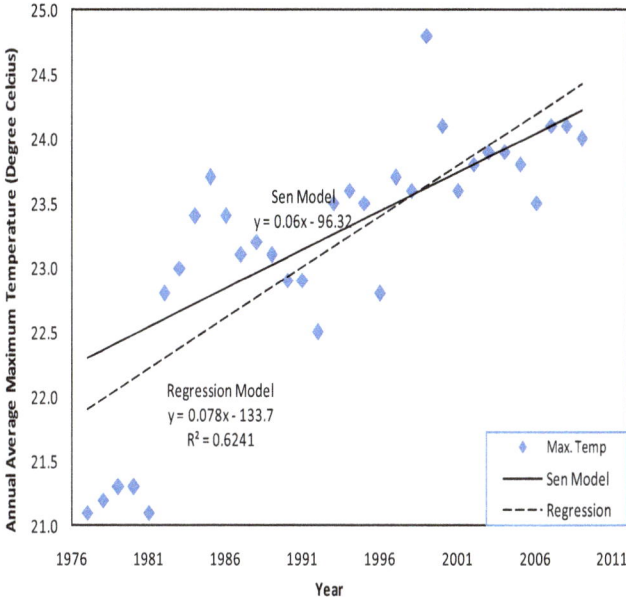

**Figure 3.** Vanderkloof Annual Average Maximum Temperature Trend (1977-2011)

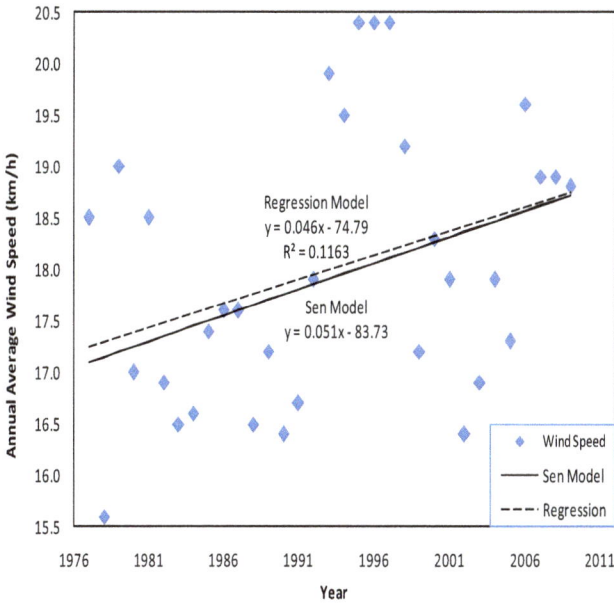

**Figure 4.** Vanderkloof Annual Average Wind Speed Trend (1977-2011)

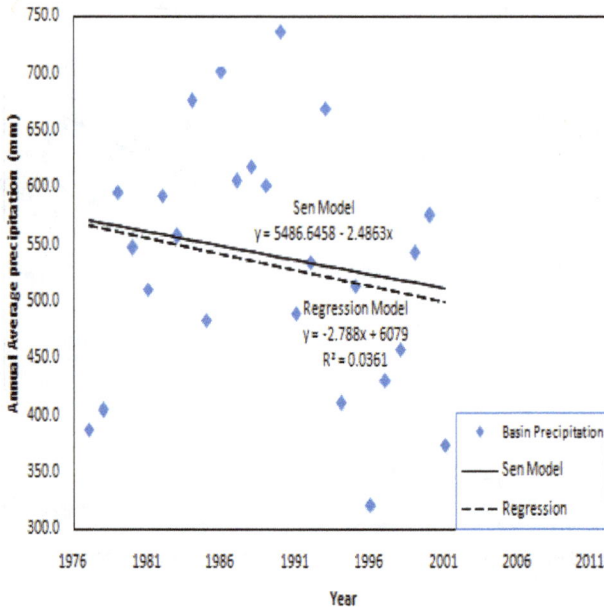

**Figure 5.** Vanderkloof Catchment Average Annual Precipitation (1977-2011)

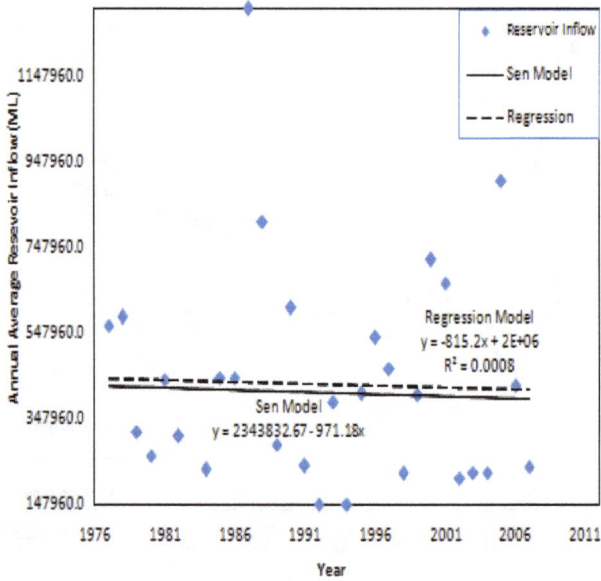

**Figure 6.** Vanderkloof Annual Average Reservoir Inflow Trend (1977-2008)

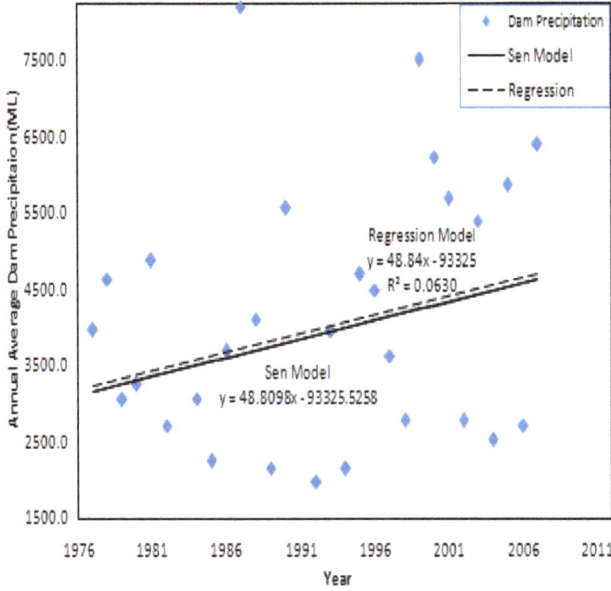

**Figure 7.** Vanderkloof Annual Average Dam Precipitation Trend (1977-2008)

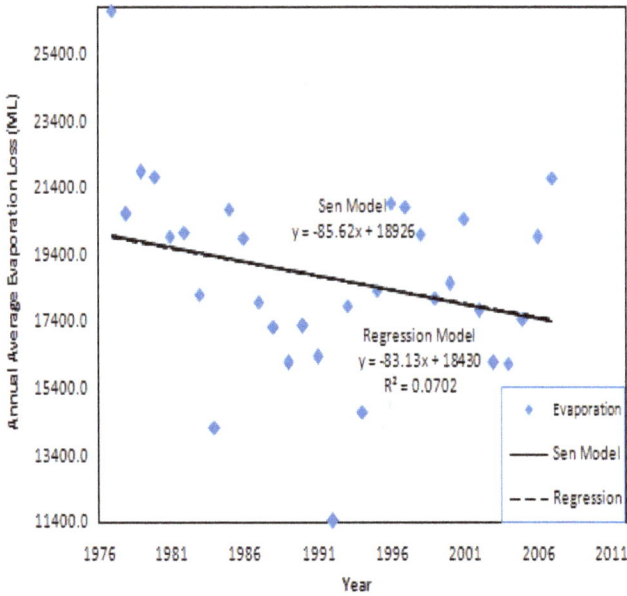

**Figure 8.** Vanderkloof Annual Average Evaporation Loss Trend (1977-2008)

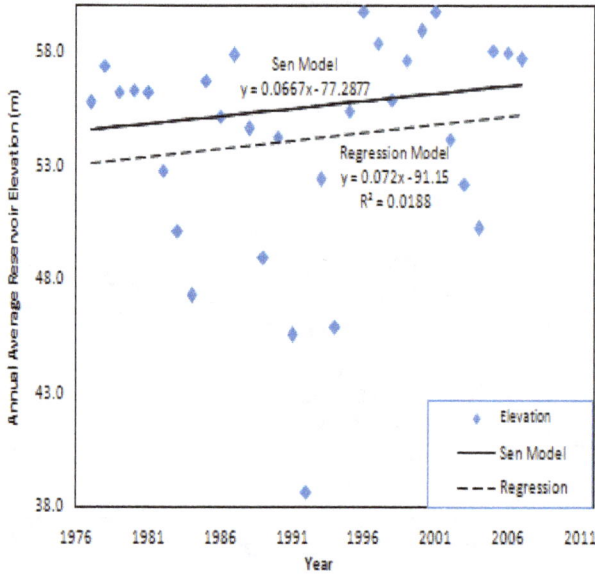

**Figure 9.** Vanderkloof Annual Average Reservoir Elevation Trend (1977-2008)

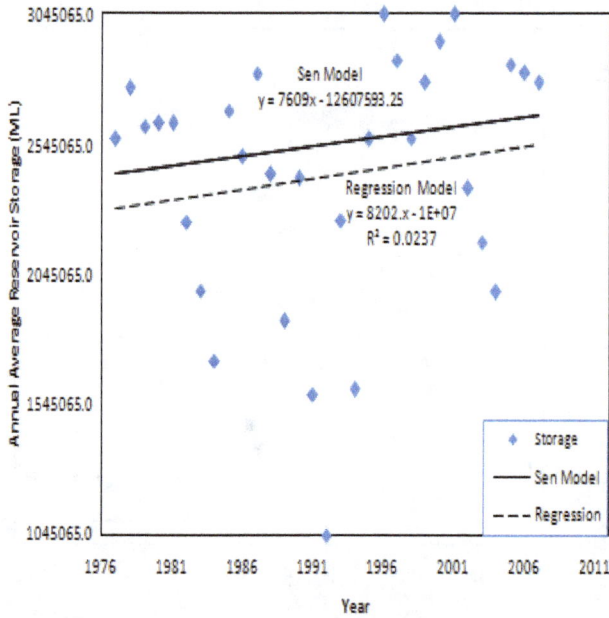

**Figure 10.** Vanderkloof Annual Average Reservoir Storage Trend (1977-2008)

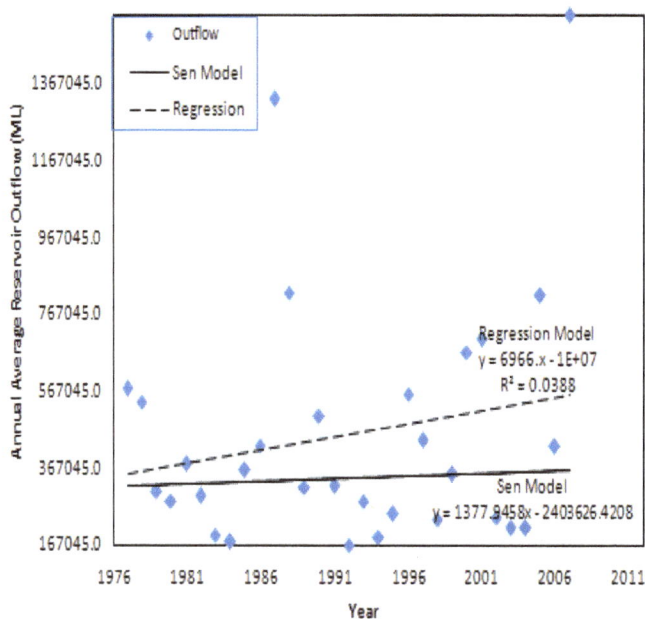

**Figure 11.** Vanderkloof Annual Average Reservoir Outflow Trend (1977-2008)

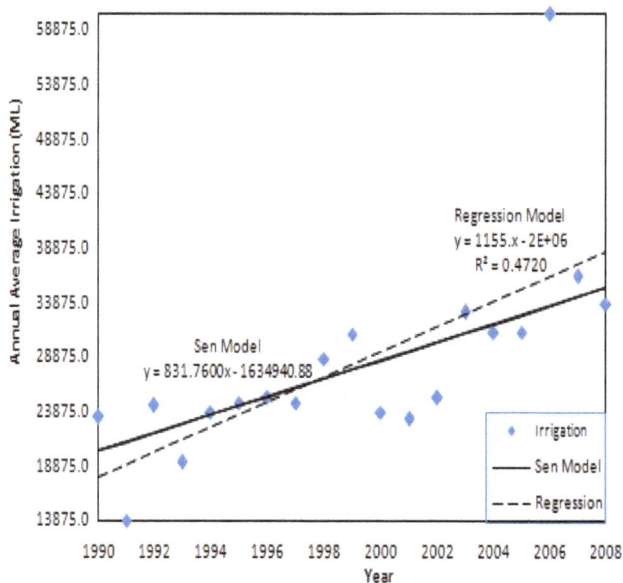

**Figure 12.** Vanderkloof Annual Average Irrigation Trend (1990-2008)

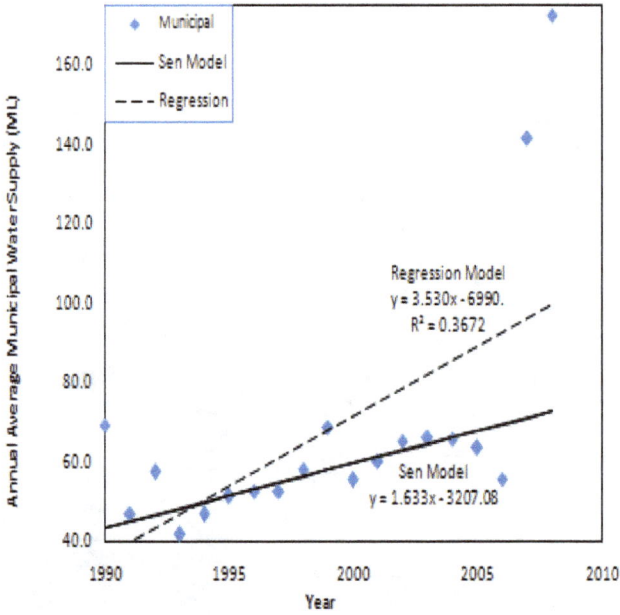

**Figure 13.** Vanderkloof Annual Average Municipal Water Supply Trend (1990-2008)

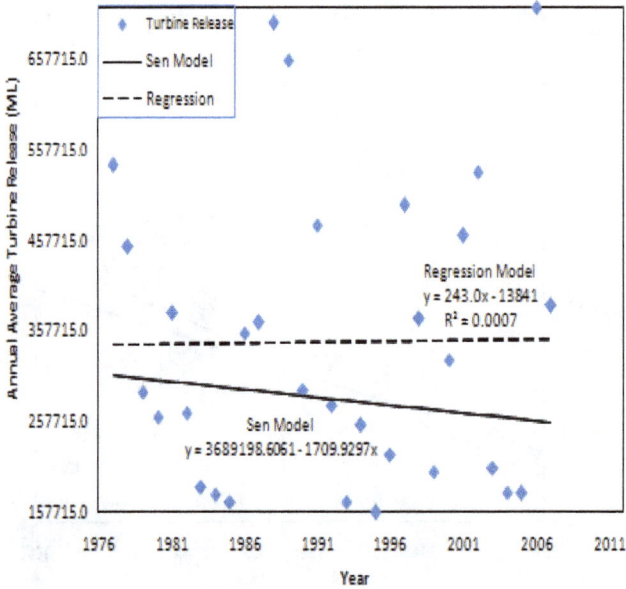

**Figure 14.** Vanderkloof Annual Average Turbine Release Trend (1977-2008)

The Mankendall analysis detected an insignificant positive trend in minimum tempratures over the 35 year period of available data (Figure 1). A Man-Kendall Statistic S = 55 and Sen slope estimate Q = 0.0085 indicate a positive trend in the timeseries. On the contrary, a regression slope coefficient a = -0.004 indicate a negative trend. [23] has noted that the time-series of hydro-meteorological random variable often exhibit a remarked skew and are usually not normally distributed. Hence in this study, the results of the non-parametric Mankendall test and Sen slope estimate are accepted. Therefore, we conclude that there is a positive trend in the time series of minimum temperature, though the trend is not significant.

For the analysis of the average temperature (Figure 2), a Man-Kendall Statistic S = 67, Sen slope estimate Q = 0.0064, and the regression slope coefficient a = 0.006 all indicate a positive trend. A correlation coefficient a = 0.164 shows a weak relationship, and the Z value of 1.0292 shows that the trend is not significant.

The analysis of the time-series of maximum temperature however, shows beyond reseaonable doubt that there is evidence of global warming in the region. The maximum temperature trend is significant at the 95% and 99% levels of significance (Table 4), thus indicating a highly significant rise in maximum temperatures (Figure 3). A Man-Kendall Statistic S = 334, Sen slope estimate Q = 0.06, and the regression slope coefficient a = 0.078 indicate a positive trend. A correlation coefficient a = 0.79 also shows a strong relationship and a Z value of 5.172 shows that the trend is highly significant.

It was found that wind speed has generally being in a uptrend. A Man-Kendall Statistic S = 113 and positive values of Sen slope estimates and regression coefficient accompanied by a Z value of 1.738 also showed that relative windspeed has being in an insignificant uptrend (Figure 4).

In order to obtain an estimate of global warming on precipitaion in the river watershed, the rainfall time series data of a nearby town was analysed. The analysis showed that rainfall in areas futher away from the water bodies has been in a down (though not significant) trend (Figure 5). Analysis of the inflow into the reservoir (Figure 6) also shows a downward insignificant trend. Thus less precipitation on the water shed has invariably resulted in less inflow from the surrounding areas. On the contrary, it was found that rainfall in the immediate vicinity of the dam, especially on the dam surface has been in an uptrend (Figure 7). Though the trends are not significant, a study of the trends however is in concordance with the speculation of [6] that areas away from water sources are likely to get less precipitation and dry out, while areas closer to water bodies will get more rainfall thus resulting in a localized climate phenomenom. Dam surface evaporation has also been in an unsignificant downtrend (Figure 8).

Analysis of reservoir elevation (Figure 9) and resrvoir storage (Figure 10) with a Z statistics of 0.8185 and 0.8838 respectively, shows that this variables have also been in an insignificant uptrend. The insignificant uptrend in the outflow from the reservoir (Figure 11), suggests that though there has been a slight drop in inflow to the reservoir due to the dry conditions of the surrounding lands, this has been more than compensated for by the increase in the dam surface precipitation and drop in reservoir evaporation.

The high increase in irrigation water supply from the reservoirs is among many other factors partly due to the impact of climate change in the surrounding area. Drying-out farmlands in the surrounding areas requires more irrigation water from the reservoir. A Man-Kendall Statistic S = 111, Sen slope estimate Q = 831.76 and a regression slope coefficient a = 1155 all indicate a strong positive trend. A correlation coefficient a = 0.687 shows a strong relationship and a Z value of 3.84 shows that the trend is significant at the 95% and 99% critical levels (Table 4). Likewise, municipal water supply to surrounding towns has been in a highly significant trend (Figure 13). Figure 14 shows that there has been a slight drop in the water supplied to the power industry.

## 4. Conclusion

Developing quantitative functional relationship between hydro-meteorological random variables and time helps provide insight into how the variables trends and provide a means of extrapolation for future climate scenarios. Quantifying the extent to which precipitation and streamflow changes are due to changes in regional climate is an important problem in hydrology. Specifically, in this study, an attempt was made to develop models that relate climatic variables to time and provide a means of estimating future climate change in quatitative terms for the Vanderkloof River basin in South Africa.

From the results of the analyses, temperatures in the vicinity of the dam have been in a significant uptrend over the years. Thus it may be concluded that there is enough evidence of global warming in the region. Global warming has produced favourable climate conditions around Vanderkloof dam as evident from the slight uptrend in the rainfall on the dam surface. The non-significant decrease in inflow to the dam has been balanced by the increased precipitation on the reservoir surface. The significant uptrend in irrigation water supply to surrounding farmlands due to dryer conditions and the significant increase in municipal water supply have resulted in a slight decrease in water supply to the power industry, though not significant. The slight uptrend in the outflow from the reservoir suggests that water supply for various uses is still sustainable under the prevailing climate condition if properly allocated. Recommendation is hereby made to the operators of the dam to optimize the release of water from the dam so as to ensure optimal operation of the dam and sustain power generation, irrigation and allocation for other uses so as to maximize the net benefit obtainable from the reservoir under the prevailing climate condition.

## Author details

Oluwatosin Olofintoye, Josiah Adeyemo* and Fred Otieno
*Departmet of Civil Engineering and Surveying,*
*Durban University of Technology, Durban, South Africa*

---

* Corresponding Author

# 5. References

[1] Academies, T.N. *America's Climate Choice*. 2008 [cited 2010; Available from: http://dels.nas.edu/basc/climate-change/index.shtml.

[2] Otieno, F.a.O., GMM *Water management tools as a means of averting a possible water scarcity in South Africa by the year 2025*. 2004, Water Institute of South Africa (WISA) Biennial Conference: Cape Town, South Africa.

[3] Hamid, A.a.K., N., *Impact of Climate Change on Hydrology, Agriculture and Natural Vegetation: A case study in an irrigation sub-division Buchiana, Punjab Pakistan*. 2003.

[4] Srinivasulu, S., and Jain, A., *A comparative analysis of training methods for artificial neural network rainfall–runoff models*. Applied Soft Computing 2006. 6: p. 295–306.

[5] Ononiwu, N.U., *Managing the effects of global climate changes in geographic information systems (GIS) environment from a river basin perspective the case of drought*. 1994, International Workshop on Impact of Global Climate Change on Energy Development: Lagos.

[6] EPA, *Climate change*. United States Environmental Protection Agency, 2009.

[7] Guo S, Y.A., *Uncertainty analysis of impact of climate change on hydrology and water resources*. Proceedings of the Rabat Symposium, 1997: p. 331 – 338.

[8] Olofintoye, O.a.A., J. , *The role of global warming in the reservoir storage drop at Kainji dam in Nigeria*. International Journal of the Physical Sciences 2011. 6(19): p. 4614-4620.

[9] Ekemezie, P.N., *Planning for variability in water availability: The hydrothermal scheduling problem*. Proceedings from the International Workshop on Impact of Global Climate Change on Energy Development March 28-30, 1994 at the Engineering Building of the Nigerian Society of Engineers in Lagos, Nigeria., 1994.

[10] P.M., A.N.A.a.A., *Exploring the impact of climate and land use changes on streamflow trends in a monsoon catchment*. Royal Meteorological Society, Published online in Wiley InterScience. (www.interscience.wiley.com), 2010.

[11] D., M., *Autocorrelation*. Notes_3, GEOS 585A, Spring 2011, 2010.

[12] McBean, E.a.M., H, *Assessment of impact of climate change on water resources: a long term analysis of the Great Lakes of North America*. Hydrol. Earth Syst. Sci., 2010. 12: p. 239-255.

[13] Olofintoye, O.O.a.S., B.F. , *Impact of Global Warming on the Rainfall of some Selected Cities in the Niger Delta of Nigeria*. 2nd Annual Civil Engineering Conference, University of Ilorin, Nigeria, 2010: p. 342 – 355.

[14] Onoz, B.a.B., M., *The Power of Statistical Tests for Trend Detection*. Turkish J. Eng. Env. Sci., 2003. 27(2003): p. 247 - 251.

[15] Adunkpe, T.L., Salami, A.W, Anwar, A.R, Jimoh, M.O and Ibiyeye, E.A, *Impact of climate change on surface water resources of Abuja*. 2nd Annual Civil Engineering Conference, University of Ilorin, Nigeria, 26 – 28 July, 2010, 2010: p. 407 – 423.

[16] Makanjuola O. R, S.A.W., Ayanshola, A.M, Aremu, S.A and Yusuf, K.O, , *Impact of climate change on surface water resources of Ilorin*. 2nd Annual Civil Engineering Conference, University of Ilorin, Nigeria, 26 – 28 July, 2010, 2010: p. 424 – 441.

[17] Khambhammettu, P., *Mann-Kendall Analysis*. Annual Groundwater Monitoring Report of HydroGeologic Inc, Fort Ord, California, 2005.

[18] R.E., W., *Introduction to Statistics*. Macmillan Publishing Co. Inc. New York. Second Edition., 1974.

[19] Oyejola, B.A., *Non-Parametric Methods and Sequential Analysis*. STA 661 lecture notes (Unpublished). Department of Statistics, University of Ilorin, Ilorin, Nigeria., 2010.

[20] Salami, A.W., Raji M.O, Sule, B.F, Abdulkareem, Y.A and Bilewu, S.O,, *Impacts of climate change on the water resources of Jebba hydropower reservoir*. 2nd Annual Civil Engineering Conference, University of Ilorin, Nigeria, 26 – 28 July, 2010, 2010: p. 442 – 461.

[21] Salmi T., M.A., Anttila P., Ruoho-Airola T., and Amnell T., *Detecting Trends of Annual Values of Atmospheric Pollutants by the Mann-Kendall Test and Sen's Slope Estimates -The Excel Template Application Makesens*. Finnish Meteorological Institute Publications on Air Quality, No. 31, Helsinki, Finland., 2002.

[22] Adamu, S.O.a.J., T.L. , *Statistics for Beginners*. Onibonoje Press and Book Industries (Nig.) Ltd. Ibadan, Nigeria. First Editiion., 1975.

[23] Viessman, W., Krapp, J.W and Harbough, T. E, *Introduction to Hydrology*. Harper and Row Publishers Inc., New York. , 1989. Third edition.

# Effects of Alternative Energy on Environment

# Chapter 9

# Energy Perspective, Security Problems and Nuclear Role Under Global Warming

Hiroshi Ujita and Fengjun Duan

Additional information is available at the end of the chapter

## 1. Introduction

It is the critical issues of the 21st century to achieve global scale 3E problems, which are keeping Environmental preservation, Energy security, and Economic growth. Recently there are several recommendations to affect national energy policy. Climate change due to carbon dioxide in atmosphere has not been fully proved, but Precautionary Principle to reduce carbon emission has been adopted internationally because it will be too late to cope with disaster after a century. It is a time to take much longer time span for energy planning to cope with future energy crisis, which seems inevitable due to apparent limit of resources.

Role and potentials of nuclear energy system in the energy options are discussed from the viewpoint of sustainable development with protecting from global warming. They are affected dramatically by different sets of energy characteristics, nuclear behavior and energy policy even under the moderate set of presumptions. Introduction of thousands of reactors in the end of the century seems inevitable for better life and cleaner earth, but it will not come without efforts and cost. The analysis suggests the need of long term planning and R&D efforts under the wisdom.

New regime establishment has been discussed toward climate change in Section II. The feasible target for new emission scenario called Z650 (Overshoot & Zero-Emission) instead of traditional concept and energy mix against global warming has been proposed. Taking the effort for energy-saving as major premise, carbon capture and storage for fossil fuel, renewable energy and nuclear energy should be altogether developed, which means energy best mix should be achieved, under the constraint of keeping $CO_2$ concentration in the atmosphere around 450ppm.

Energy security problems and nuclear role have been also discussed in Section III. The basic overview of energy security and method of evaluation are reorganized. Energy security which had wide conception is indicated by method that compares the energy security level in each country. The role and potential of nuclear power from the viewpoint of the energy security is an important point to examine the direction and the role of the nuclear power industry in the future. It is understood that the energy security is severely affected by the case without nuclear energy.

Energy issue and nuclear energy role after the Fukushima Daiichi Accident has been further examined in Section IV. The root causes and countermeasures of the Fukushima Daiichi Accident are described and the direction for energy and nuclear power after the accident is discussed.

The path and key issues for "Sustainable development" has been summarized in Section V. Nuclear power and renewable energy should be two wheels towards low carbon societies against global warming with economic growth and with avoiding energy security problem.

## 2. Toward a new climate change regime establishment

### 2.1. Climate change history

In order to address climate change, most countries joined the United Nations Framework Convention on Climate Change (UNFCCC) in early 1990s to examine how to reduce global warming. The Third Conference of Parties (COP3) took place in 1997, and adopted the principle update to the treaty, the Kyoto Protocol. In the protocol, industrialized countries and economies in transition (Annex I countries) committed to reduce their aggregate greenhouse gas emissions by about 5.2% during the period of 2008-2012 (so called the first commitment period) compared to 1990 emission levels. With the coming up of the expiration date of the first commitment period, the post-2012 climate regime has been examined during recent years. According to the Bali Action Plan adopted at the COP13 in 2007, intensive negotiations aimed at urgently enhancing the implementation of the Convention up to and beyond 2012 have been conducted during 2008 and 2009. However, the COP15 taking place in Copenhagen in December 2009 failed to produce an international agreement involving binding greenhouse gas emissions reduction targets. From then on, the international negotiation on climate change fell into a chaotic state. The COP16 held in Cancun in the end of 2010 and the COP17 held in Durban in the end of 2011 adopted the Cancun Agreements and the Durban Agreements that consist of significant decisions by the international community. The agreements represent key steps forward in capturing plans to reduce GHG emissions and to help developing countries protect themselves from climate impacts. However, the framework of the climate regime has not been clarified.

Fruitless negotiations on international binding scheme of GHG emissions reduction illustrate that the absence of a common vision become the biggest obstacle of combating global warming. It is time for us to go back to the beginning of the issue to consider what kind of world we can share.

## 2.2. Global emission pathway

In general, the base of the climate regime combating global warming is that it is necessary to limit the global surface temperature to 2°C compared to pre-industrial levels (so called "2°C target"). In the Copenhagen Accord and following COP Agreements, this target was reconfirmed. Based on the target and the fourth assessment report of IPCC, the G8 Summit (Declaration 2007, 2008 and 2009) argued that the worldwide greenhouse gas emissions must be reduced by at least 50% in 2050 compared to the levels of 1990 or recent years. However, the ambitious argument failed to get global consensus due to the strong opposition by most developing countries who claimed that the reduction plan did not have sufficient scientific background and did not leave enough space for their economic growth. Therefore, it is necessary to reexamine the scientific analyses of the climate change for developing a reliable emission pathway which can be accepted worldwide.

Employing the schemes of zero emission and overshoot, a research group developed a new stabilization concept named "Zero-emission Stabilization (Z-Stabilization)" instead of the traditional equilibrium stabilization. Their researches (Matsuno et al., in [1]) documented that the Z-Stabilization could avoid long-term risks while meeting short term need of relatively large emissions. Based on the new concept of stabilization and the 2°C target, a global GHG emission scenario named Z650 was proposed (Figure. 1). The scenario was designed based on two assumptions, one is that the amount of cumulative $CO_2$ emissions in the $21^{st}$ century would be 650GtC equivalent, the other is that the zero-emission would be achieved in 2160. Some recent researches (e.g., UKCCC, 2008 [2]; Allen et al., 2009 [3]) also employed the concept of zero emission or near zero emission for seeking best options of climate change mitigation. These researches suggest, from practical viewpoint, that a functional form with a peak within several decades following by monotonic decrease to approach to zero is necessary for a reliable emission pathway.

The performance of the designed Z650 scenario was examined, along with a typical 450ppm equilibrium stabilization scenario (E450), though projection experiment by using a simplified climate system model. Figure 2 shows the emission pathways (a), the $CO_2$ concentrations (b), the global temperature rises from the pre-industrial period (c), and the sea level rises due to thermal expansion (d) of the two scenarios. The $CO_2$ concentration under the Z650 scenario increases more rapidly, exceeds 450ppm in about 2030, and goes to its peak of about 480ppm around 2070 due to the lager amount of emissions during the early period of $21^{st}$ century. It declines thereafter because the emission will be less than the natural absorption, crosses the 450ppm line around 2160, and goes down steadily. In contrast the concentration under E450 scenario stays below 450ppm, and increases steadily to approach the final equilibrium state. As a result, the maximum temperature rise under Z650 scenario is 1.8°C at around 2100 (if all GHG was taken into account, the peak value would be 2.3°C). The peak will last only several decades, and then the temperature will decrease to a stable state (1.7°C higher than the pre-industrial level). At meanwhile, almost no significant difference of sea level rise occurs between the two scenarios. These results obtained through the projection experiment indicate that the proposed Z650 scenario could be a new solution on combating climate change given by science. According to the Z650 scenario, the global CO2 emissions will peak between 2020 and 2030 with a ratio of approximate 1.3 and decrease to around 0.75 in 2050 compared to 2005 level.

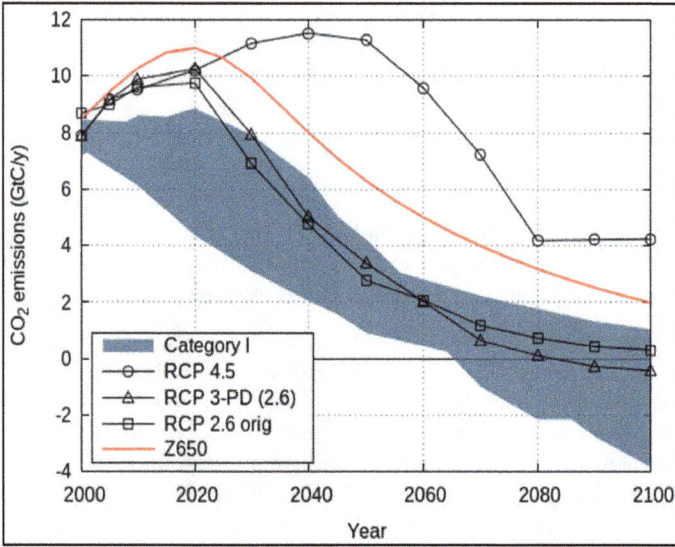

**Figure 1.** CO2 emission pathways: RCPs and Z650 (Matsuno et al., in [1])

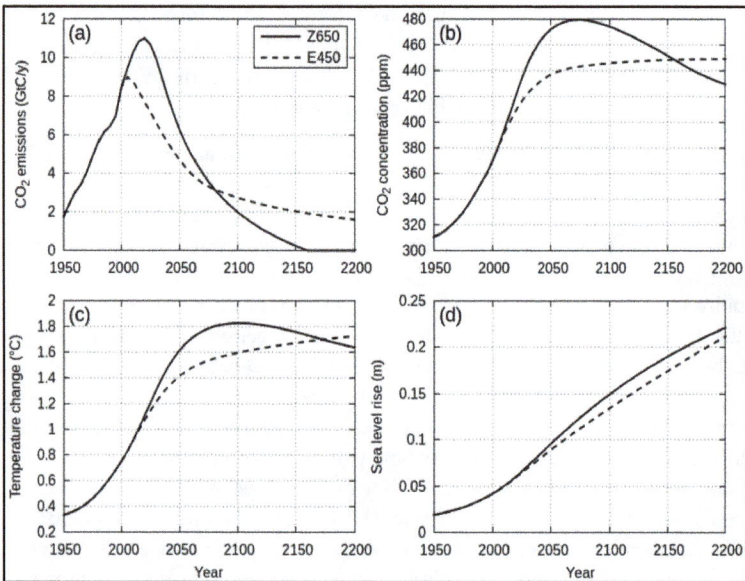

**Figure 2.** Comparison between Z650 and E450 during short to midterm (Matsuno et al., in [1])

## 2.3. Optimal way toward the global vision

In order to examine the technical feasibility of the Z650 scenario and investigate the optimal way to realize it, numerical experiments of global energy system optimization using GRAPE (Global Relationship Assessment to Protect the Environment) model (Kurosawa et al., 1999 in [4]) were conducted. Fifteen regions were set in the model to cover the global aggregate, those are: United States, Western Europe, Japan, Canada, Oceania, Russia, Central Europe, East Europe, China, India, ASEAN countries, Middle East and Northern Africa, Southern Africa, Brazil, and Latin America. The former 8 regions were defined as industrialized countries, and the rest regions were defined as developing countries. The final energy demands for every region were assumed based on population and economic growth, while the technology assumptions were examined based on previous researches. The cost minimization of global energy system was carried out to optimize the global and regional energy supply.

Three main scenarios were analyzed for the period of 2000 to 2150. BAU (Business as Usual), which is the baseline scenario of CO2 emissions, assumed no changes of current the energy and environmental policies in the future. It is very similar with the Reference Scenario of IEA (IEA, 2010 in [5]) (Figure.3). REF (Reference), which is the reference scenario of economic assessment, assumed that energy conservation would be promoted according to regional capacities and conditions but no CO2 reduction policy. It has a similar performance with the New Policy Scenario of IEA (IEA, 2011 in [6]) (Figure.4). Z650, which is the mitigation scenario, assumed a global CO2 emission cap based on scientific Z650 scenario described above (Figure. 2(a)).

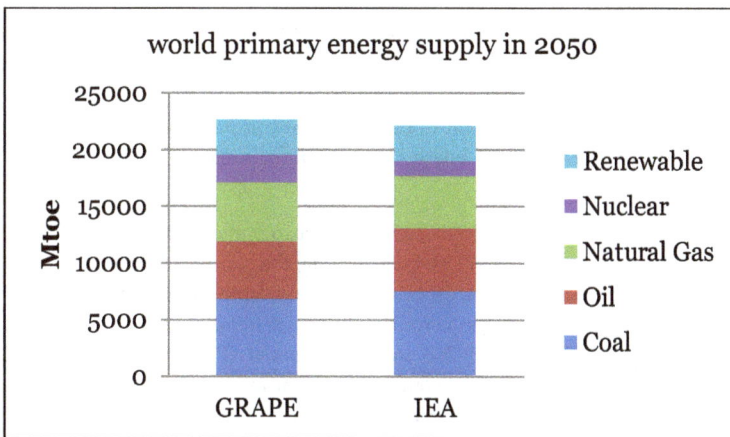

**Figure 3.** Comparison between BAU in this study with the Reference Scenario of IEA

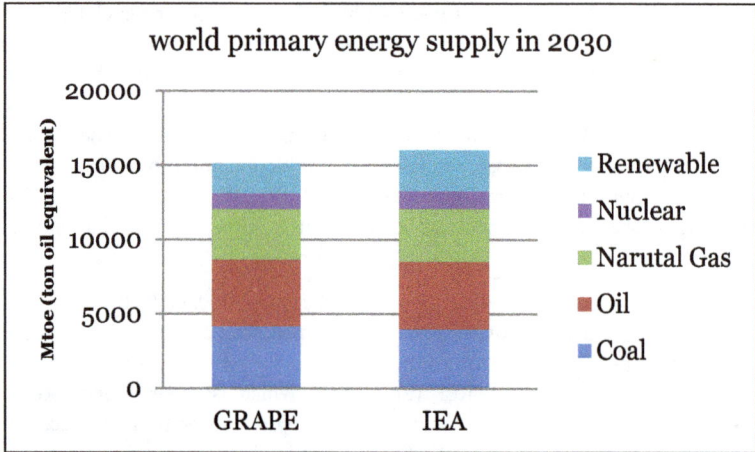

**Figure 4.** Comparison between REF in this study with the New Policy Scenario of IEA

### 2.3.1. Long-term energy vision

The simulated global total primary energy supply (TPES) for the three scenarios is shown in Figure 5. Under BAU, the TPES with a large portion of fossil fuel increases substantially, triples in 2100 compared with the 2000 level. The TPES of REF increases slightly during the later stage, almost doubles in 2100 compared with the 2000 level, due to the influence of the regional energy conservation measures. However, the main component is still the fossil fuel.

On the other hand, the resulted TPES of Z650 is the cleaner in combination despite the same amount with REF. In order to prevent global warming, the consumption of fossil energy will peak at 2030, and the clean energies, especially the renewable energy will play an essential role during the second half of the century. As the results, portion of Fossil: Nuclear: Renewable is 5: 2: 3 in 2050, while 3: 2: 5, in 2100. Regional TPES for Z650 is also examined as shown in Figure 6. In industrialized countries, total primary energy is almost constant up to 2100, where share of fossil fuel gradually decreases and share of renewable energy mainly increases alternatively. In developing countries, total primary energy continuously increases up to 2100, where peak of fossil fuel consumption is around 2040, and both nuclear and renewable energy increase remarkably.

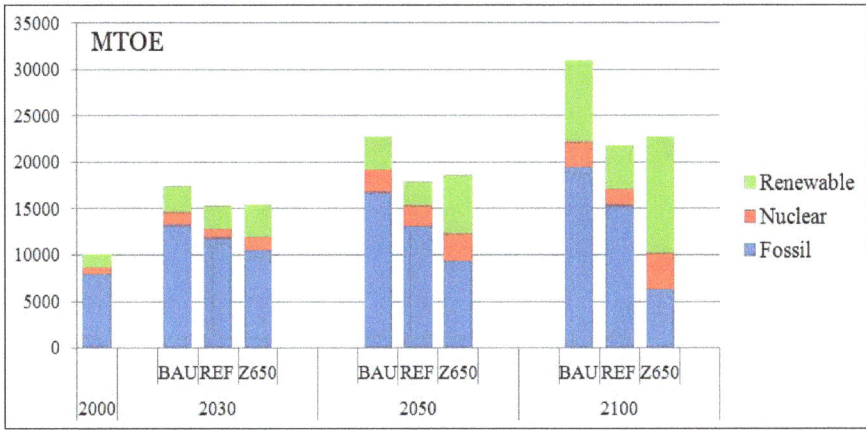

**Figure 5.** Global total primary energy supplies for the three scenarios

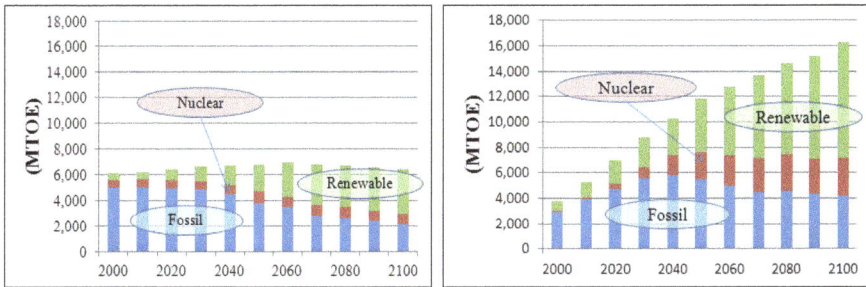

**Figure 6.** Regional total primary energy supply for Z650 (left: industrialized countries; right: developing countries)

The results for global power generation trends for Z650 are shown in Figure 7. In general, decarbonization and pluralization processes will be improved together. The fossil fuel will play essential role during early stage, but will decrease after the peak in 2040. During the second half of the century, it will cover less than 15% of the total power generation. The nuclear energy will increase constantly during the first half, and provide approximate 30% of the global power. As to renewable energies, large scale utilization of the wind power will start from 2020, while that of the solar photovoltaic power will start from 2050. Both of them increase steadily till the end of the century. Together with the stable hydro power and increasing biomass power generation, the renewable energies will cover almost 40% of global power generation in 2050, and a portion of about 60% in 2100.

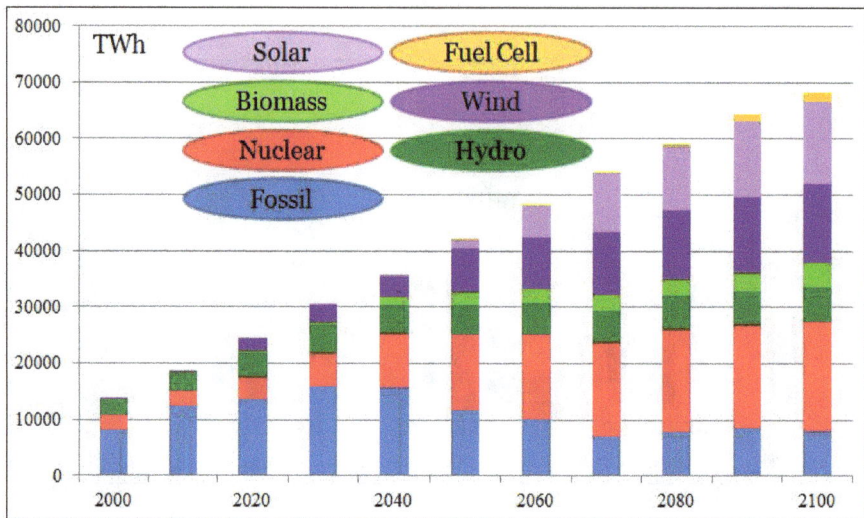

**Figure 7.** Global power generation for Z650

## 2.3.2. Energy related $CO_2$ Emissions

Based on the global $CO_2$ emission cap of Z650, the global energy system optimization projected regional $CO_2$ emissions (Figure. 8). Emissions of industrialized countries peak in 2010 and emissions in 2050 will be reduced by 50% compared to 2005 levels. On the other hand, emissions by developing countries will peak in 2030 at 1.6 times 2005 emissions and decline to 1.1 times 2005 emissions in 2050.

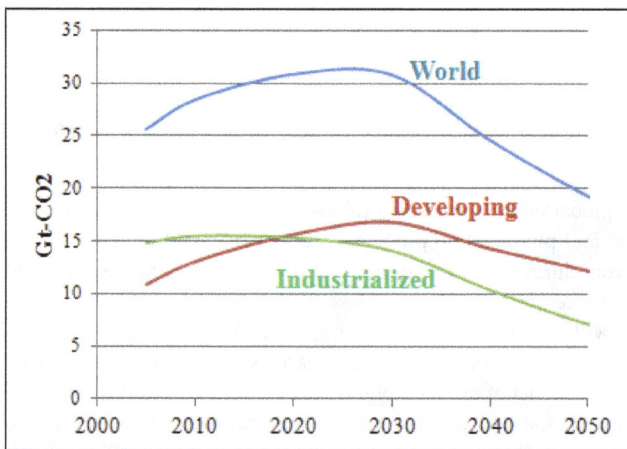

**Figure 8.** Global and regional CO2 emissions of Z650

From the standpoint of regional equitability, per capita emissions in the industrialized nations will approach that of the developing nations by 2100 and the $CO_2$ emissions per GDP of the developing nations will approach that of the industrialized nations (Figure. 9). The results for industrialized nations show that $CO_2$ emissions per capita and $CO_2$ emissions per GDP will converge around 2050. Global emissions in 2030 will be 1.6 times that of 1990 (1.2 times that of 2005) and will be about 1990 levels. Compared to the REF scenario without $CO_2$ constraint, the ratio for global emissions in 2030 for Z650 is 0.82. For industrialized nation the ratio is similar at 0.89. For 2050, the ratio to the REF scenario for industrialized nations of 0.48 is similar to the global ratio of 0.46 (Table 1). As the reduction potential is higher for developing nations, the effect is larger. In general, the resulted regional emission curves reflect the differences of financial and technical capability among areas. These results provide useful information for global harmony.

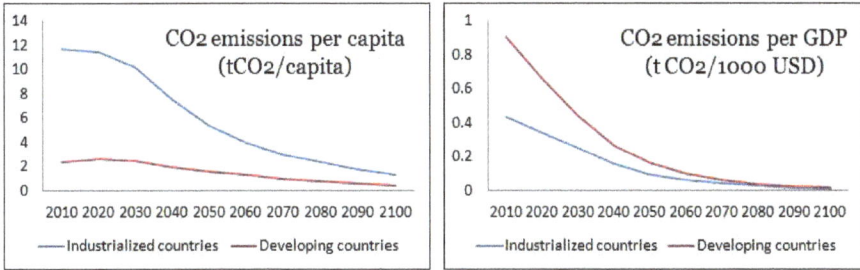

Figure 9. Two $CO_2$ emission index based on simulation of Z650

| Region | CO₂ emissions | | | | | |
|---|---|---|---|---|---|---|
| | 2030 | | | 2050 | | |
| | Ratio to 1990 levels | Ratio to 2005 levels | Ratio to REF of 2030 | Ratio to 1990 levels | Ratio to 2005 levels | Ratio to REF of 2050 |
| World | 1.60 | 1.20 | 0.82 | 1.00 | 0.75 | 0.46 |
| Industrialized countries | 1.05 | 0.95 | 0.89 | 0.53 | 0.48 | 0.48 |
| USA | 1.16 | 0.96 | 0.90 | 0.57 | 0.47 | 0.47 |
| EU15 | 0.89 | 0.86 | 0.91 | 0.46 | 0.45 | 0.53 |
| Japan | 0.93 | 0.79 | 0.90 | 0.55 | 0.47 | 0.66 |
| Developing countries | 2.82 | 1.54 | 0.77 | 2.05 | 1.12 | 0.45 |
| China | 2.77 | 1.48 | 0.74 | 1.53 | 0.82 | 0.37 |
| India | 3.42 | 1.91 | 0.72 | 2.83 | 1.57 | 0.37 |
| ASEAN | 3.74 | 1.64 | 0.80 | 3.41 | 1.50 | 0.57 |

Table 1. Global and regional emissions in major industrialized and developing countries

Compared with BAU, emission reductions by region and sector till 2050 in Z650 are investigated. Among the regional emission reductions, that of the developing countries with substantial economic growth in the future occupies more than two thirds in the following 40 years (Figure. 10). Especially the reductions in China, India and ASEAN countries contribute 31%, 13% and 8% of the total reduction in 2050 respectively. It means that the decarbonization in the regions with substantially increasing energy demands will hold the key to combat global warming. Among the industrialized countries, the United States will contribute the most. Its reduction occupies 14% of global reduction in 2050. While reduction in Japan only contributes approximate 1% of global reduction until 2050.

The results of analysis of $CO_2$ reductions by sector show that energy conservation contributes the most during the whole period (Figure. 11), occupies 42% and 32% of all reduction in 2030 and 2050 respectively. The second contributive sector is the power generation. It will contribute 25% and 27% of all reduction in 2030 and 2050 respectively. The carbon capture and storage (CCS) will play an increasing role in later stage, contribute 27% of all reduction in 2050.

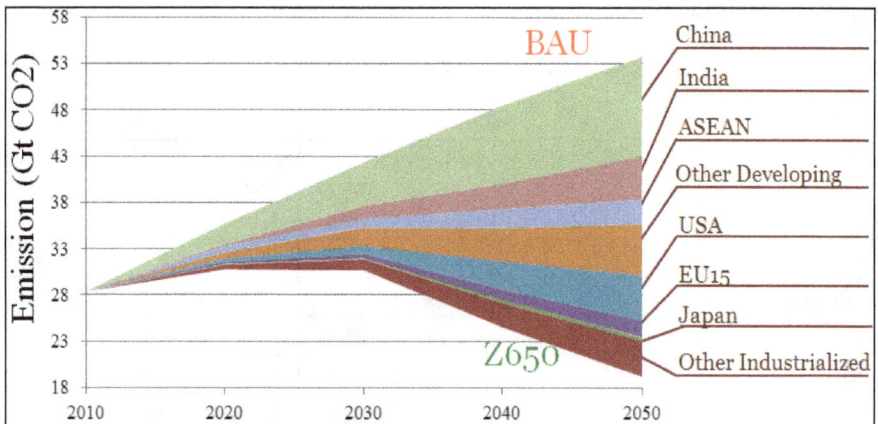

**Figure 10.** Simulated regional emission reductions in Z650 compared with BAU

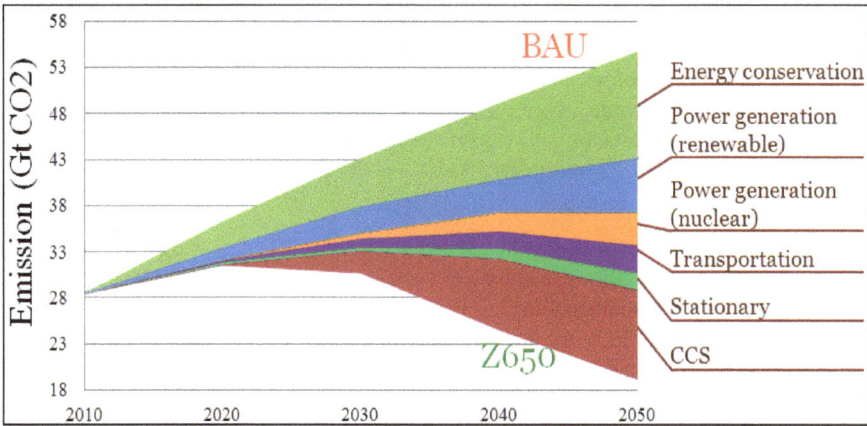

**Figure 11.** Contributions of each sector to CO₂ emission reduction based on simulations

## 2.3.3. Economic assessment

An economic assessment was conducted based the analysis of necessary additional investments and the fossil fuel saving. The analysis is based on the accumulative statistics during 2010-2050. In REF, the world will emit 1,462 Gt CO2 during the 40 years, in which 622 Gt generated in industrialized countries while 840 Gt generated in developing countries. At the meanwhile, total energy system costs will be 232 trillion USD (in 2005 value) in the world with almost the same portions in industrialized and developing countries (Table 2).

| REF | CO2 Emissions (ratios to 2005 levels) | | Acc. Emissions GtCO2 (2010-2050) | Acc. Costs T$ (2010-2050) | |
|---|---|---|---|---|---|
| | 2030 | 2050 | | | |
| World | 1.5 | 1.6 | 1462 | 323 | |
| Industrialized countries | 1.1 | 1.0 | 622 | 154 | |
| Developing countries | 2.0 | 2.5 | 840 | 169 | |
| Z650 (Z650+) | CO2 Emissions (ratios to 2005 levels) | | Acc. Reductions GtCO2 (2010-2050) | Additional Investments T$ (2010-2050) | Fuel Saving T$ (2010-2050) |
| | 2030 | 2050 | | | |
| World | 1.2 | 0.75 | 362 | 11 (42) | 14 (10) |
| Industrialized countries | 1.0 (0.7) | 0.5 (0.2) | 114 (256) | 4 (37) | 5 (10) |
| Developing countries | 1.5 (1.9) | 1.1 (1.5) | 248 (106) | 7 (5) | 9 (0) |

**Table 2.** Global and regional economic analysis based on the simulations

In order to achieve the Z650 vision against global warming, an accumulative emission reduction of 362 Gt CO2 is to be carried out, one third in industrialized countries and two thirds in developing countries. For the purpose, total additional investments of 11 trillion USD are necessary worldwide, which is equivalent to 0.28% of the global accumulative GDP in the same period. The data for industrialized and developing countries are 4 and 7 trillion USD, 0.18% and 0.43%, respectively. Most of the investments are distributed in transportation and power sectors.

At meanwhile, the additional investment will yield significant savings in fossil fuel consumption. The total fuel savings in the Z650 compared to the REF are 57 Gtoe of coal and 32 Gtoe of oil. However, additional 26 Gtoe of natural gas will be consumed. Calculated using current prices of the fossil fuels, the undiscounted value of these fuel saving is 14 trillion USD, 5 in industrialized countries and 9 in developing countries. Thus, in this case the additional investments could be covered by the fuel savings during the following 40 years both globally and regionally. There would be a good balance between benefit and investment from the optimal energy mix. This assumes the technologies to be used by 2050 are those technologies that currently appear to be feasible and are expected to be widely deployed by 2030.

In order to evaluate the economic performance further, an additional scenario analysis was conducted. In the new scenario, which is so called Z650+, the emission cap for industrialized countries is added to constraint conditions according to the G8 Summit Declaration. That is the industrialized countries will reduce their emission by 80% in 2050 compared with 2005 levels. The projection results are also shown in Table 2.

The accumulative emission reduction is the same, but one third will be carried out in developing countries and two thirds in industrialized countries. Due to the lower reduction potential of low cost in industrialized countries, the total necessary additional investments jump to 42 trillion USD, which is equivalent to 1.09% of the global accumulative GDP in the same period. The data for industrialized and developing countries change to 37 and 5 trillion USD, 1.66% and 0.31%, respectively. At the same time, the fuel savings will be less than Z650, and will mainly distribute in industrialized countries. As a result, the good balance between additional investments and fuel saving is destroyed. In addition, the high cost in industrialized countries would not bring benefits to developing countries.

## 2.4. Role of nuclear energy

As mentioned above, nuclear energy will play an important role to achieve the proposed vision against global warming. Its share in global TPES will increase steadily during the first half of the 21st century, from approximate 10% in 2030 to almost 20% in 2050, and will keep the level in the second half of the century. It will contribute more in power generation sector. Approximate 20% of global electricity in 2030 and more than 30% in and after 2050 will be generated by nuclear energy.

In order to evaluate the role of nuclear energy, the analysis on two sub-scenarios based on Z650 were carried out. One is NuPO, in which nuclear energy will be phased out with considering

Fukushima Daiichi Accident affect, that is no new plant will be built from 2020 and the current plants will be closed according to designed life time. The other is NoFBR, which means the technology of Fast Breeder Reactors (FBR) will not be utilized. In usual case such as Z650 scenario, we assumed that the FBR technology will be available and introduced from 2050.

### 2.4.1. Impact to TPES

The global TPES of the Z650, NuPO and NoFBR are shown in Figure 12. In the case of phasing out nuclear energy, natural gas including that from unconventional resources will be the main alternative during the first half of the period. However, large-scale introduction of renewable energies, especially the offshore wind energy, occurs during the second half of the period due to the limitation of natural gas resources. On the other hand, the absence of breeder technology does not cause significant influence to TPES during the early stage. But the increase of nuclear energy utilization will be limited by the uranium resources thereby more natural gas will be introduced during the middle stage. Within the end stage of the period, similar to the characteristics in NuPO, large-scale of renewable energy will be introduced.

Anyway, Z650 scenario shows Light Water Reactor (LWR) will play important role in the first half century, while FBR, latter half.

### 2.4.2. Impact to power generation

Figure 13 illustrates the projection results of the global power generation for the Z650, NuPO and NoFBR. In general, similar portfolio is necessary for both NuPO and NoFBR compared with Z650. The natural gas, biomass and wind energy will be the main alternatives to nuclear energy during the early stage. While natural gas with CCS, solar energy and fuel cell will be the main alternatives during the late stage. However, the scales of introducing these technologies are smaller in NoFBR compared with NuPO due to the availability of LWR. And the more coal can be used through the technology of IGCC with CCS during the middle stage. According to the technology portfolio, the global average costs for power generation in NuPO are much higher than in Z650 during the whole period and will be almost twice in 2100 (Figure. 14). On the other hand, the global average costs for power generation in NoFBR are not significantly different with those in Z650 till around 2060. However, it will increase rapidly during the end stage in the case of NoFBR, and will be approximately 50% higher than in Z650 (Figure. 14).

### 2.4.3. Economic impact

The same economic assessments as for Z650 are performed for NuPO. Compared to the Z650 scenario, global total additional investment through 2050 would increase from 11 trillion USD to 17 trillion USD while benefits from fuel saving would decline from 14 trillion USD to 9 trillion USD. The additional investment and fuel savings are 6 trillion USD and 5 trillion USD for industrialized countries, 11 trillion USD and 4 trillion USD for developing countries. These results indicate that the more negative impacts will happen in developing countries. There is no significant difference between the economic performance of NoFBR and Z650 till 2050.

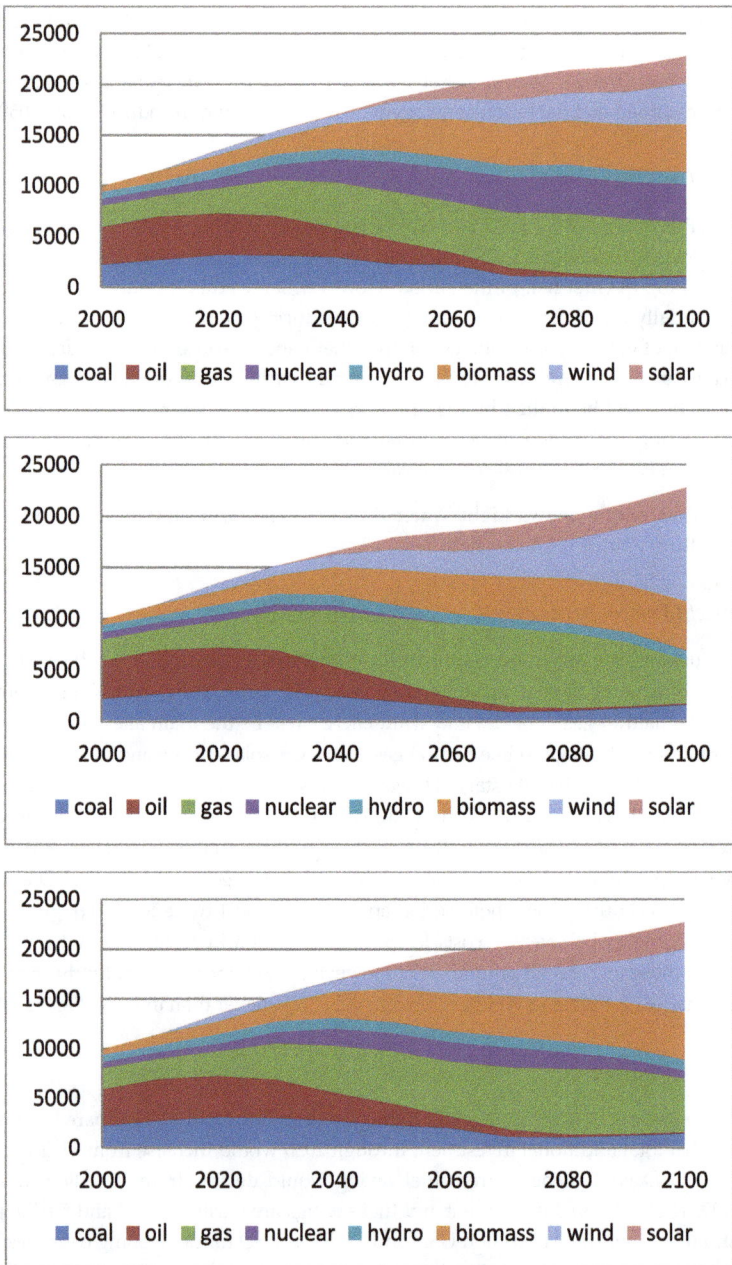

**Figure 12.** Projected TPES (upper: Z650; middle: NuPO; lower: NoFBR) (unit: Mtoe)

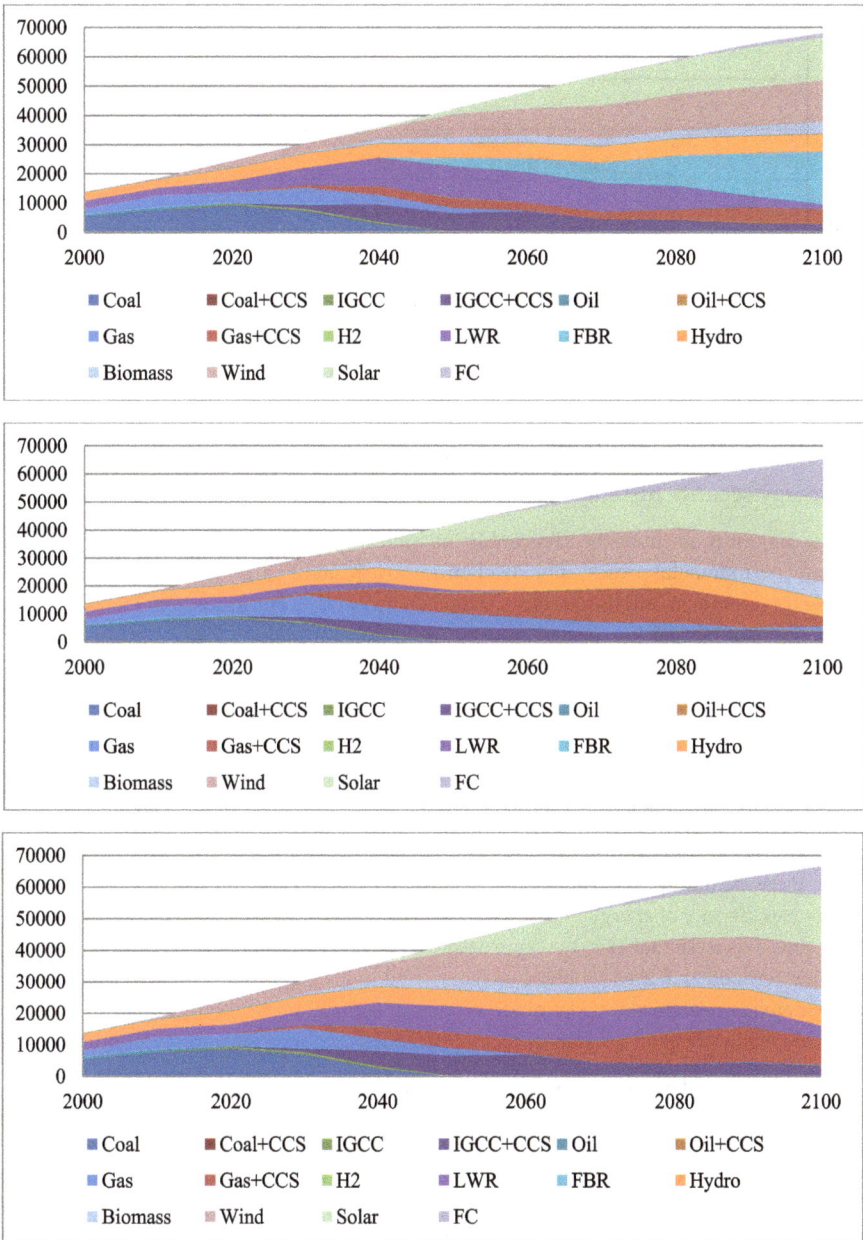

**Figure 13.** Figure 13 Projected global power generation (upper: Z650; middle: NuPO; lower: NoFBR) (unit: TWh)

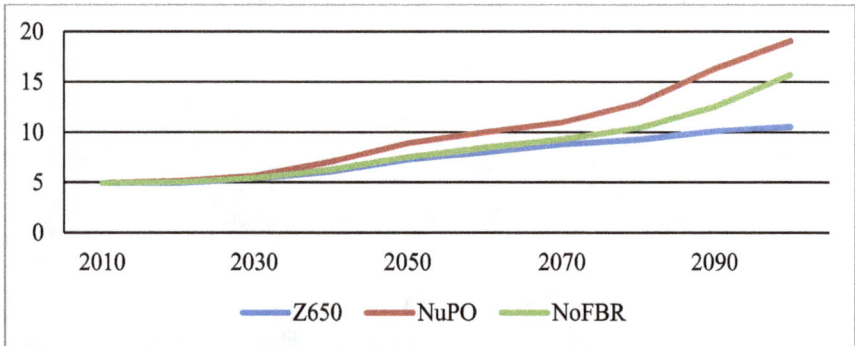

**Figure 14.** Projected global power generation costs (unit: cents/kWh)

## 3. Energy security problems

### 3.1. Energy security history

It is the biggest problem of the 21st century to achieve global scale 3E problems, which are keeping Environmental preservation, Economic growth, and Energy security. Recently there are several recommendations to affect national energy policy. For instance in Japan, the role and progress of nuclear is expected to solve the global warming up, by Council for Science and Technology Policy, the Atomic Energy Commission and the Agency of Natural Resources and Energy, and etc.

Expert meeting on "Nuclear Energy and Security of Supply" was held for OECD/NEA/NDC, during Dec. 2007- Dec. 2008, in Paris. One of the authors has attended the meeting for Japanese Expert. The explanation, examples, and proposals are based on the meeting discussion [7].

In fact, there are multiple concepts for energy security, due to the differences of the quantity of resources, density of energy network, or the needs of the times [7-9]. As broad definitions, under time axis or spatial axis, two approaches, divided to the long term on global problem and short term on each area problem, are advocated in the paper. The problem of short term on each area is recognized as definition of energy security of narrow meaning in general, which are further categorized into incidental (temporal) problem and structural problem.

The model is popular for poor resource countries to evaluate security risk based on the imported energy resources portfolio methodology which targets to energy best mix. Several evaluations of energy securities such in Europe and Japan are discussed [7-9]. As specific example of evaluation, the evaluation process is shown how to evaluate the security risk from the five points of view, energy efficiency, diversification index of energy resource portfolio, energy resource dependency from Middle East and Russia, self-sufficiency ratio in the primary energy supply, and $CO_2$ emission index. The comparison results are also indicated in seven developed countries belong to OECD. Furthermore, the study of nuclear role from the viewpoint of different results of nuclear existence or not is discussed [8]. The Scenario Planning analysis of "Two China in 2015" is also introduced [10].

## 3.2. Concept of energy security

### 3.2.1. Energy resources

Figure 15 shows evolution of primary energy structure, shares of oil and gas, coal, and non-fossil sources, in percent, historical development from 1850 to 1990 (triangles) and in scenarios to 2020 (open circles), 2050 (diamonds), and 2100 (closed circles) [11]. Three cases as follows are indicated in the figure;

Case A includes three high-growth scenarios,

Case B has a single middle-course scenario, and

Case C is the most challenging.

The primary energy had changed to coal of fossil fuels from firewood which is originated by solar energy since the Industrial Revolution, shift to oil occurred in the 20th century, and then it has come to nuclear power and fossil fuels in general in 2000. In the future, to meet the challenges of energy resource depletion and global warming, it will be migrated to renewable energy and nuclear energy in any scenario.

Table 3 shows energy intensity (electric power generation) for each electric power source in [12]. Looking at the energy density of various types of power, nuclear and coal-fired power plants are large and overwhelmingly 1 GWh/m2/year, while renewable energy is very small about 10kWh/m2/year, renewable energy significant expansion in the primary energy ratio would be difficult to expect. It is expected to have the division of roles and complement each other, nuclear power as a backbone power source, and renewable energy as distributed one.

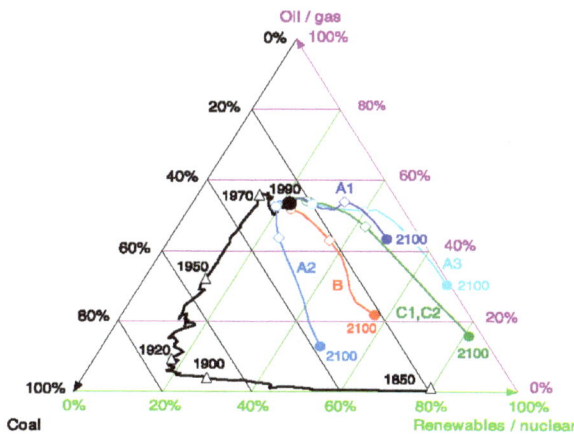

**Figure 15.** Evolution of primary energy structure.

| Candidate | Power density per square meters  [kWh/m$^2$ · year] | Remarks |
|---|---|---|
| Electrical needs in house | 35 | Detached home (160sq.m.  40A) |
| Electric needs in office | 400 | Eight-story (architectural area 3,000sq.m.) |
| Biomass power | 2 | Poplar plantation (6years-cycle) Generating efficiency  34% |
| Wind power | 21 | Tehachapi (U.S.A.) C.F.20% |
| Solar power | 24 | Roof of detached house (160sq.m. 3kW, equipment  availability15%) |
| Hydro power | 100 | Average of 100 hydro power plants in Japan |
| Coal-fired power | 9,560 | Hekinan coal-fired power plant (2.1million kW) |
| Nuclear power | 12,400 | Kashiwazaki-Kariha (8.212million  kW) |

**Table 3.**  Energy intensity (electric power generation) for each electric power source.

Discovery, production, and projection of oil and gas with CO2 emission is shown in Figure 16 [13]. Looking at the fossil fuel resources, whereas the amount of discovery peaked in 2000 on the border, as shown in the figure, because the peak in demand is still ahead, it is expected to accelerate the decrease in supply.

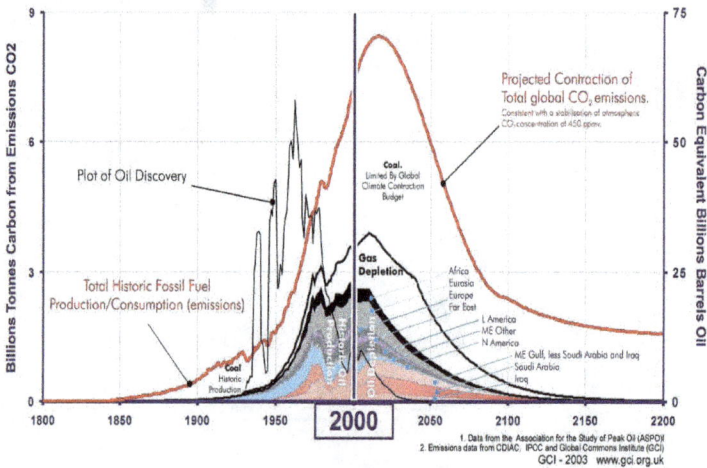

**Figure 16.**  Discovery, production, and projection of oil and gas with CO2 emission.

As understood from Figure 17 which shows fossil fuel resources per capita [14], the oil and natural gas unevenly distributed in the Middle East and Russia, on the other hand coal is a large amount of endowment in the world mainly in North America and Russia. The challenge is anticipated to be significant that there has less abundance of fossil fuels, while increase in demand in Asia. It is growing awareness of energy security in the countries of East Asia led by China is a matter of course.

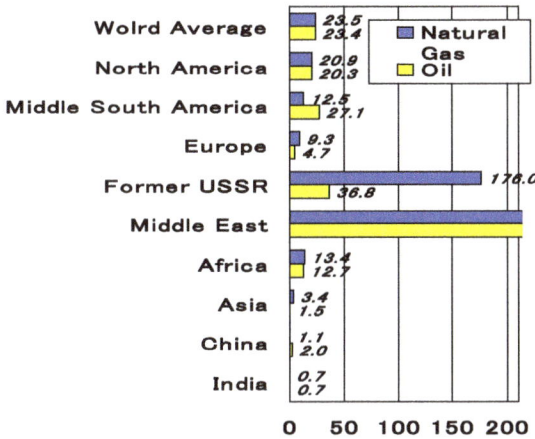

(a) Proven oil and natural gas reserves (Ton per capita)

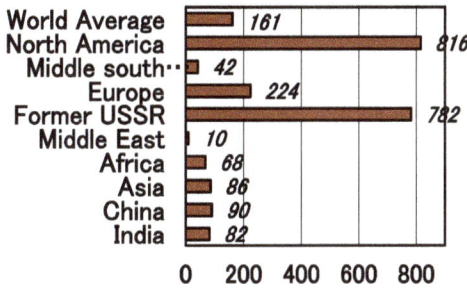

(b) Proven coal reserves (Ton per capita)

**Figure 17.** Fossil fuel resources per capita.

Uranium, on the other hand, look at the next 50 years, initially is large in consumption in developed countries, from 2020 consumption in developing countries will increase in the supply and estimated additional resources to the resource confirmation become severe by 2050 that also somehow with the addition of promising high-cost resources [15]. According to the simulation results of the authors shown in Figure 18, Uranium is consumed in industrialized countries first, while consumption in developing countries increased after

2020. The uranium resources are used significantly in nuclear power, by the time 2100 are also likely to depletion [8]. For this reason, the introduction of fast breeder reactors can be required as countermeasures as soon as possible, considering long Plutonium breeding time.

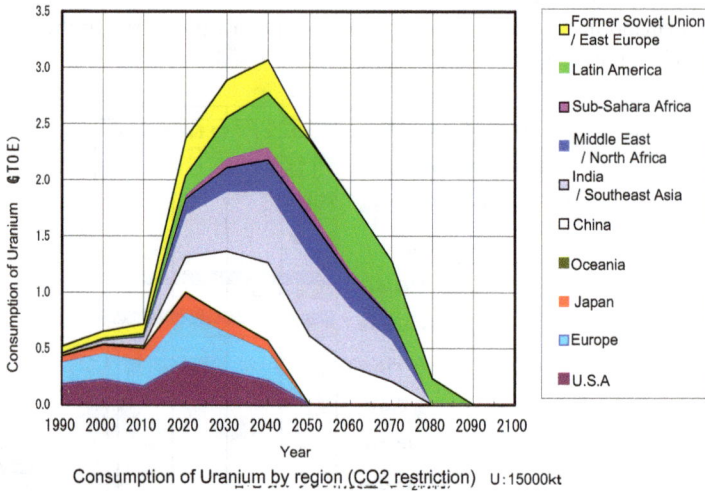

Figure 18. Uranium usages in the World.

### 3.2.2. Major factors threatening energy security

In a familiar concept, safety issue is a measure of the risk factors that occur in good faith in an organization act in basically. In contrast, security issue, as seen in the information security issue, is a measure to risk factors due to external attacks or malicious action in the organization.

As a broad concept, there is a national security. The underlying is to ensure the national interests for the people as a nation in power relations between nations. As they say, energy security and food security accounts for the foundation of national security. Apart from this, there are also domestic issues such as counterterrorism.

Energy security, in a situation no one knows what will happen (risk factors), is to ensure sufficient energy source as a nation. World War II was said to be a scramble for oil. Another example is that the population to be able to survive is determined by the amount of supplied energy. In this way, energy is the backbone of the nation, and energy policy is also considered as a measure of energy security.

In the OECD / NEA expert meeting, "security of energy supply and role of nuclear energy " held in 2008, the security of energy supply was discussed [7]. At the meeting, "the economics of imported energy, social, political and technical problems" were discussed. OECD / NEA has first announced Nuclear Outlook 2008 in 2009, in which the idea " nuclear power is alternative resource, and can be supplied by the countries that political stability is important for energy security" was also showed [16].

### 3.2.3. Definition of energy security

There are different energy security concepts in Europe and U.S.A. with Japan. The important issues in Europe are electric power network in the community (EU), and the prevention of large scale black out, and fuel supply (Gas and Uranium shortage etc.). On the other side, the important issues in Japan are the improvement of self-sufficiency rate for energy import and making good portfolio of energy resources to be caused by few natural resources.[9] They are not opposing concepts between the Europe concept (stability of supply network) and Japanese one (self-sufficiency and market power). The diversity of defined energy security is to be indicated by nation or entity, for instance of U.S.A. which has electric network vulnerability same as Europe. Another example, regardless China locates in the Continent of Asia, China is regarded to have similar concept to Japan of the island country. For these reasons, it is searched for various and hierarchical definition to approach accurate analysis of security risk, and that the risk is also examined on time and spatial axes. These multilateral considerations are the essence to measure security risk.

The broad sense definition of energy security risk is classified by time axis and spatial axis in Table 4 [8]. The short term energy security risk is narrow definition of the energy security as it is called. It means energy security risk happened in term around 10 years, and can be categorize in nation, area or global under spatial axis. The problem of energy resource supply from other countries, especially the approach to energy resources best mix, is one of the most important problem for isolated and few natural resources country, like Japan,. Expanding for use of energy by developing countries like BRICs is serious matter not only it causes energy resources conflict for other countries but it also brings out strong demand for keeping energy resources for themselves.

The other hand, on long term around 100 year problem, it is the global energy problem which are shortage of fossil fuel energy and global warming. They are the most important issues in recent energy environment problems.

| Range | Area | Content |
|---|---|---|
| **Narrow meaning SoS**<br><br>**Short Term ~10y**<br>**Energy Crisis** | •Country<br>•Region<br><br>•World | •Energy supply- Best mix<br>•Fuel supply- U problem<br>•Electricity supply- Network<br>•Developing countries usage (China, India, etc.)- Best mix |
| **Long Term ~100y**<br>**Energy Problem** | •Global | •Fossil Fuel Exhaustion<br>•Global Warming |

**Table 4.** Wide meaning energy security - short term regional crisis vs. long term global problem.

In short term energy crisis can classify into incidental (temporal) crisis like accident or terrorism and structural crisis like Middle East instability or expanding of energy demand in Asia, as shown Table 5 [8]. The measures are different in these crises, immediate action as typified by oil reserve and long term political solution as typified by resource development.

| | Cause | Consequence | Countermeasure |
|---|---|---|---|
| Contingent Crisis | • Conflict, <br> • Accident, <br> • Terrorism | • Energy supply chain (Sea-lane) interruption | • Petroleum reserve <br> • International and Regional Corporation <br> • Anti-Accident <br> • Anti-Terrorism |
| Structural Crisis | • Middle East instability, <br> • Energy demand increase in Asia, <br> • Technology development stagnation, <br> • Environmental problem | • Price fluctuations, <br> • Supply shortage, <br> • Resource straggle, <br> • Weak consumer | • Energy Technology Development <br> • Energy Policy <br> • Foreign policy <br> • Defense policy |

**Table 5.** Short term regional energy crisis.

## 3.3. Evaluation method for energy security

### 3.3.1. Indexes for energy security

The most basic index of energy security in island country is self-sufficiency ratio in the primary energy supply. The self-sufficiency ratio is 96% in England and 140% in Canada, while 50% with nuclear, 8% without nuclear in France. In Japan, it is 14% with nuclear power, without nuclear power, only 8%. Diversified index of primary energy is indicated in Figure 19 [17]. Balanced energy supply country can get low number, which means that they have achieved Best-Mix in energy resource. Canada has the most balanced energy portfolio. While China is indicated the highest number 0.55, because of China relied heavily on coal-fired thermal power. Japan is also indicate higher number 0.31 than the average of OECD or World (0.27), because Japan largely relies on oil.

The Basic Act on Energy Policy in Japan points out the necessary of energy resources diversification, that is one of the course to keep the steady supply of energy. Table 6 shows that oil dependency in Japan on primary energy supply placed in high level as 77% when it is at oil crisis in 1970's. It takes still in higher level index (50%) compere with the global average (40%), now. On the other hand, Japan continues to make effort to reduce oil dependency. It is understood that the rest three items on the table become low level, such as the ratio of oil proportion to total imports [18].

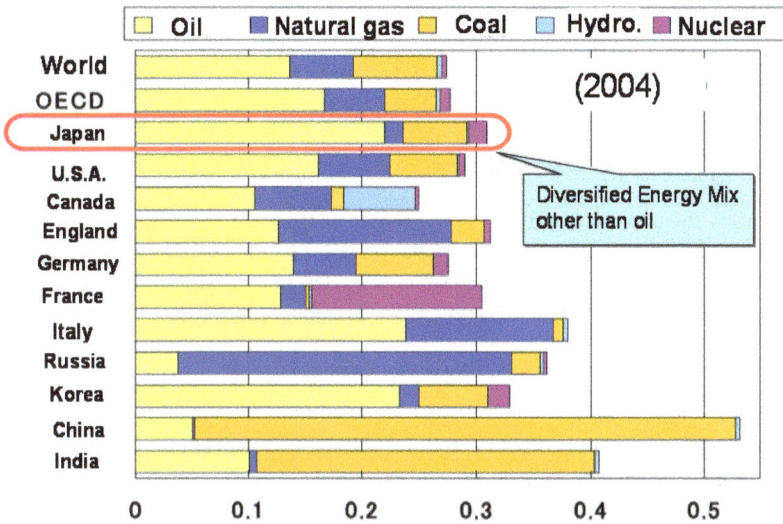

**Figure 19.** Diversified index of primary energy.

| | Oil crisis (73.10~74.8) | Second Oil crisis (78.10~82.2) | Gulf War (90.8~91.2) | Iraq War (03.3~) | Present |
|---|---|---|---|---|---|
| Ratio to primary energy | 77% (73.04~74.03) | 72% (79.04~80.03) | 58% (90.04~91.03) | 50% (92.04~93.03) | 50% (93.04~94.03) |
| Dependence on the Middle East | 78% (73.04~74.03) | 76% (79.04~80.03) | 71% (90.04~91.03) | 86% (92.04~93.03) | 89% (94.04~95.03) |
| Thermal power ratio to total electric capacity | 73% (73.04~74.03) | 53% (79.04~80.03) | 29% (90.04~91.03) | 10% (92.04~93.03) | 10% (94.04~95.03) |
| Oil proportion to total imports | 30% (74.04~75.03) | 30% (79.04~80.03) | 13% (90.04~91.03) | 11% (92.04~93.03) | 12% (94.04~95.03) |
| Oil imports to total GDP | 4. 1% (74.04~75.03) | 4. 8% (81.04~82.03) | 1. 0% (90.04~91.03) | 0. 9% (92.04~93.03) | 1. 3% (94.04~95.03) |

**Table 6.** Energy security index of oil dependency in Japan.

On diversified index of region of crude oil import, Japan and East Asia dependent on Middle East heavily (0.7-0.9), while the index is 0.2 in U.S.A., 0.3 in Europe and China. It means that these nation import from multi region and keep good balanced portfolio [18]. Basic Act on Energy Policy said, "As reducing excessive dependence on specific geographic regions for the import of primary energy sources". But 90% of oil has imported from Middle

East as shown the table which is the energy security index concerning oil, in Japan. In U.S.A. it is also to become big problem which the diversified index to Middle East is rising 20% today. On the other hand, it is the problem that Europe depend 30% supply of crude oil and 20% supply of natural gas on Russia.

### 3.3.2. Share index models to prove procurement stability of energy sources

According to the Kainou, energy security consists of structural risk and individual risk of the conversion, production, and transportation for each phase. Structural risk can be evaluated by the variance of its configuration [18]. Herfindahl Index of the formula (1) is a typical evaluation formula and is also referred to as stable supply risk.

$$H = \sum W_i^2 \text{(Wi: Share of each risk factor)} \tag{1}$$

The method is shown in Figure 20 to evaluate a comprehensive risk matrix which reflects importing region is unevenly distributed or where energy source is supplied and so on. It is thought that this evaluation index is the most comprehensive energy security. According to this evaluation, "whereas the highest risk of oil energy sources, coal has the least variance and risk, and nuclear power is an energy source that has the next least risk and has minus co-variance (small connection to other energy sources)" [19]. Judging from the energy security (without taking into account the environmental issues), it is the best mix for Japan to reduce greatly the dependence on oil imports, to increase the ratio of coal drastically, and then to increase the ratio of nuclear power on the structure of primary energy, which can lead to minimization of risk.

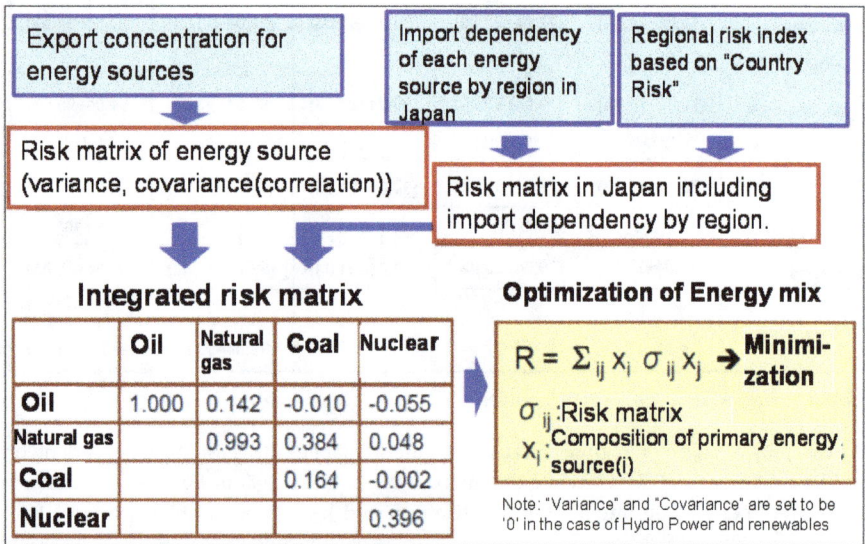

```
Export concentration for        Import dependency      Regional risk index
energy sources                  of each energy         based on "Country
                                source by region in    Risk"
                                Japan

Risk matrix of energy source
(variance, covariance(correlation))    Risk matrix in Japan including
                                       import dependency by region.
```

**Integrated risk matrix**

|            | Oil   | Natural gas | Coal   | Nuclear |
|------------|-------|-------------|--------|---------|
| **Oil**    | 1.000 | 0.142       | -0.010 | -0.055  |
| **Natural gas** |  | 0.993       | 0.384  | 0.048   |
| **Coal**   |       |             | 0.164  | -0.002  |
| **Nuclear**|       |             |        | 0.396   |

**Optimization of Energy mix**

$$R = \sum_{ij} x_i \, \sigma_{ij} \, x_j \rightarrow \text{Minimization}$$

$\sigma_{ij}$ :Risk matrix
$x_i$ :Composition of primary energy source(i)

Note: "Variance" and "Covariance" are set to be '0' in the case of Hydro Power and renewables

**Figure 20.** Risk based best energy composition. (METI, 2001)

Since uranium resources are distributed over countries that are socio-politically stable and adequate diversity of supply is maintained (absence of apparent over-concentration of market power to specific countries/ regions), nuclear fuel is understood as less risky in terms of procurement than a number of fossil fuels (specifically gas and oil). This can be measured primarily by calculating share indices. The method is shown in IEA (2007) which is based on Herfindahl-Hirschman Index (HHI), which is defined as the sum of square of share of all supply options with certain modification to reflect different socio-political risks. Putting domestic fuel supply as risk-free, HHI is modified as ESMC (which stands for energy security market concentration), with:

$$ESMC = \sum S_{if}^2, \text{ where Sif: share of import of fuel f from country/region i.}$$

Since there are different degrees of socio-political stableness across countries/regions, ESMC is expanded to:

$$ESMCpol = \sum (r_i * S_{if}^2), \text{ where } r_i: \text{political risk associated with exporting country/region i}$$

CRIEPI applied a similar methodology using HHI and risk premium [19]. Under the Japanese context, inter-temporal evaluation of Japan's primary energy mix was conducted, whose result is shown in Figure 21. The Risk Index in the figure is a procurement stability index that reflects instability of energy mix induced from global resource distribution, global trade share, Japan import structure of each energy resource, as well as socio-political risks of countries with resource deposits or exports. It has its maximum value 1 when all the primary energy needs are met solely by imported oil (in the same composition in the reference year, which in this case was set at 2005), while its minimum value 0 when the energy supply is preoccupied by domestic risk-free sources, such as hydro and other renewables. It revealed that the primary energy structure has evolved with remarkable improvements in its robustness since the period of oil crises in 1970s, through efforts to substitute oil with alternatives such as nuclear energy and natural gas, also shown in the background of figure.

### 3.3.3. Multiple indexes model to evaluate energy security level

#### 3.3.3.1. Method

In this section, the method is explained which was used in a comparison study of security of supply using five parameters among seven OECD countries using OECD 2005 year data [20]. Japan energy policies have stressed three targets: energy security, lower energy prices, and environmental protection. In response to the recent structural imbalance of oil supply and demand, Japan has placed energy security at the top agenda of its energy policies. Review of the energy security level has importance in formulating and steering the energy policies. Although energy security meant basically national energy security that puts the main priority on a stable energy supply, it is required to consider energy security from wider viewpoints of global energy security, which includes environment, nuclear concerns, international relations and others as its priority aspects.

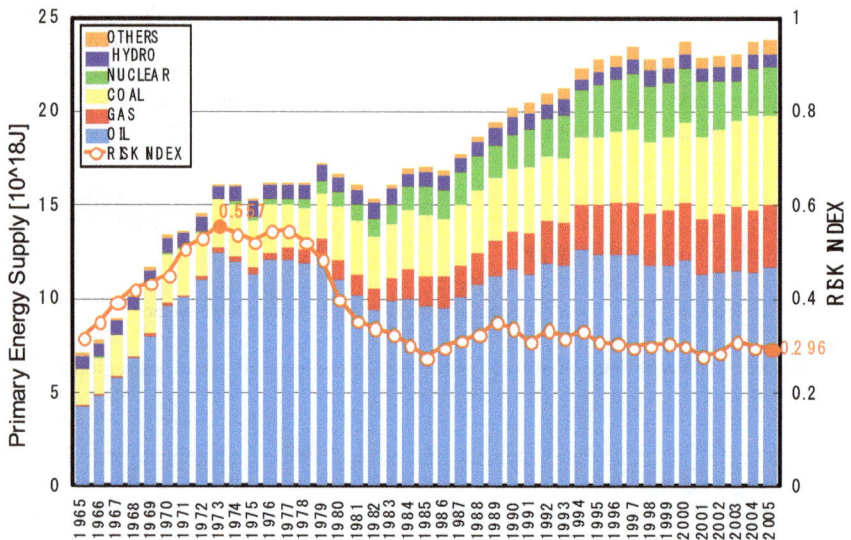

**Figure 21.** Historical evolution of primary energy mix and its procurement stability index in Japan.

The estimation is shown on energy security levels of OECD's G7 Summit member countries, which are Canada, France, Germany, Italy, Japan, UK and US. These nations occupy 81% of GDP, and 76% of primary energy supply in OECD member countries. The energy security levels are shown by the scores of standard deviation of the following factors, that is, are compared relatively for each country. Those scores of energy security levels for each country are estimated by the following process:

1st step: Select factors and indexes concerned on energy security.

The factors include as follows:

1.  Ratio of self-sufficiency of energy supply, for index of energy supply independence, or resource amount rich or not.
2.  Share of energy imported from specified areas, such as the Middle East and Russia, for index of stable energy supply, or of import risk by two big threats of political condition instability and marketing power.
3.  Diversity of energy supply, for index of energy best mix, which is calculated by $\sum i$ $Wij^2$,

    Wij: Share of respective energy i (Coal, Oil, Gas, Nuclear, and Renewable) in energy supply for each country j.

4.  Energy consumption per unit GDP, for index of energy usage efficiency.
5.  $CO_2$ emission ratio, for index of global environmental problem measure, which is calculated by $\sum i$ $Wij \times Ci$,

    Ci: $CO_2$ emission per unit energy consumption of respective energy for each country.

2nd step: Estimate the deviation of data of each nation on each factor

A normal distribution of data is assumed. The lower point for each factor indicates the better performance, that is the higher score of Yij, from the viewpoint of energy security.

$$Yij = 50 + 10 \times Xij - Ave\ Xi\ /\ Std\ Xi$$

Ave Xi: Average of Xij for each factor
Std Xi: Standard deflection of Xij for each factor
Xij: Data on each factor i of each nation j
Yij: Deviation of data on each factor i of each nation j

3rd step: Sum up the above estimated scores of the adopted factors

Average score of Zj indicates the relative energy security level of seven countries.

$$Zj = Ave\ Yij$$

Ave Yij: Average of Yij for each nation j

The estimation is made by the data of OECD/IEA energy statistics [8,11,14].

### 3.3.3.2. Results

The scores on energy security levels of seven OECD nations for 2005 were calculated. The scores of Canada, US, and UK, resource-supplying countries, are relatively higher scores, compared with other nations, poor resource countries. Among the poor resource countries, Germany shows good diversification and France has good self-sufficient rate due to its high nuclear production capability, while Japan has low scores for factors except energy usage efficiency.

Figure 22 shows the trend of the scores on energy security levels of seven OECD nations for about 30 years (presented by five points) [8]. The method estimates energy security levels based on relative comparison. With a view on energy security levels of seven OECD member countries, the scores estimated by this method show that Japan is now placed at a lower level than most major OECD member countries.

### 3.3.3.3. Survey of nuclear contribution on security of energy supply

Nuclear energy has a great potential to improve energy security. Here, to clarify the nuclear contribution on energy security, a virtual world of 'without nuclear energy' is considered. Nuclear energy contribution on energy security is identified by comparing the levels between two cases, one is with nuclear energy and the other is without nuclear energy.

This study excludes energy usage efficiency as an index for the comparison which is equal in both cases. Nuclear energy contribution is considered to be zero and then nuclear energy is allotted to fossil energy sources in proportion to their respective portions in energy supply of the 2005 year data, for other indexes, diversity of energy supply, self-sufficiency ratio, share of energy imported from the Middle East and Russia, and $CO_2$ emission ratio.

As the method mentioned in the previous section is used for comparison purpose among countries, only order is meaningful. To evaluate the nuclear energy contribution, relative value is used. Each index except energy usage efficiency, which is not affected existence of nuclear, is normalized, where best one is unity while worst one is zero in two cases.

The comparison results are shown in Figure 23 [8]. Normalized scores of France with a great nuclear portion in energy supply become worse drastically, and those of Japan with scare energy resource and having nuclear energy promotion program, becomes worse also. Italy is not apparently affected by with or without nuclear. Other nations possessing nuclear energy with rich energy resources would not be affected seriously.

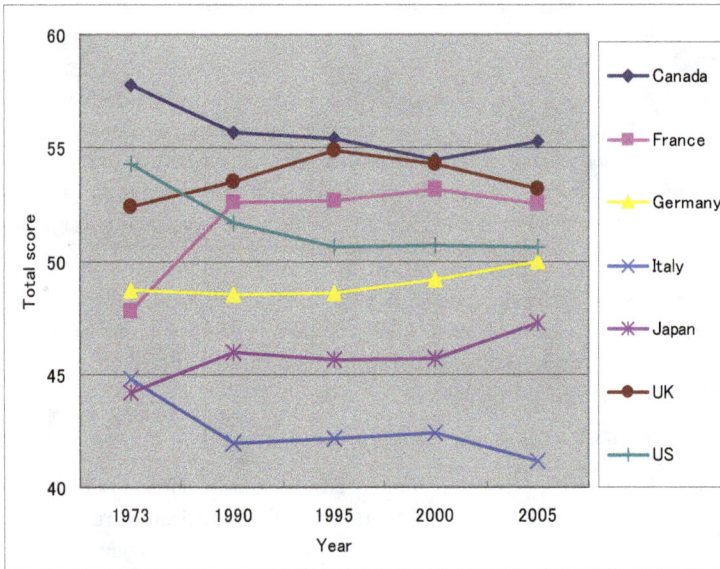

**Figure 22.** Trend of total scores of energy security levels of seven OECD countries.

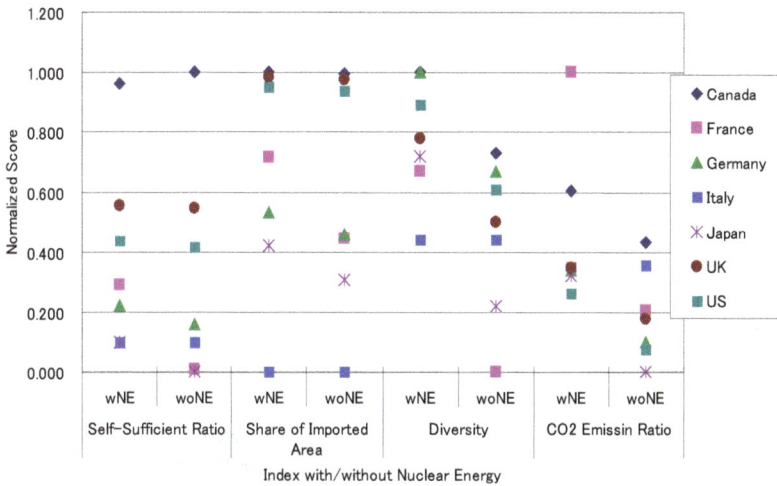

**Figure 23.** Comparison results of four normalized indexes with (wNE) and without(woNE) nuclear energy.

### 3.3.4. Emergent scenario of two China by scenario planning in 2015

The scenario-planning-based-approach has been proposed for the development of science and technology strategy through the analysis of energy crises in East Asia [10]. The method, with the discussion of experts of various fields, enables the comprehensive understanding of the problem to be considered, the development of a robust science and technology strategy for uncertain future, and the evaluation of individual research and development theme from various aspects. This is an example of a structural crisis in Table 5.

#### 3.3.4.1. Master plan: Shared awareness of the issues

The research team conducts comprehensive analysis of risk factors, development of two scenarios with emphasis on 'China's future'. Investigation of strategic viewpoints needed on science and technology and evaluation of individual research and development theme for each of the scenarios are also performed.

#### 3.3.4.2. Extraction of risk factors and determination of the scenario structure

The fragility of the energy system of East Asia, with aggravating further from now on, has a large possibility to bring national power decline and resource protectionism in the countries in East Asia. Such situation is a threat onto security of Japan, while it can also serve as an opportunity to growth national power and international presence of Japan, by advanced technology development and its technology transfer.

The two China images of 'Sovereign Right China' and 'Open China' were built and the 'Resource scramble scenario' and the 'Japan isolation scenario' were created from each in this research. As foresight which China image becomes dominant from now on is difficult, Japan needs to build a technology strategy with consideration of both possibilities.

In extracting risk elements as the components of the crisis scenario, eight risk categories are set as follows:

1.   China resource protectionism,
2.   China science and technology organization,
3.   geopolitics and international relations,
4.   energy infrastructure,
5.   motorization,
6.   electricity crisis,
7.   nuclear accident and nuclear proliferation, and
8.   environmental problem.

Although the elements mentioned in this stage was 30 items or more, as a result of scrutinizing these further, to realization of a master plan, and 18 items were listed for an element with an uncertain prospect in this time as shown in Table 7. It classifies into four categories for convenience.

| China | • Market Mechanism in Energy Sector?<br>• Foreign Resources required ?<br>• Technology  and Political System ?<br>• Prosperity and Political Stability?<br>• Electricity Generation increase?<br>• GHGs Control ? |
| --- | --- |
| Geographical features | • Korean Peninsula?<br>• Russian Resources?<br>• US–China Trading Friction ?<br>• Sea Lane? |
| Infrastructure, Terrorism | • Oil shipment corporation?<br>• Nuclear Accident?<br>• South –East Asia Terrorism ?<br>• Nuclear Safeguard in South Asia? |
| Technology, Environment | • Japanese Technology Superiority ?<br>• Automobile Efficiency?<br>• Energy Saving Mind?<br>• Post Kyoto Protocol? |

**Table 7.** Asia crisis- scenario planning uncertainty factors to 2015. (MEXT, 2005)

*3.3.4.3. Two China - "Open China" and "Sovereign Right China"*

Based on the work described until now, examination intensive about the future image of China used as the base of each scenario was performed. Consequently, the two China images were formed as shown in Figure 24. Although it will probably be common but still hypothesis that it is shifting to "Open China" from "National power China" as a trend, and it cannot predict which "China" becomes dominant till around 2020. Japan is required to construct strategy based on both possibilities. In any case, nuclear power is dominant for energy technology.

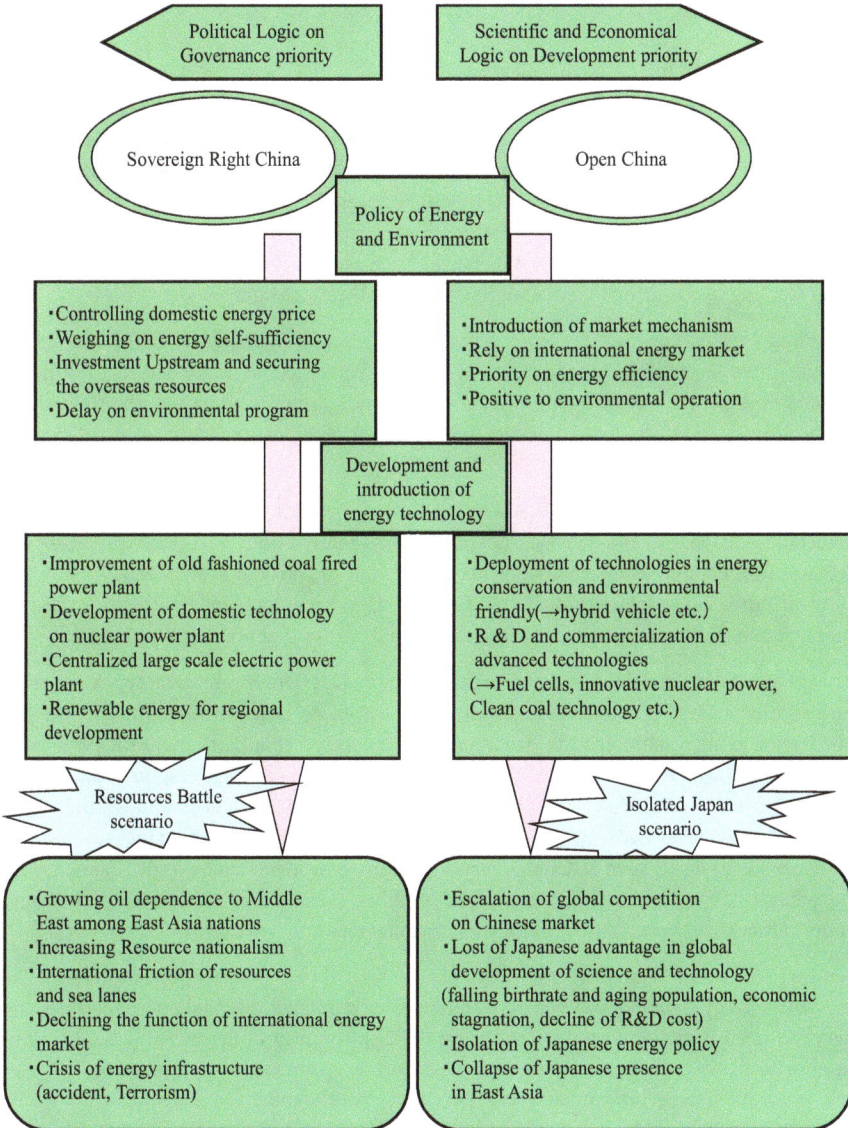

**Figure 24.** Emergent scenario of two China by scenario planning.

## 4. Energy issue and nuclear energy role after the Fukushima Daiichi accident

The Fukushima Daiichi Accident raised a new challenge of securing the safety of utilization. National nuclear policies of many countries are being reexamined along with the safety evaluation.

Safety design principle is "Defense in Depth" concept, which should be further reconsidered reflecting the accident causes. Usual systems focus on the forefront function, such as preventing damage, expansion mitigation, or incident prevention, while safety critical systems increases attention to back-up functions such as incident expansion mitigation or environmental effects mitigation, if it has a large enough impact on the environment. Common Mode Failure of External Initiating Event such as Earthquake or Tsunami, which is usually Rare Event, or auxiliary systems failure such as Off-site Power, EDG, Buttery, or Sea Water Cooling loss is difficult to install to Defense in Depth design.

According to the "Defense in Depth" concept reflecting Fukushima accident, we should consider three level safety functions; usual normal system, usual safety system, and newly installed emergency system including external support function. Anyway the diversity is significantly required for not only future reactor concept but also existing plant back-fit activities.

Swiss Cheese Model proposed by Reason, J indicates operational problem other than design problem [21]. Fallacy of the defense in depth has frequently occurred recently because plant system is safe enough as operators becomes easily not to consider system safety. And then safety culture degradation would be happened, whose incident will easily become organizational accident. Such situation requires final barrier that is Crisis Management.

Concept of "Soft Barrier" has been proposed here [22]. There are two types of safety barriers, one is Hard Barrier that is simply represented by Defense in Depth. The other is Soft Barrier, which maintains the hard barrier as expected condition, makes it perform as expected function. Even when the Hard Barrier does not perform its function, human activity to prevent hazardous effect and its support functions, such as manuals, rules, laws, organization, social system, etc. Soft Barrier can be further divided to two measures; one is "Software for design", such as Common mode failure treatment, Safety logic, Usability, etc. The other is "Humanware for operation", such as operator or maintenance personnel actions, Emergency Procedure, organization, management, Safety Culture, etc.

Premise here is that "Global warming and energy security are the invariant problems". The long-term energy demand and supply simulation to minimize the total energy system cost was conducted for energy prediction during the 21st Century in the world [22-23]. Taking the effort for energy-saving as major premise, carbon capture and sequestration for fossil fuel, renewable energy and nuclear energy should be altogether developed, which means energy best mix should be achieved, under the CO2 constraint around 450ppm atmosphere [24]. Nuclear phase-out scenario, in which new nuclear plant construction is prohibited, is possible from the simulation even considering the issue of global warming, with the following problems; increasing energy costs, little room for countermeasure, and large uncertainties of technology. The role of nuclear is also examined to understand energy security is severely affected by the case without nuclear energy. Therefore, rational use of nuclear power is requested, that is each country should make decision, Japan and several European countries will be also phase out, while China, India and ASEAN countries will continue to be introduced. If the accident happens again anywhere in, it will become the global phase-out. Therefore, rational unified safety standards (organizational structure, design and operation, regulations) should be reviewed based on the Fukushima Daiichi Problem world-wide analysis and established in the world.

## 5. The path and key issues for "sustainable development"

Figure 25 shows the path and key issues for "Sustainable development" [8], which is the target of the 21st century.

- Stabilization of world population.
- Reduction of the south and north economic difference.
- Preservation of global environment.
- Preservation and effective use for rare resources.

It is necessary to improve economic development and the living standard in the developing countries, as a fundamental solution for the population growth in the world and for the reduction of "Income gap between North and South" also. It is necessary to secure energy that is low-cost and resource restriction free, to support economic in the developing countries. Moreover, great control of carbon-dioxide emissions is necessary to mitigate the climate change influence. It is necessary to achieve a worldwide spread of technical improvement/ recycling society system that aims at the efficiency improvement of the resource use.

- The hydrocarbon resources such as oil and natural gases are to use for the recycling as the raw material.
- Drastically conversion into non-fossil energy is indispensable.
- Expansion of nuclear power energy use: It is precondition to secure the durability by constructing the fuel cycle with the viewpoint of the resource and waste.
- Use of renewable energy source: It is important to improvement the technology that aims at cost reduction.

Because Japan has the feature as the following energy systems, it is considered to be possible to deal with the energy crisis enough if the policy is correctly set.

- Development and usage of highly effective energy conversion technology.
- Usage of nuclear power generation.
- Renewable energy technology development; especially, the world is led in the Photovoltaic technology.

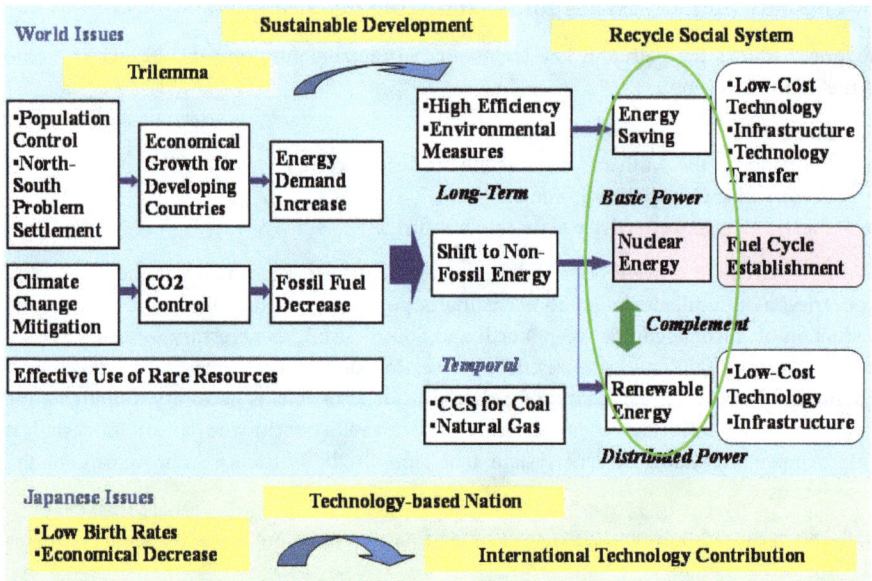

**Figure 25.** Global Environmental and Energy Resource Problems, and Energy Perspectives

## 6. Conclusions

In order to address the biggest challenge to global sustainable development caused be global warming, a new post-2012 climate regime was examined to be scientifically sound, economically and technologically rational. The key findings are as the following.

1.  Instead of the traditional 450ppm equilibrium stabilization of IPCC, a new scenario based on zero-emission and overshoot schemes was proposed recently. The essential limitation is that the total emission during the 21$^{st}$ century should be lower than 650GtC. The scientific examinations demonstrated that the so called Z650 scenario could avoid long-term risks. At the meanwhile it could meet short term need of relatively large emissions. The proposal improves the possibility of international agreement compared with the G8 Summit proposal, which argued that the worldwide greenhouse gas emissions must be reduced by at least 50% in 2050 compared to the 1990 or recent year levels.
2.  A numerical experiment of global energy system optimization shows the technical feasibility of the Z650 scenario not only globally but also regionally. The obtained time series total primary energy mixes suggest that the consumption of fossil energy will peak at 2030, and the clean energies, especially the renewable energy will play an essential role during the second half of the century. The resulted regional emission curves reflect the differences of financial and technical capability among areas. The industrialized countries will reduce their emissions by 50% in 2050 compared with 2005 levels, while the emissions of developing countries will increase by 10% at the same time. The results of individual industrialized countries fit with the national targets well.
3.  The cost-effective analysis shows that the Z650 scenario is economically rational. Compared with the reference case, the additional investments in Z650 scenario could be covered by the fuel savings during the following 40 years (2010-50) both globally and regionally.
4.  Nuclear energy will play an important role for achieving the vision against global warming. Large-scale introductions of the more expensive renewable energies during early stage are necessary without nuclear energy or next generation nuclear technology. As a result, the power generation cost will increase rapidly thereby the negative economic impact will be significant especially in developing countries. Therefore, rational use of nuclear power is requested to combat global warming.

Compared with the threat from global warming, energy security is the more traditional key issue for global and regional sustainable development. Based on the overview of energy security concepts and existing evaluation methods, we proposed a new integrated index to evaluate national energy security from the wide conception. Case studies employing the index for OECD countries and China were conducted to evaluate the role of nuclear energy. The key findings are the following.

1.  From the viewpoint of self-sufficient ratio, nuclear energy affects security index largely in the energy importing nations but slightly in the resources nations.
2.  From the viewpoint of energy diversity and CO2 emissions, the absence of nuclear energy decreases the security index significantly by which influences the sustainability of national economic growth.
3.  The nuclear policies of China will influence not only the domestic economic growth but also the energy situations in the world, especially the surrounding nations.

Nuclear energy will play an important role from the necessity of mitigating climate change, as well as improve energy security. However, the Fukushima Daiichi Accident raised a new challenge of securing the safety of utilization. Following the safety design principle of "Defense in Depth", three level safety functions should be considered for the hardware. Those are, the usual normal system, usual safety system, and emergency system including external support function. On the other hand, software for design including common mode failure treatment, safety logic, and usability should be improved together with the human-ware for operation including personnel actions, emergency procedure, organization, management, and safety culture.

Sustainable development is the final target for human society. The energy related environmental issues and energy issue are the main challenges during the 21$^{st}$ century. Although the energy conservation is the most important issue in the energy policy, the utilization of nuclear energy is also essential to maintain the global environment and energy security together with the improvement of the renewable energy and the development of the carbon dioxide isolation technology for the fossil fuel. Therefore, it is necessary to continue technological development so as to demonstrate each potential as for the basic energy in 21st century.

## Author details

Hiroshi Ujita
*Tokyo Institute of Technology, Department of Nuclear Engineering, Japan*
*The Canon Institute for Global Studies, Japan*

Fengjun Duan
*The Canon Institute for Global Studies, Japan*
*The University of Tokyo, Japan*

## Acknowledgement

The global warming discussion is based on a research project launched by the Canon Institute for Global Studies. We'd like to thank the project members, Dr. Tetsuo Yuhara, Mr. Masanori Tashimo, Dr, Takahisa Yokoyama, Ms Yuriko Aoyanagi, Mr. Kazuaki Matsui, Dr. Toshikazu Shindo, Dr. Kazuhiro Tsuzuki, Dr. Atsushi Kurosawa, Mr. Ken Oyama, Dr. Yasumasa Fujii, and Dr. Ryo Komiyama. We also thank Prof. Matsuno, Mr. Toshihiko

Fukui, Mr. Ryozo Hayashi, Mr. Kazumasa Kusaka, and Mr. Akihiro Sawa for their kindly advices on the research.

The energy security problem analysis is based on the discussion within OECD/NEA Expert Meeting on "Nuclear Energy and Security of Supply". We'd like to thank the experts of OECD/NEA. We also thank Mr. Kazuaki Matsui and Dr. Eiji Yamada for their fruitful discussion on the research.

# 7. References

[1]   Matsuno et al. Equilibrium stabilization of the atmospheric carbon dioxide via zero emissions - An alternative way to stable global environment, Proc. Japan Academy, Ser. B, in press.
[2]   UKCCC: The 2050 target, Building a low-carbon economy – the UK's contribution to tackling climate change, 2008.
[3]   Allen, M. R., Frame, D. J., Hemingford, C., Jones, C. D., Love, C. D., Meinshausen, M., and Meinshousen, N.: Warming caused by cumulative carbon emissions towards the trillionth tone, Nature 458, 1163-1166, 2009.
[4]   Kurosawa et al., Analysis of carbon emission stabilization targets and adaptation by assessment model, The Energy Journal, Vol. 20 (Special I), 157-176, 1999.
[5]   IEA, World Energy Outlook 2010.
[6]   IEA, Energy Technology Perspective 2010.
[7]   OECD/NEA, The Security of Energy Supply and the Contribution of Nuclear Energy, NEA6358, 2010.
[8]   H. Ujita, K. Matsui, E. Yamada, Proposal on Concept of Security of Energy Supply with Nuclear Energy, ICAPP '09, Tokyo, Japan, May 10-14, 2009.
[9]   H. Ujita, A Study on Energy Security and Nuclear Energy Role, JNST, Vol.10, No.1, 2011 (in Japanese).
[10] R. Omori, H. Horii, Analysis of Energy Crises in East Asia using Scenario Planning method and It's Implications for Japan's Science and Technology Policy, Sociotechnica, pp.1-10, Sociotechnology Research Network, 2005.
[11] IIASA/WEC, Global Energy Perspective 1998.
[12] Y. Uchiyama, et al., Design of optimal power of our country from the viewpoint of risk, economic efficiency, and security, Electric Economic Research, No.20, 1986 (in Japanese).
[13] Global Commons Institute: "www.gci.org.uk", 2003.
[14] EDMC/energy/ economy sumarry2004, REEJ.
[15] IAEA, "Analysis of Uranium Supply to 2050", 2001.
[16] Nuclear Energy Outlook (NEO) 2008, OECD/NEA (2008).
[17] BP Statistical Review of World Energy June 2005.
[18] METI, Recent Energy Situation and Our Policy Trends of 2005 (in Japanese).
[19] K. Nagano et al., A Valuation Study of Fuel Supply Stability of Nuclear Energy, CRIEPI Socio-economic Research Center, Y07008, 2008 (in Japanese).

[20]  IEA, Energy Statistics of OECD Countries (2006).

[21]  J. Reason, Managing The Risk of organizational Accidents, Ashgate Publishing Limited., 1997.

[22]  H. Ujita, Panel discussion: "Nuclear energy: is Fukushima the end of a paradigm?", The MEDays Forum, Tangier, Morocco, November 16-19, 2011.

[23]  T. Yuhara, H. Ujita, International Seminars on Planetary Emergencies and Associated Meetings, 44th Session, The Role of Science in the Third Millennium, Erice, Italy, 20 August 2011.

[24]  T. Yuhara et al., "Towards the harmony - Principles for the new climate regime-", The 2nd CIGS Symposium, Sep. 16, 2011.

# Chapter 10

# Efforts to Curb NOx from Greenhouse Gases by the Application of Energy Crops and Vegetation Filters

Zsuzsa A. Mayer, Andreas Apfelbacher and Andreas Hornung

Additional information is available at the end of the chapter

## 1. Introduction

Nitrogen, in a similar way to carbon, has a complex and fragile global cycle. Anthropogenic activities from the beginning of the 20th century have interfered with this fine nitrogen-balance by capturing $N_2$ from the atmosphere for fertiliser production. When stable N from the atmosphere enters the forage crop production and stock-raising cycle it returns to the environment as waste and in more reactive forms.

During the combustion of energy crops the fuel-bound N forms greenhouse gases which are liberated to the atmosphere, therefore both fertiliser applications and biomass combustion can be directly linked to nitrogen related environmental problems.

Short-rotation plantations irrigated with effluent have both high nitrogen uptake capacity [1] and also enhance growth characteristics without the application of fertilisers or competition with fresh water usage [2, 3]. Furthermore wastewater irrigation[1] reduces the cost of wastewater treatment while crops cultivated on the land can provide solution for the increasing energy demand of rural areas without destroying existing forestry [2].

In order to choose appropriate feedstock and design a biomass-to-energy conversion technology both the economical and environmental aspect of a project should be considered. Biomass pyrolysis, which is the thermal degradation of the biomass in an inert atmosphere, provides an advanced liquid fuel. Pyrolysis liquid (or bio-oil) is the subject of intense research and investigations for direct energy applications to provide green electric power with highest efficiency [4].

---

[1]Throughout this chapter, the term *wastewater* will refer either to treated wastewater (effluent) or untreated (raw) wastewater. *Wastewater irrigation* can refer to both flood irrigation, spray irrigation, subsurface drains and other applications.

This chapter introduces the use of energy crops into the global nitrogen-cycle by following nitrogen from wastewater irrigation via energy conversion (biomass pyrolysis) and finally back to soil in a stable form to close the circle (Fig. 1).

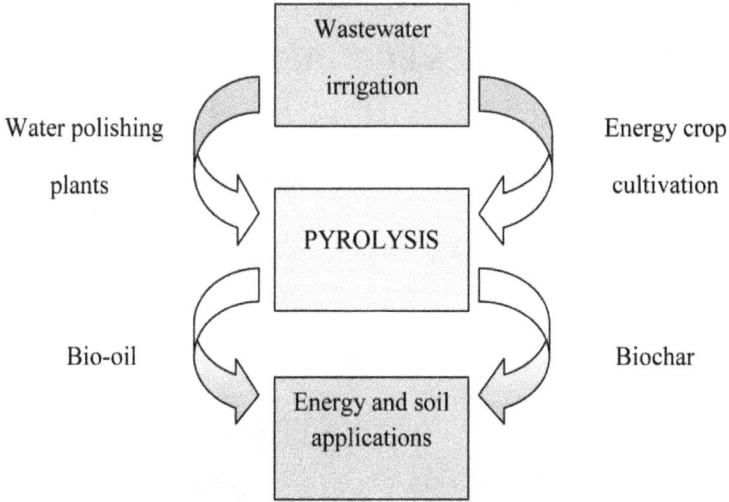

**Figure 1.** General scheme showing the conversion of wastewater irrigated vegetation filters to energy and to soil amendment

## 2. Wastewater and wastewater treatments

### 2.1. Nitrogen in municipal wastewater

Nitrogen in domestic wastewater is present in both inorganic and organic forms. Organic nitrogen from human diet and metabolism is transformed into free ammonia ($NH_3$) and ammonium cation ($NH_4^+$) by microorganism [5, 6] The $NH_3$ to $NH_4^+$ ratio in water is depending on temperature and pH. The presence of free $NH_3$ above the concentration of 0.002 mg/L is toxic for the ecosystem [7]. Ammonia is also the source of inorganic nitrate and nitrite ($NO_3^-$, $NO_2^-$) nitrogen in wastewater [6]. Inorganic nitrogen is an essential plant nutrient. However, high concentrations in water cause *eutrophication*; an extreme bloom in the population of plants with an enhanced growth period followed by the necrosis of the biomass. The degradation of dead plant tissues increases oxygen demand of fresh water, therefore, eutrophication leads to oxygen scarcity and decreased self-cleaning ability of the biomass system [8]. The presence of nitrate and nitrite anions in drinking water is blamed for causing cyanotic conditions like shortness of breath, methemoglobinemia and blue-baby syndrome [9, 10].

To protect human health and aquatic life the nitrogenous contaminants of wastewater must be controlled. Table 1 contains some requirements set up by different governments and some typical nitrogen values in different types of wastewater.

| | Form of nitrogen | Concentration (mg N/L) | Source |
|---|---|---|---|
| Typical nitrogen concentration in *grey wastewater* | TN[a] | 0.6–74 | [11] |
| Typical nitrogen concentrations in domestic *raw wastewater* | TN | 20–80 | [11] |
| Requirement of the European Council for urban *wastewater* treatment | TN | 10 | [12] |
| Primary standards of the *National Primary Drinking Water Regulations* by US EPA[b] | Nitrate-N | 10 | [13] |
| Health value of the *Australian Drinking Water Guidelines* | Nitrate-N | 11.3 | [14] |

[a]TN: *Total Nitrogen*; Sum of organic nitrogen, ammoniacal nitrogen, nitrate-nitrogen and nitrite-nitrogen
[b]EPA: *Environmental Protection Agency*, U.S.

**Table 1.** Typical nitrogen values and requirements in water and wastewater

## 2.2. Biological wastewater treatment

The physicochemical removal of nitrogen from wastewater is possible, however, biological methods have proved to be more effective and less expensive treatments [15].

The biological removal of nitrogen is based on the mixed populations of live bacteria naturally present in wastewater which are able to convert nitrogen compounds to other chemical forms. The mineralization (consecutive steps of *ammonification, nitrification* and *denitrification)* of the wastewater-derived organic matter provides oxygen, nitrogen and energy for the bacteria to produce new cells [16].

The *activated sludge* formed by these living microorganisms is the core of modern industrial wastewater treatment technologies. To ensure the most suitable environmental conditions for the microorganisms (e.g. aerobic zone for nitrification and anoxic zone for denitrification) several industrial processes have been designed like the *Bio-Denitro process,* modified *Ludzack-Ettinger process, Bardenpho process,* etc. [15].

When these conventional wastewater treatment facilities are not available – mostly in developing countries – stabilization ponds are the most widely used municipal wastewater treatment systems [17]. Even if the climate favours microbial activity these stabilization ponds cannot reduce the concentration of nitrogen satisfactorily [18].

## 2.3. Vegetation filters

If the high cost of the commercial technologies discounts the use of sufficient wastewater treatment, the unregulated or poorly regulated water turns to a potential risk factor to human health and environment [19, 20]. To eliminate this risk it is crucial to reduce the concentration of nitrogen and other pollutants before any effluent reaches the environment.

The application of biological filter systems like soil and *vegetation filters* represents an alternative on-site wastewater treatment. While the first pilot tests were carried out by big companies to treat cannery effluents, the treatment of municipal water receives more and more attention now in developing countries [21, 22]. This type of wastewater management is able to reduce the concentration of organic and inorganic contaminants in the water and remove 73-97% of the *total nitrogen* [23]. This low-cost treatment also assimilates nitrogen as plant nutrients back into the environment while pathogens from the wastewater cannot compete with the natural microbial population of the soil [24, 25].

## 3. Nitrogen, the essential plant nutrient

### 3.1. Nitrogen in soil

The role of soil in the biological-cycle is to store and supply nitrogen and other essential nutrients for plants. The average amount of organic nitrogen in soil is 3300 kg/ha, however, the available nitrogen for plants is less than 1 % of the above volume as vegetations are not able to uptake any kind of forms of soil nitrogen [26].

### 3.2. Nitrogen uptake in plants

The synthesis of plant cell components (e.g. amino acids, nucleic acids, enzymes, chlorophyll etc.) is unachievable without nitrogen; nitrogen deficiency in plants causes slow growth which can be recognized by the pale green colour of the leaves. Without available nitrogen there are no processes in plants [26]. For the formation of new cells, plants uptake nitrogen – along with water – in the form of $NH_4^+$ or $NO_3^-$ during their growing period (*assimilation*), or store extra nitrogen (*immobilization*) [27].

Nitrogen is being absorbed from soil during the whole life of the plants but the nitrogen use efficiency of plants varies according to the stage of maturity, seasons, environmental conditions of the site and the fertility status of the soil as well [26, 28]. The latter factor is particularly important in terms of crop yield as nitrogen supply is a main limitation factor to plant growth [29].

### 3.3. Synthetic nitrogenous fertilisers

If the nitrogen supply within the soil is not sufficient, land productivity can be improved by organic and inorganic (also known as synthetic) macronutrient plant fertilisers. The most widely used synthetic fertilisers are ammonia-based products [30]. The source of nitrogen in these fertilisers is the atmosphere containing molecular nitrogen in 78 %. The direct reaction of molecular nitrogen and molecular hydrogen to $NH_3$ is the base of the widely applied Haber-Bosch process [31] which provides more than 140 million tonnes of ammonia to farmers around the world every year [32].

Modern soil fertility management in the 20th century has made a significant contribution to the growth of Earth's population which has almost quadrupled since 1900s. To sustain this growing population industry produces millions of tons of fertiliser which is responsible for more than 1% of the world's energy consumption. Since hydrocarbon combustion is the main energy source of ammonia production, the fertiliser industry is a major contributor to greenhouse gas emission [33]. In addition to the energy consumed during production transportation of the fertilisers is also contributing to the world's greenhouse gas emission with 37 Tg $CO_2$-eq per year [34]. There is also an estimated 2.5-4.5 Tg N emitted from the nitrogen-fertilised soil to the atmosphere each year [35].

## 3.4. Nitrogen uptake in effluent-irrigated short-rotation crops

If the cost or availability of the technology does not make it possible to apply inorganic fertilisers, alternative – and possibly more sustainable – nitrogen sources should be considered to increase the productivity of agricultural land.

Similar to inorganic fertilisers, wastewater is a source of supplemental nitrogen. According to studies, nitrogen uptake of rain-fed Eucalyptus in New Zealand is in the range of 30-80 kg/ha/year while the uptake in effluent-irrigated plantations is one magnitude higher [1]. During wastewater irrigation, plants uptake nitrogen for their growth and polish the water. The absorbed N nutrients are converted to amino acids and stored in wood [36] or transferred from roots to shoots for protein synthesis [37]. Research results have also proved that plants have enhanced growing characteristics as a result of wastewater, grey water or effluent irrigation [38-40]. Table 2 shows the increments in storage and transport amino acid concentrations due to wastewater irrigation.

| Free amino acid ($\mu$g / mg) | Arginine | Asparagine | Aspartic acid | Glutamine | Glutamic acid |
|---|---|---|---|---|---|
| Control willow | 0.054 | 0.141 | 0.066 | 0.002 | 0.048 |
| Wastewater irrigated willow | 0.404 | 0.177 | 0.102 | 0.013 | 0.103 |

Sample: Willow (*Salix*) from the bioremediation programme of Agri-Food & Biosciences Institute (ABFI, Hillsborough, N. Ireland). Trees were in their second year of re-growth after coppicing and plantations were irrigated with farm wastewater (TN: 100 mg N/L); Source: Chapter authors

**Table 2.** Free amino acid content of willow from wastewater irrigated plot and from a control plot

## 3.5. Nitrate-leaching

Even though vegetation has the potential to store wastewater-derived nitrogen, nutrient uptake is not the only limitation factor of the land applications of wastewater.

Due to the metabolism of microorganisms, nitrogen in soil and wastewater is predominantly present in the form of $NO_3^-$ and $NH_4^+$, which are readily available plant nutrients. The surface charge of clay minerals in soil is negative which attaches the wastewater derived ammonium ion to soil matrix, but ions with negative charge are carried by water [41]. Due to heavy rains or improper agricultural activities nitrate nitrogen can leach below the root system of plants into the groundwater with a negative effect both on the environment and drinking water quality. Nitrate concentration in groundwater can reach extremely high values; one of the reported Indian examples was 1500 mg nitrate in one litre of water, 150 times higher than the permitted value by the WHO [42].

Nitrate is a primary pollutant of groundwater. Although chemical reduction, biological denitrification and other in-situ treatments of groundwater are feasible [43], nitrate leaching is still the main limitation factor of wastewater irrigation; treatments cannot prevent the formation of groundwater contamination or solve the problem of nutrient loss of the soil. Without an effective prevention system the only groundwater protection is source control which means the limitation of wastewater loading.

## 4. Energy from biomass

### 4.1. Heating values

Treating contaminated water by vegetation filters require fast-growing plants, like willow [44]. Willow is also a widely cultivated fuelwood for energy applications with an annual yield of 9-13 t/ha in Europe [45].

An important feature of fuelwood and other energy crops is their composition which determines their heating (or calorific) value [46, 47]. The higher heating value (HHV) is the energy available from the fuel and it is generally given in units of energy per unit of weight (cal/g; J/g or Btu/lb). Table 3 contains some typical heating values of fuelwood and other solid fuels. Energy crops can displace approximately 0.44 tonnes of oil equivalent when converted to electricity [48] and contribute to the reduction of greenhouse gas emission by 100-2070 Mt $CO_2$-eq/year [49].

The quality characteristics of the biomass have a significant effect on the yield of energy during a biochemical or thermochemical conversion process [50]. For example high oxygen and carbon content favours combustion and increases the heating value [51] while the general model of heating values predicts a slight decrease in HHV when nitrogen content of biomass increases [52].

|  |  | HHV (MJ/kg) | Source |
|---|---|---|---|
| **Fuelwood** | | | |
| | Softwood (average) | 20.0 | [53] |
| | Hardwood (average) | 18.8 | [53] |
| | Straw (maize silage) | 20.0 | [49] |
| **Charcoals** | | | |
| | Charcoal from rice husk | 17–18 | [54] |
| | "High quality" charcoal | 28–33 | [55] |
| **Fossil fuels** | | | |
| | General purpose coal | 32–42 | [56] |
| | Petrol | 45–47 | [56] |

**Table 3.** Heating values of energy crops, charcoals and fossil fuels

## 4.2. Biomass combustion and nitrogen liberation

Nitrogen content of trees ranges between 0.3 and 1 % [57]. Nitrogen in short-rotation plants is generally higher and significant differences can be found between species. Short-rotation plants represent a cheap and renewable energy source with high energy potential. The combustion of these plants is also a $CO_2$ neutral energy conversion technology, however, combustion converts fuel-nitrogen to nitric oxides ($NOx = NO + NO_2$) and nitrous oxide ($N_2O$) [58-63] which are contributors to acid rain formation [64]. $N_2O$ is also a greenhouse gas with a *global warming potential* (GWP) of 289 where 1 unit represents the global warming potential of $CO_2$ over 20 years [65]. The emission of NOx contributes to acidification and it also causes eutrophication and ground-level ozone formation [66].

Increased nitrogen content in the biomass also means increased emission of NOx during combustion [67]. The estimated emission of NOx from biomass combustion was 5-5.9 TgN in 2000 [35] and based on the fact that the energy demand and the biomass fuel consumption are increasing [68], this NOx emission must be even more significant now and need to be decreased drastically.

To control the harmful effects of combustion plants' pollutants, organisations like Environmental Protection Agency of the United States (US EPA) or the Intergovernmental Panel on Climate Change (IPCC) have elaborated their guidelines and emission criteria [69, 70]. The most common way to fulfil these regulations is the application of flue gas cleaning systems (primary reduction with excess air, secondary catalytic reduction, etc) [67] but these technologies add cost, particularly in small bioenergy facilities. Another effective way to reduce the environment impact of biomass-derived NOx pollution is the application of alternative energy conversion technologies with better emission characteristics.

# 5. Pyrolysis

## 5.1. Biomass conversion to solid, liquid and gas products

Pyrolysis is a thermochemical process where the biomass (e.g. energy crop) is being converted into more effective energy sources. During the pyrolysis process the macromolecules and biopolymers of the biomass undergo a thermal degradation in the absence of oxygen, which leads to solid, liquid and gaseous products.

The thermal decomposition and conversion can be interpreted as the independent degradation of the three main organic woody biomass compounds, cellulose, hemicellulose and lignin [71, 72] which have an average ratio of 45/24/28 wt % in softwood and 45/31/21 wt % in hardwood, respectively [53]. The few parentage of wood inorganics remains in the solid product of pyrolysis while the lignocellulosic compounds undergo thermal degradation.

The biomass conversion at different pyrolysis temperatures can be followed by the thermal degradation and the weight loss of the main wood compounds on Fig. 2. The ratio of the gases, vapours and solid products depend on the temperature, residence time and heating rate of pyrolysis [73, 74]. Increasing the highest treatment temperature of pyrolysis increases the liquid and gas yields and decreases char yield (Fig 3). Due to secondary reactions of vapours liquid yield has a maximum which is followed by a reduction at higher temperatures and the gas yield increases at the expense of biochar yield [73, 75].

In terms of nitrogen oxide emission, pyrolysis is a more desirable energy conversion technology than combustion; while biomass combustion releases fuel-nitrogen in the form of NOx, the inert atmosphere of pyrolysis does not favour to the formation of these or any other oxidized pollutants [76].

## 5.2. Pyrolysis liquids (bio-oil)

Pyrolysis has the ability to generate highly energetic bio-oil which represents most of the energy content of wood (Fig. 4) with the additional benefit that it can be easily pumped or transported. Another advantage of the bio-oil from energy crops and vegetation filters is the lack of jeopardy to the security of food supply, unlike the dangers of sugar-, starch- and vegetable oil-based conventional bio-fuels –which conquer valuable agriculture lands [77].

Bio-oil is still a relatively new energy source and its energy applications are still developing, but its combustion in boilers, turbines and engines has been successfully used for heat and electricity production [78, 79]. Table 4 contains some typical power output values.

| Hot water generation | |
| --- | --- |
| Boiler fuelled with pyrolysis oil (BTG Biomass Technology Group BV, The Netherlands) | 150 kW |
| **Electric power generation** | |
| Pyrolysis liquid combustion in diesel engine (VTT Energy, Finland) | 84 kW |
| Pyrolysis liquid combustion in diesel engine (Wärtsilä Diesel International, Taiwan) | 1.5 MW |
| Pyrolysis liquid combustion in gas turbine (University of Rostock, Germany) | 75 kW |
| **Combine heat and power generation (CHP)** | |
| Pyrolysis liquid combustion in a Stirling CHP unit (ZSW, Germany) | 10-25 kW$_{th}$, 4-9 kW$_e$ |

(Source: Czernik, 2004)

**Table 4.** Power outputs from bio-oil combustion

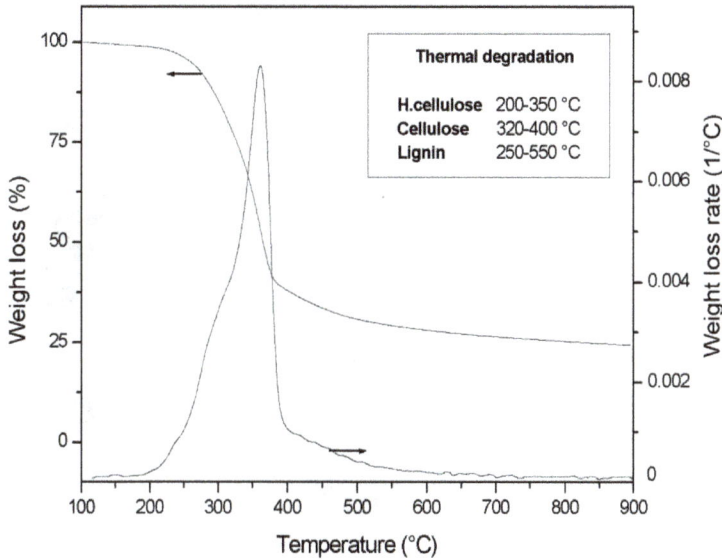

Sample: 10 mg grinded willow (*Salix*) pyrolysed in a Mettler TGA/DSC 1 Star System.
Heating rate: 20 °C/min. Purging gas: He (Source: Chapter authors)

**Figure 2.** Typical thermal degradation curves of wood pyrolysis

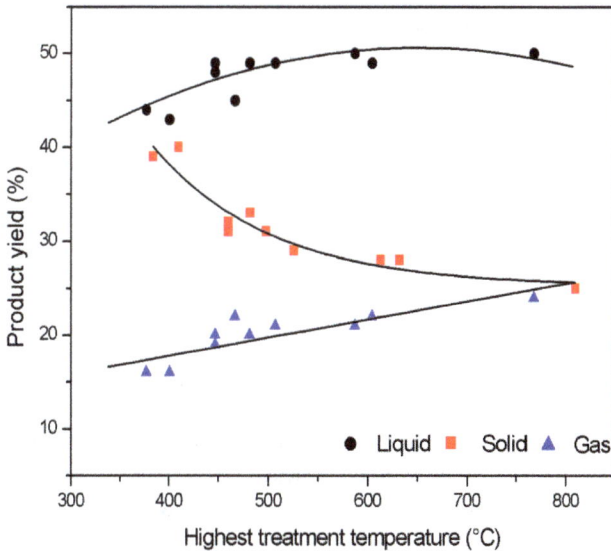

Sample: 100 g chipped willow (*Salix*) pyrolysed in a fixed bed reactor;
Heating rate: 30 °C/min, Purging gas: N₂ (Source: Chapter authors)

**Figure 3.** Effect of pyrolysis temperature on product distribution

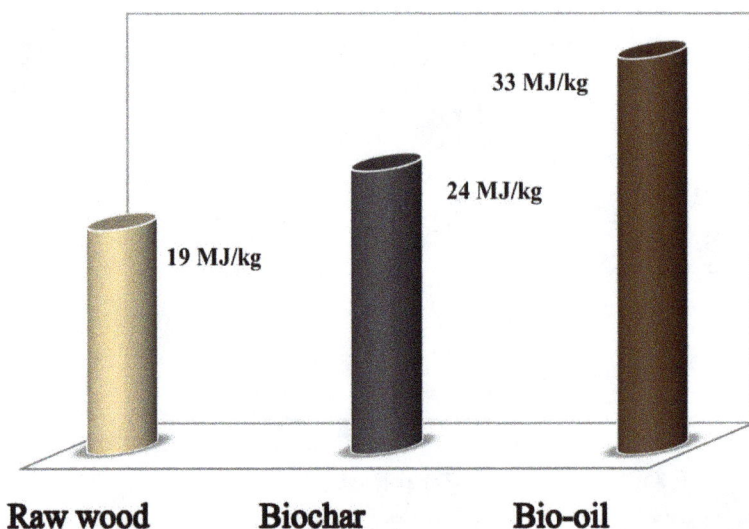

Raw material: 100 g chipped willow (*Salix*) pyrolysed in a fixed bed reactor
Highest treatment temperature: 460 ⁰C, Purging gas: N₂ ; Source: Chapter authors

**Figure 4.** Energy content of raw wood and its solid and liquid pyrolysis products

## 5.3. Nitrogen in biochar and in pyrolysis gases

Biochar and biomass char are the solid co-products of the pyrolysis process. They are mainly made of carbon and the ash content of the biomass. Despite of their similarities, historical definitions distinguish *biochar* from *biomass char* which is also known as *charcoal*. While the latter has been produced and used as fuel for heat for centuries, the former belongs to a new concept of soil management and carbon sequestration [80]. Other names like *black carbon*, *dark earth* (terra preta) or *agrichar* can be also fined in literature.

Enrichment of the fuel-bound nitrogen of biomass occurs in the biochar independently from the applied pyrolysis technique [81]. Nitrogenous gases (e.g. ammonia, hydrogen cyanide and isocyanic acid) are released during pyrolysis, but only at high temperature. The ration of these main gaseous nitrogen products depending on both the type of biomass and the conditions of the pyrolysis process [76]. Nitrogen-free gases leave the system when pyrolysis temperature is increased which results nitrogen depletion in char at high temperatures [81]. However, low pyrolysis temperature does not favour the liberation of fuel nitrogen therefore most of the nitrogen (approximately 60-75 % at 500 °C) remains captured in the char [72, 76, 82].

The nitrogen functionalities in biochar are *pyrrolic*-N, *pyridinic*-N, *quaternary*-N and *amines* [83-85] and incubation tests evidenced that these stable nitrogen forms with low bioavailability [86].

# 6. Biochar properties

## 6.1. Biochar as a fuel

Due to its high fixed carbon content biochar is a renewable energy source with a heating value up to 30-35 MJ/kg [87]. Biomass char has higher energy density and better combustibility properties than traditional biomass, and higher reactivity than coal due to its oxygen content [88] and its incoherent carbon structure [89]. The combustion of biomass char is able to displace traditional fuels, however, the combustion of biochar recycles atmospheric $CO_2$ and liberates the char-bond N in the form of NOx without the benefit of carbon or nitrogen sequestration.

According to different estimations biomass pyrolysis with soil applications of the biochar has a negative greenhouse gas emission – with a $CO_2$ *equivalent* ranging from few hundred kg up to a few tonnes of $CO_2eq$ $t^{-1}$ *dry biomass* – with a positive net energy [90-92]. A detailed calculation and complete life cycle assessment of biochar can be found in the work of Roberts at et al [90].

## 6.2. Biochar as a soil amendment

The most widely acknowledged benefit of biochar's soil applications is its long-term carbon sequestration potential [90, 93]. Other potentials of biochar is stimulation of $N_2$ fixation and the biological transformation of nitrogen in soil [94-96].

Biochar is also known for the ability to contribute to soil properties by changing its physical and chemical characteristics. The most important physicochemical properties of biochar are directly related to the type of the biomass used for char production and the applied temperature of pyrolysis [55, 97] therefore biochar contribution to soil quality factors can be both positive and negative [98, 99]. By selecting the right feedstock, setting the right pyrolysis conditions and elaborately characterising the physicochemical properties, char can be applied to soil as an amendment.

The pyrolysis temperature related structural changes of biochar can be seen on the infrared absorption spectra of Fig. 5. Comparing these spectra it can be seen that char gradually loses its structural complexity at higher pyrolysis temperatures as wood carbonisation becomes more completed. Char samples prepared at 300 °C and 400 °C show dramatic decreases in intensity in almost all functional groups; this is the temperature range in which the majority of the pyrolysis mass loss of wood occurs due to the degradation of cellulose, hemicellulose and lignin. Hemicellulose peak at 1736 $cm^{-1}$ becomes undetectable in char prepared at 400 °C but the O – H (3413 $cm^{-1}$) and $CH_n$ related vibrations (2956, 2924, 2851 $cm^{-1}$) show dramatic decreases in char prepared at higher temperatures (400 – 600 °C) where the thermal degradation of cellulose is already completed. Pyrogenic char (prepared at 700 °C or over) has no measured transmittance due to the lack of organic functionalities and the disordered carbon structure.

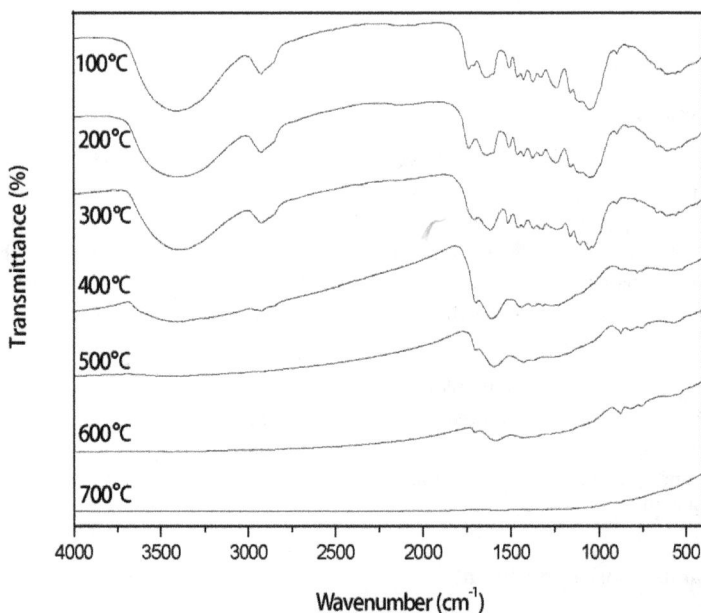

The first strong broad band between 3700 and 3000 cm⁻¹ of dried willow (100 and 200 °C) is the stretching vibrations of O – H functional groups. In the region of 2975-2840 cm⁻¹ the unresolved group of medium weak bands is related to C – H stretching vibrations of $CH_n$ groups. The peak at 1736 cm⁻¹ is assigned to the absorption of free carbonyl groups, therefore it is a typical hemicellulose marker [112, 113]. Bands around 1600 and 1500 cm⁻¹ are generally considered as lignin markers as this is the region of the skeletal vibrations of aromatic rings [114].
Sample: 2 mg grinded wood or biochar blended with 200 mg KBr and pressed into pellet
Spectra: recorded on a Perkin Elmer Spectrum 100 FT-IR spectrometer, 4 scans per experiments, resolution: 4 cm⁻¹
Source: Chapter authors

**Figure 5.** Changes in the infrared spectra of biochar obtained at different pyrolysis temperatures

The functional groups on biochar surface determine the pH and the cation exchange capacity and the nutrient retention in soil [100, 101]. The pH also has an impact on the mobility of ions and affects soil microbial activity [102].

As well as the changes of biochar surface, the increasing pyrolysis treatment temperature also increases C content, decreases H and O content and increases the ash content in char [103]. These changes in char composition increase hydrophobicity [99] and aromaticity [103]. Hydrophobicity and aromaticity play a major role in the future stability of biochar in soil [103] and the estimated half-life of char with O/C over 0.2 is 100-1000 year and greater than 1000 years in case of char when O/C is smaller than 0.2 [104].

The composition changes in the carbonised char is also accompanied by changes in the physical appearance of the biochar; Pyrolysis vapours can develop pores in biochar [105]

The increased porosity affects the water-holding capacity of soil and the surface area – a shelter for microorganism [97]; bulk density, which affects the pore size distribution of soil and the conditions for gas exchange [106]; and total dissolved solids, which give an estimation on the amount of the mobile charged ions, migrating from char to soil [80].

Due to the high specific surface and adsorbent capacity, biochar can increase the water and nutrient retention capacity of the soil [107, 108] while a biochar buffer layer in soil can reduce both nitrate leaching and gaseous loss of soil nitrogen [107, 109]. Improved nitrogen recovery in soil will directly result in increased plant growth.

Biochar properties are strongly affected by the pyrolysis temperature [97, 110] which makes possible to design biochar, remediate specific soil issues and realise a new type of soil management [99, 111].

## 7. Conclusions

Nitrogen always has been the "weakest link" in the food chain and agriculture. Without additional nitrogen the present capacity of Earth's topsoil is not able to satisfy our hunger for biomass for food or energy.

Wastewater is a valuable source of nitrogen but nitrate leaching is harmful for groundwaters and results in nutrient lost from the soil. Plants cultivated for wastewater treatment can be considered as energy crops and bring land back into economic use.

To obtain an economically attractive feedstock for energy conversion applications, efforts should be made to maximise the utilisation of the sources (land, irrigation water etc) and the energy gained from the biomass with a minimum environmental impact. Pyrolysis of wastewater irrigated energy crops offers the advantages in both fields, therefore it is an excellent candidate to supply green energy for rural areas in developing countries while the soil application of biochar can retain and assimilate the wastewater derived nitrogen back into the environment.

- In terms of the nitrogen-cycle, biomass combustion liberates 5-5.9 Tg of NOx-N each year into the atmosphere. However, the cultivation of wastewater irrigated energy crops and the pyrolysis of the vegetation filters have the potential to reduce the emission of NOx-N and other greenhouse gases the following ways:
- Vegetation filters reduce the concentration of water contaminants and lower nitrogen content by 97%. Wastewater can provide nitrogen and nutrients for plants and increase biomass yield without the application of inorganic soil fertilisers.
- Energy crops can uptake wastewater derived nitrogen and double the concentration of the storage amino acids.
- Compare to traditional combustion the pyrolysis of energy crops does not favour the formation of NOx.
- Pyrolysis captures 60-75% of the biomass derived nitrogen in the biochar. The soil applications of the biochar provide a long-term nitrogen sequestration and reduce the amount of the reactive nitrogen forms which accompany the traditional water treatment processes.

## Author details

Zsuzsa A. Mayer, Andreas Apfelbacher and Andreas Hornung

*European Bioenergy Research Institute (EBRI), Aston University, Birmingham, United Kingdom*

## Acknowledgement

Financial support has been provided by the Science Bridge project (EP/G039992/1), the initiative of the Research Councils (RCUK) and Department of Science and Technology (DST). The authors are thankful for the permission to partly reproduce our paper, originally published in Z. A. Mayer, A. Apfelbacher and A. Hornung, "Nitrogen cycle of effluent-irrigated energy crop plantations from wastewater treatment to the thermochemical conversion processes", Journal of Scientific and Industrial Research, 70, pages 675-682, Copyright 2011 NISCAIR.

## 8. References

[1] Nicholas I (2003) Nitrogen Uptake in New Zealand short Rotation Crops. Short Rotation Crops for Bioenergy: New Zealand. Crops 1985: 235-240.

[2] Pandey A, Srivastava R K (2010) Role of Dendropower in Wastewater Treatment and Sustaining Economy. Journal of Cleaner Production 18: 1113-1117.

[3] Vasudevan P, Thapliyal A, Srivastava R K, Pandey A, Dastidar M G, Davies P (2010) Fertigation Potential of Domestic Wastewater for Tree Plantations. Journal of Scientific & Industrial Research 69: 146-150.

[4] Czernik S, Bridgwater A V (2004) Overview of Applications of Biomass Fast Pyrolysis Oil. Energy & Fuels 18: 590-598.

[5] Patterson R A (2003) Nitrogen in Wastewater and its Role in Constraining On-site Planning in Future Directions for On-site Systems: Best Management Practice. Proceedings of On-site '03 Conference. University of New England, Armidale 30 Sept-2 Oct 2003. Published by Lanfax Laboratories Armidale. ISBN 0-9579438-1-4. pp 313-320.

[6] Abeliovich A (1992) Transformations of Ammonia and the Environmental Impact of Nitrifying Bacteria. Biodegradation 3: 255-264.

[7] US Environmental Protection Agency (1985). Ambient Water Quality Criteria for Ammonia. (440/5-85-001). EPA

[8] Oglesby R T, Edmondson W T (1966) Control of Eutrophication. Water Pollution Control Federation 38: 1452-1460.

[9] Knobeloch L, Salna B, Hogan A, Postle J, Anderson H (2000) Blue Babies and Nitrate-Contaminated Well Water. Environmental Health Perspectives 108: 675-678.

[10] L'Hirondel J, L'Hirondel J L (2002) Nitrate and Man: Toxic, Harmless or Beneficial. Wallingford: CABI UK Publishing.

[11] Eriksson E, Auffarth K, Henze M, Ledin A (2002) Characteristics of Grey Wastewater. Urban Water 4: 85-104.

[12] European Commission (1991). Council directive of 21 May 1991 concerning urban waste water treatment. (91/271/EEC) EEC. Available from: http://eur-lex.europa.eu/

[13] US Environmental Protection Agency National Primary Drinking Water Regulations. EPA Available from: http://water.epa.gov/drink/contaminants/

[14] Australian Government National Health and Medical Research Council (1996). Australian Drinking Water Guidelines (ADWG). Available from: http://www.nhmrc.gov.au/

[15] Cooper P, Day M, Thomas V (1994) Process Options for Phosphorus and Nitrogen Removal from Wastewater. Water and Environment 8: 84-92.

[16] Verstraete W, Philips S (1998) Nitrification-Denitrification Processes and Technologies in New Contexts. Environmental Pollution 102: 717-726.

[17] von Sperling M (1996) Comparison Among the Most Frequently Used Systems for Wastewater Treatment in Developing Countries. Water Science and Technology 33: 59-72.

[18] Kivaisi A K (2001) The Potential for Constructed Wetlands for Wastewater Treatment and Reuse in Developing Countries: A Review. Ecological Engineering 16: 545-560.

[19] Corcoran E, C. N, Baker E, Bos R, Osborn, H. S (2010) Sick Water? The Central Role of Wastewater Management in Sustainable Development. A Rapid Response Assessment. United Nations Environment Programme, UN-HABITAT (Birkeland Trykkeri AS, Norway) 2010.

[20] Dixon A M, Butler D, Fewkes A (1999) Guidelines for Greywater Re-Use: Health Issues. Journal of the Chartered Institution of Water and Environmental Management 13: 322-326.

[21] Gilde L C, Kester A S, Law J P, Neeley C H, Parmelee D M (1971) A spray Irrigation System for Treatment of Cannery Wastes. Journal of the Water Pollution Control Federation 43: 2011-2025.

[22] Bendixen T W, Hill R D, Dubyne F T, Robeck G G (1969) Cannery Waste Treatment by Spray Irrigation-Runoff. Water Pollution Control Federation 41: 385-391.

[23] Delgado A N, Periago E L, Viqueira F D-F (1995) Vegetated Filter Strips for Wastewater Purification: A Review. Bioresource Technology 51: 13-22.

[24] Bogosian G, Sammons L E, Morris P J L, Oneil J P, Heitkamp M A, Weber D B (1996) Death of the Escherichia Coli K-12 Strain W3110 in Soil and Water. Applied and Environmental Microbiology 62: 4114-4120.

[25] Toze S (1997) Microbial Pathogens in Wastewater. Literature Review for Urban Water Systems Multi-Divisional Research Program. Technical Report 1/97. CSIRO Land and Water, (Australia)

[26] Troeh, F. R.; Thompson, L. M. (2005) Soils and Soil Fertility, Oxford: Wiley-Blackwell

[27] Masclaux-Daubresse C, Daniel-Vedele F, Dechorgnat J, Chardon F, Gaufichon L, Suzuki A (2010) Nitrogen Uptake, Assimilation and Remobilization in Plants: Challenges for Sustainable and Productive Agriculture. Annals of Botany 105: 1141-1157.

[28] Gomez A, Leschber R, P. L H (1986) Sampling Problems for the Chemical Analysis of Sludge, Soils, and Plants. London and New York: Elsevier Applied Science Publishers.

[29] Vitousek P M, Howarth R W (1991) Nitrogen Limitation on Land and in the Sea: How Can It Occur? Biogeochemistry 13: 87-115.

[30] European Fertilizer Manufacturers' Association. Production of Urea and Urea Ammonium Nitrate. EFMA. Brussels, 2000. Available from: http://www.efma.org

[31] Haber F, Le Rossignol R (1913) Über Die Technische Darstellung Von Ammoniak Aus Den Elementen. Zeitschrift für Elektrochemie und angewandte physikalische Chemie 19: 53-72.

[32] Manchester K L (2002) Man of Destiny: The Life and Work of Fritz Haber. Endeavour 26: 64-69.

[33] Wood S, Cowie, A, A Review of Greenhouse Gas Emission Factors for Fertiliser Production. In For IEA Bioenergy Task 38, International Energy Agency: 2004. Available from: http://www.ieabioenergy-task38.org

[34] International Fertilizer Industry Association Annual Production and International Trade Statistics. In: Series of Statistical Reports on 2007 Productio Capacity, Production and International Trade of Key Fertilizers, Raw Materials and Intermediates. IFA (2008) Paris, France

[35] Jaeglé L, Steinberger L, Martin R V, Chance K (2005) Global Partitioning of NOx Sources Using Satellite Observations: Relative Roles of Fossil Fuel Combustion, Biomass Burning and Soil Emissions. Faraday Discuss 130: 407-423.

[36] Haynes R, Goh K (1978) Ammonium and Nitrate Nutrition of Plants. Biological Reviews 53: 465-510.

[37] Dickson R (1989) Carbon and Nitrogen Allocation in Trees. Annales des Sciences Forestieres 46: 631-647.

[38] Chen G Z, Miao S Y, Tam N F Y, Wong Y S, Li S H, Lan C Y (1995) Effect of Synthethic Waster on Young Kandelia Candel Plants Growing under Greenhouse Conditions. Hydrobiologia 295: 263-273.

[39] Mohammad Rusan M J, Hinnawi S, Rousan L (2007) Long Term Effect of Wastewater Irrigation of Forage Crops on Soil and Plant Quality Parameters. Desalination 215: 143-152.

[40] Mohammad Rusan M J, Mazahreh N (2003) Changes in Soil Fertility Parameters in Response to Irrigation of Forage Crops with Secondary Treated Wastewater. Communications in Soil Science and Plant Analysis 34: 1281 - 1294.

[41] Carroll D (1959) Ion Exchange in Clays and Other Minerals. Geological Society of America Bulletin 70: 749-779.

[42] Jacks G, Sharma V P (1983) Nitrogen Circulation and Nitrate in Groundwater in an Agricultural Catchment in Southern India. Environmental Geology 5: 61-64.

[43] Della Rocca C, Belgiorno V, Meriç S (2007) Overview of In-situ Applicable Nitrate Removal Processes. Desalination 204: 46-62.

[44] Meers E, Vandecasteele B, Ruttens A, Vangronsveld J, Tack F (2007) Potential of Five Willow Species (Salix Spp.) for Phytoextraction of Heavy Metals. Environmental and Experimental Botany 60: 57-68.

[45] Ericsson K, Rosenqvist H, Nilsson L J (2009) Energy Crop Production Costs in the EU. Biomass and Bioenergy 33: 1577-1586.

[46] Bech N, Jensen P A, Dam-Johansen K (2009) Determining the Elemental Composition of Fuels by Bomb Calorimetry and the Inverse Correlation of HHV with Elemental Composition. Biomass and Bioenergy 33: 534-537.

[47] Thipkhunthod P, Meeyoo V, Rangsunvigit P, Kitiyanan B, Siemanond K, Rirksomboon T (2005) Predicting the Heating Value of Sewage Sludges in Thailand from Proximate and Ultimate Analyses. Fuel 84: 849-857.

[48] Aylott M J, Casella E, Tubby I, Street N, Smith P, Taylor G (2008) Yield and Spatial Supply of Bioenergy Poplar and Willow Short-Rotation Coppice in the UK. New Phytologist 178: 358-370.

[49] Sims R E H, Hastings A, Schlamadinger B, Taylor G, Smith P (2006) Energy Crops: Current Status and Future Prospects. Global Change Biology 12: 2054-2076.

[50] Kenney W A, Sennerby-Forsse L, Layton P (1990) A Review of Biomass Quality Research Relevant to the Use of Poplar and Willow for Energy Conversion. Biomass 21: 163-188.

[51] Gaur S, Reed T B (1998) An Atlas of The thermal Data For biomass and Other Fuels, Final Subcontract Report, New York: Marcel Dekker

[52] Channiwala S A, Parikh P P (2002) A Unified Correlation for Estimating HHV of Solid, Liquid and Gaseous Fuels. Fuel 81: 1051-1063.

[53] Demirbas A (1997) Calculation of Higher Heating Values of Biomass Fuels. Fuel 76: 431-434.

[54] Demirbas A (1999) Properties of Charcoal Derived from Hazelnut Shell and the Production of Briquettes Using Pyrolytic Oil. Energy 24: 141-150.

[55] Antal M J, Grønli M (2003) The Art, Science, and Technology of Charcoal productionon. Industrial & Engineering Chemistry Research 42: 1619-1640.

[56] Rose J W, Cooper J R (1977) Technical Data on Fuel. 7th edn London: British National Committee, World Energy Conference.

[57] Leppälahti J, Koljonen T (1995) Nitrogen Evolution from Coal, Peat and Wood During Gasification: Literature Review. Fuel Processing Technology 43: 1-45.

[58] Thompson D, Brown T D, Beér J M (1972) NOx Formation in Combustion. Combustion and Flame 19: 69-79.

[59] Fenimore C P (1971) Formation of Nitric Oxide in Premixed Hydrocarbon Flames. Symposium (International) on Combustion 13: 373-380.

[60] Kuo K K-Y (1986) Principles of Combustion. New York: Wiley.

[61] Nelson H F (1976) Nitric-Oxide Formation in combustion. AIAA Journal 14: 1177-1182.

[62] Leonard P A, Plee S L, Mellor A M (1976) Nitric Oxide Formation from Fuel and Atmospheric Nitrogen Combustion Science and Technology 14: 183-193.

[63] Hayhurst A N, Vince I M (1983) The Origin and Nature of "Prompt" Nitric Oxide in Flames. Combustion and Flame 50: 41-57.

[64] Likens G E, Bormann F H, Johnson N M (1972) Acid Rain. Environment: Science and Policy for Sustainable Development 14: 33-40.

[65] Elrod M J (1999) Greenhouse Warming Potentials from the Infrared Spectroscopy of Atmospheric Gases. Journal of Chemical Education 76: 1702-1705.

[66] Pitts J N (1993) Anthropogenic Ozone, Acids and Mutagens: Half a Century of Pandora's NOx. Research on Chemical Intermediates 19: 251-298.

[67] Nussbaumer T (1997) Primary and Secondary Measures for the Reduction of Nitric Oxide Emissions from Biomass Combustion. In: Bridgwater A V, Boocock D G B,

editors. Development and Thermochemical Biomass Conversion. London: Chapman and Hall.

[68] Victor, N. M.; Victor, D. G., Macro Patterns in the Use of Traditional Biomass Fuels, in Stanford/TERI workshop on "Rural Energy Transitions", (New Delhi) 2002.

[69] US Environmental Protection Agency (1982). Air Quality Criteria for Oxides of Nitrogen. (600/8-82-026). EPA

[70] Intergovernmental Panel on Climate Change (2006). Guidelines for National Greenhouse Gas Inventories. IPCC Available from: http://www.ipcc-nggip.iges.or.jp

[71] Antal M J, Jr., Várhegyi G (1995) Cellulose Pyrolysis Kinetics: The Current State of Knowledge. Industrial & Engineering Chemistry Research 34: 703-717.

[72] Skodras G, Natas P, Basinas P, Sakellaropoulos G P (2006) Effects of Pyrolysis Temperature, Residence Time on the Reactivity of Clean Coals Produced From Poor Quality Coals. Global NEST Journal 8: 89-94.

[73] Bridgwater A V, Meier D, Radlein D (1999) An Overview of Fast Pyrolysis of Biomass. Organic Geochemistry 30: 1479-1493.

[74] Bridgwater A V (1999) Principles and Practice of Biomass Fast Pyrolysis Processes for Liquids. Journal of Analytical and Applied Pyrolysis 51: 3-22.

[75] Di Blasi C, Branca C, D'Errico G (2000) Degradation Characteristics of Straw and Washed Straw. Thermochimica Acta 364: 133-142.

[76] Hansson K-M, Samuelsson J, Tullin C, Åmand L-E (2004) Formation of HNCO, HCN, and NH3 from the Pyrolysis of Bark and Nitrogen-Containing Model Compounds. Combustion and Flame 137: 265-277.

[77] Müller A, Schmidhuber J, Hoogeveen J, Steduto P (2008) Some Insights in the Effect of Growing Bio-Energy Demand on Global Food Security and Natural Resources. Water Policy 10: 83-94.

[78] Hornung A, Apfelbacher A, Richter F, Schneider D, Schöner J, Seifert H, Tumiatti V, Franchi P, Lenzi F, Haloclean - Intermediate Pyrolysis - Power Generation From Rape, 16th European Biomass Conference & Exhibition, Valencia, Spain, 2008.

[79] Hossain A K, Davies P A (2010) Use of Pyrolysis Oil for CHP Application: Difficulties and Prospects. In: Proceedings of the World Renewable Energy Congress (WRECXI), Abu Dhabi, UAE, September 25-30, pp. 121-126.

[80] Joseph S, Peacocke C, Lehmann J, Munroe P (2009) Developing a Biochar Classification and Test Methods. In: Lehmann J, Joseph S, editors. Biochar for Environmental Management: Science and Technology. London: Earthscan. pp. 107-126.

[81] Johnsson J E (1994) Formation and Reduction of Nitrogen Oxides in Fluidized-Bed Combustion. Fuel 73: 1398-1415.

[82] Tan L L, Li C-Z (2000) Formation of NOx and SOx Precursors During the Pyrolysis of Coal and Biomass. Part I. Effects of Reactor Configuration on the Determined Yields of HCN and NH3 During Pyrolysis. Fuel 79: 1883-1889.

[83] Pels J R, Kapteijn F, Moulijn J A, Zhu Q, Thomas K M (1995) Evolution of Nitrogen Functionalities in Carbonaceous Materials During Pyrolysis. Carbon 33: 1641-1653.

[84] Zhu Q, Money S L, Russell A E, Thomas K M (1997) Determination of the Fate of Nitrogen Functionality in Carbonaceous Materials During Pyrolysis and Combustion Using X-Ray Absorption Near Edge Structure Spectroscopy. Langmuir 13: 2149-2157.

[85] Gong B, Buckley A N, Lamb R N, Nelson P F (1999) XPS Determination of the Forms of Nitrogen in Coal Pyrolysis Chars. Surface and Interface Analysis 28: 126-130.

[86] Bridle T, Pritchard D (2004) Energy and Nutrient Recovery from Sewage Sludge Via Pyrolysis. Water Science and Technology 50: 169-175.

[87] Sharifi H (1999) A Methodology for Achieving Agility in Manufacturing Organisations: An Introduction. International Journal of Production Economics 62: 7-22.

[88] Backreedy R I, Jones J M, Pourkashanian M, Williams A Burn-out of Pulverised Coal and Biomass Chars. Fuel 82: 2097-2105.

[89] Henrich E, Bürkle S, Meza-Renken Z I, Rumpel S (1999) Combustion and Gasification Kinetics of Pyrolysis Chars from Waste and Biomass. Journal of Analytical and Applied Pyrolysis 49: 221-241.

[90] Roberts K G, Gloy B A, Joseph S, Scott N R, Lehmann J (2009) Life Cycle Assessment of Biochar Systems: Estimating the Energetic, Economic, and Climate Change Potential. Environmental Science & Technology 44: 827-833.

[91] Gaunt J, Cowie A (2009) Greenhouse-gas Accounting and Emissions Trading. In: Lehmann J, Joseph S, editors. Biochar for Environmental Management: Science and Technology. London: Earthscan. pp. 317-340.

[92] Brownsort P, Carter S, Cook J, Cunningham C, Gaunt J, Hammond J, Ibarrola R, Sims K, Thornley P  An assessment of  the Benefits and Issues Associated  with  the Application of Biochar to Soil. Available from: http://www.biochar.org.uk

[93] Lehmann J, Czimczik C, Laird D, Sohi S (2009) Stability of Biochar in Soil. In: Lehmann J, Joseph S, editors. Biochar for Environmental Management: Science and Technology. London: Earthscan. pp. 183– 206.

[94] Sohi S P, Krull E, Lopez-Capel E, Bol R (2010) A Review of Biochar and Its Use and Function in Soil. In: Donald L S, editor. Advances in Agronomy. Academic Press. pp. 47-82.

[95] Lehmann J, da Silva J P, Steiner C, Nehls T, Zech W, Glaser B (2003) Nutrient Availability and Leaching in an Archaeological Anthrosol and a Ferralsol of the Central Amazon Basin: Fertilizer, Manure and Charcoal Amendments. Plant and Soil 249: 343-357.

[96] DeLuca T H, MacKenzie M D, Gundale M J (2009) Biochar Effects on Soil Nutrient Transformations. In: Lehmann J, Joseph S, editors. Biochar for Environmental Management: Science and Technology. London: Earthscan. pp. 251-270.

[97] Downie A, Crosky A, Munroe P (2009) Physical Properties of Biochar. In: Lehmann J, Joseph S, editors. Biochar for Environmental Management: Science and Technology. London: Earthscan. pp. 13-32.

[98] Shackley S, Sohi S, Brownsort P, Carter S, Cook J, Cunningham C, Gaunt J, Hammond J, Ibarrola R, Mašek O (2010) An Assessment of the Benefits and Issues Associated with the Application of Biochar to Soil. Department for Environment, Food and Rural Affairs (DEFRA), London, pp. 132.

[99]  Novak J M, Lima I, Xing B, Gaskin J W, Steiner C, Das K, Ahmedna M, Rehrah D, Watts D W, Busscher W J (2009) Characterization of Designer Biochar Produced at Different Temperatures and Their Effects on a Loamy Sand. Annals of Environmental Science 3: 195-206.

[100] Liang B L, Solomon J, Kinyangi D, Grossman J, O'Neill J, Skjemstad B, Thies J, Luizão J, Petersen F, Neves J (2006) Black Carbon Increases Cation Exchange Capacity in Soils. Soil Science Society of America 70: 1719-1730.

[101] Chan K Y, Xu Z (2009) Biochar: Nutrient Properties and Their Enhancement. In: Lehmann J, Joseph S, editors. Biochar for Environmental Management: Science and Technology. London: Earthscan. pp. 67-84.

[102] Thies J E, Rillig M C (2009) Characteristics of Biochar: Biological Properties. In: Lehmann J, Joseph S, editors. Biochar for Environmental Management: Science and Technology. London: Earthscan. pp. 85-106.

[103] Krull E S, Baldock J A, Skjemstad J O, Smernik R J (2009) Characteristics of Biochar: Organo-Chemical Properties. In: Lehmann J, Joseph S, editors. Biochar for Environmental Management: Science and Technology. London: Earthscan. pp. 53-65.

[104] Spokas K A (2010) Review of the Stability of Biochar in Soils: Predictability of O: C Molar Ratios. Carbon 1: 289-303.

[105] Cetin E, Moghtaderi B, Gupta R, Wall T F (2004) Influence of Pyrolysis Conditions on the Structure and Gasification Reactivity of Biomass Chars. Fuel 83: 2139-2150.

[106] Schjønning P, Thomsen I K, Petersen S O, Kristensen K, Christensen B T (2011) Relating Soil Microbial Activity to Water Content and Tillage-Induced Differences in Soil Structure. Geoderma 163: 256-264.

[107] Laird D, Fleming P, Wang B, Horton R, Karlen D (2010) Biochar Impact on Nutrient Leaching from a Midwestern Agricultural Soil. Geoderma 158: 436-442.

[108] Steiner C, Glaser B, Geraldes Teixeira W, Lehmann J, Blum W E, Zech W (2008) Nitrogen Retention and Plant Uptake on a Highly Weathered Central Amazonian Ferralsol Amended with Compost and Charcoal. Journal of Plant Nutrition and Soil Science 171: 893-899.

[109] Clough T J, Condron L M (2010) Biochar and the Nitrogen Cycle: Introduction. Journal of Environmental Quality 39: 1218-1223.

[110] Brewer C E, Schmidt-Rohr K, Satrio J A, Brown R C (2009) Characterization of Biochar from Fast Pyrolysis and Gasification Systems. Environmental Progress & Sustainable Energy 28: 386-396.

[111] Mukherjee A, Zimmerman A R, Harris W (2011) Surface Chemistry Variations among a Series of Laboratory-Produced Biochars. Geoderma 163: 247-255.

[112] Owen N, Thomas D (1989) Infrared Studies of "Hard" and "Soft" Woods. Applied spectroscopy 43: 451-455.

[113] Yang H, Yan R, Chen H, Lee D H, Zheng C (2007) Characteristics of Hemicellulose, Cellulose and Lignin Pyrolysis. Fuel 86: 1781-1788.

[114] Schwanninger M, Rodrigues J C, Pereira H, Hinterstoisser B (2004) Effects of Short-Time Vibratory Ball Milling on the Shape of FT-IR Spectra of Wood and Cellulose. Vibrational Spectroscopy 36: 23-40.

# Chapter 11

# Impact of Uses of 3-Dimensonal Electronics IC Devices and Computing Systems on the Power Consumptions and Global Warming Issues

Karl Cheng, Bharat Raj Singh and Alan Cheng

Additional information is available at the end of the chapter

## 1. Introduction

According to the American Energy Information Administration (EIA) and to the International Energy Agency (IEA), the world-wide energy consumption will on average continue to increase by 2% per year. A yearly increase by 2% leads to a doubling of the energy consumption every 35 years (*Michaelbluejay, 2012*). This means the world-wide energy consumption is predicted to be twice as high in the year 2047 compared to today (2012) which is about 2500 GW of capacity 35 years from now. Current global consumer electronics devices (TV and PC etc.) account for roughly 15% residential electricity usage annually and are growing. Nearly 185 GW of capacity is consumed. About 370 medium size (500 MW) power plants are required. Estimated 650 megaton's of $CO_2$ emissions per year today and could surpass 1000 megaton's by 2030 without policy intervention. With the switch and design architecture changed from 2-D to 3-D for all TV and PC electronics devices the energy consumption could be reduced to the tenth of the current consumed capacity which is about 18 GW of capacity today only (*Cheng, K., 2011*). This could save about 600 megaton's of $CO_2$ emissions per year today. 37 medium size (500 MW) power plants are needed only instead of 370 power plants (*Massoud Pedram, 2009; Noah Horowitz, 2011*).

From the recent study, it is learnt that 3-dimensional electronics integrated circuit device (3-D IC) and 3-dimensional computing systems (3-D Computer) will be the ultimate architecture for all our daily used electronics equipments such as TV, PC, cell phone, PDA and GPS devices etc. because of the 3 dimensional integrated circuit has the benefit of small size, light weight, high speed and less power consumption to meet the global warming regulation. No PCB, no package and even no cable connector are needed in the future 3 dimensional computer system. It has been proved in the laboratory that the 3 dimensional computing systems consumed only about one seventh to one tenth of that of the 2 dimensional counterparts. In the next few years to come each of us will have a mobile device such as i-phone or tablet computer to carry with all the time. This means the number of the device needed is about the population of the world. Therefore, it is vital urgent to turn all the current 2 dimensional computing system to the 3 dimensional architecture. This will lead to the seven to ten times of power saving and will benefit to our environment by using less carbon and meet the global warming regulation.

The era of cloud computing is near and becoming more reality for us. More business organizations are taking the instant tremendous large volume of data from cloud computing to analyze and direct their business orientation for profit consideration. For the private section to guide their investment for instant profit is becoming more reality by manipulating the instant data information provided from the cloud computing. The advantage of the cloud computing technology is the tremendous large instant data base available for public. Handling this large amount of instant data into your local need is one of the biggest challenges of this new era. In order to quickly process the data obtained from the cloud, a local smart terminal computing system is required. An example of a cloud computing terminal is an advanced GPS device. Live traffic information can be processed in real time such that alternate routes can be derived. Such smart terminals required for this application could be further advanced by applying 3 dimensional computing systems. This means in the near future era of cloud computing these smart terminals are needed almost every corner of the world. Again, it is vital urgent to turn all the current 2 dimensional computing system to the 3 dimensional architecture which will cut down the residential power requirement of electronic gadgets from 15% to 1.5%-2.1% and would help in reduction of power requirement substantially and approximately 600 megaton's of $CO_2$ emissions per year could be saved on the basis of current data. It would thus protect our environment by using less carbon and meet the global warming regulation if all the electronics equipments are switched to 3-D. **Source:** (*Latest industry research & analysis; Prediction of energy consumption world-wide*).

## 2. Construction features of 2-d to 3-d integrated circuit devices

A 2-dimensional (2-D) system chip or device, for instance, may consist of 5 elements as shown in the **Fig.** 1-a. This system can be partitioned and then stacked up to form a 3-dimensional (3-D) system chip or device with the same functionality as shown in **Fig.** 1-c. **Fig.** 1-d is also a 3-D but sometime it is named 2.5-D because only small portion of the whole system are stacked up to form a 3-D. As can be seen is that the global wirings (wirings between elements) shown in orange color in **Fig.** 1-b are much longer than the global wirings shown in **Fig.** 1-d, i.e. the wirings in 3-D system are much shorter than the 2-D's. With the shorter wirings the system may functioning more efficiency. Besides, it consumes less power too.

In general, the power for IC circuit can be expressed as in the following equation:

$$\text{System Power consumption} \doteq C \bullet V \bullet V \bullet f \tag{1}$$

where C=capacitance, V=voltage swing, f=frequency of operation.

So the advantage of consumed less power in 3-D comes from the smaller capacitance due to the reduced size and length of the interconnection among the system.

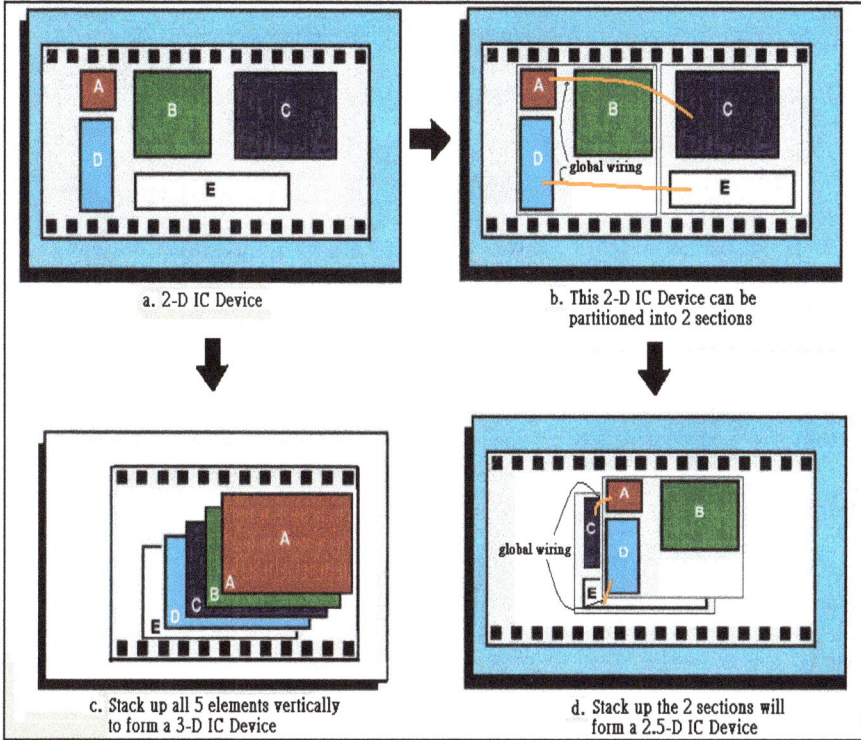

a. 2-D IC Device

b. This 2-D IC Device can be partitioned into 2 sections

c. Stack up all 5 elements vertically to form a 3-D IC Device

d. Stack up the 2 sections will form a 2.5-D IC Device

**Figure 1.** 2-D IC Device converted to 2.5-D or 3-D IC Device

2-D IC system as described in **Fig.1**-a can be partitioned more and form a totally 3-D IC system device or chip (**Fig. 1**-c) with all 5 elements stacked vertically into 5 layers. With the same token, stacking up more of the same or different functioning 3-D IC chips vertically will finally become a complete close-up system, a 3D-Computer as shown in **Fig. 2** below. The form factor (size and weight) of the 3-D computer system is much compacted and smaller than its 2-D counterpart.

**Fig. 2** shows the 3 systems; 2-D IC, 3-D IC and the 3-D Computer. According to the lab experimental results indicated that the 2.5-D IC  with only portion of the system stacked to form a 3-D IC device like the one in **Fig. 1**-d, the power consumption is about 3 quarter of that 2-D IC counterpart. With more detail partitioning and stacking into more layers as shown in Fig. 2, the 3-D Computer, the power consumed by this 3-D Computer system is only one seventh or one tenth of the 2-D counterpart. The experimental results [1] have been performed in the lab as shown in **Table 1** below:

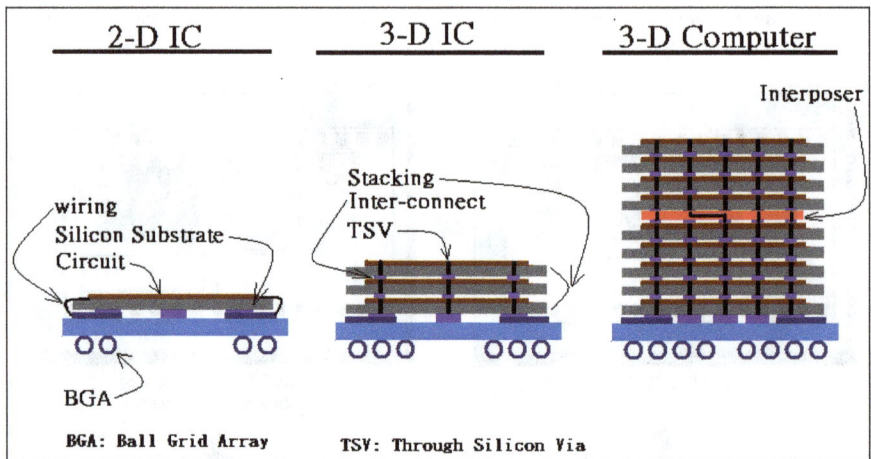

**Figure 2.** 2-D IC vurses 3-D IC and 3-D Computer

| Technologies | Size (M³) | Weight (Kg) | Power (KW) |
|---|---|---|---|
| System with all MSI/LSI only | ~1.000 | ~125 | ~0.6 |
| System with Conventional VLSI | ~0.038 | ~16 | ~1.27 |
| System with Wafer Scale VLSI | ~0.012 | ~6 | ~0.95 |
| System with Multi-Wafers Stacked VLSI (3-D Computer) | ~0.003 | ~1 | ~0.14 |

**Table 1.** Power Consumption Comparison between 2-D and 3-D Computer system

From the above **Table 1**, the size, weight and power of 3-D Computer are at most (0.003/0.012=)1/4 in size, 1/6 in weight and (0.14/0.95=)1/7 in power as compared with the other 2-D Computer counterparts.

The system power consumption as stated in the above equation (1) is also proportional to the voltage swing, V. With the system stacked in a very compact form like that in 3-D Computer system, the voltage swing actually can be reduced to a point where the logic states, "1" and "0" is still valid. Supposed the voltage swing of the 3-D Computer can be reduced to half then the power consumption will be reduced to the factor of 4. Therefore, the power consumed by the 3-D computing system could reduce almost to one 4x7=thirtieth and beyond. This is a very optimal way of saving the power. This concluded that the system must go 3-D in order to save power consumption and meet the global warming regulation.

## 3. Signal, Information processing and Super Computer Architecture of 3-D IC

In this section the power of 3-D IC signal and information processing is described by some sample applications. 3-D IC's will have the advantage of parallel processing over its 2-D IC counterpart which will lead to better image, signal and data processing (*Little, M.J.,1989; Toborg, S.T., 1990; Boguslaw C. et al., 2009; Orion Jones, 2012; Sarah Perrin, 2012*).

The main technologies for integrating and fabricating the 3D-IC device and system are:

i. Through Silicon Via (TSV) technology for vertical communication;
ii. Interconnecting technology to accomplish the up and down signal flow between wafer layers and
iii. Stacking technology to complete the final assembly of the 3D-IC device and system.

The 3-D IC fabrication technology and its manufacturing processes are becoming more mature now-a-days. The TSV and the multi-layer wafers stacking technologies are going to be more prevalent in the manufacturing industry. These advancing techniques are now providing an opportunity for the integrated circuit industry to advance in an optimal circuit design using 3-D IC's. The advantages of the 3-D IC design are what we always strive for in

IC manufacturing, that is, smaller sizes, lighter weights and less power consumption.

Most of the micro-processor (μP) currently used in all computing system is 2-D. With the advancement of the TSV, wafer to wafer interconnection and stacking technologies in 3-D IC manufacturing, it is possible to produce the 3-D μP at relatively easy and low cost. What will be the different between the 2-D and 3-D μP is that the processing algorithm and the circuit design architecture. With the implementation of SIMA (Single Instruction Multiple Data) and BSWP (Bit Serial Word Parallel) processing technologies, a 3-D μP or super computer can be designed to execute the data signal paralleling while the conventional 2-D is sequentially. Therefore, the power of the 3-D signal processing could be tremendously better than that of the 2-D version (*Singh, A. D., 1985*).

The 2-D conventional microprocessor is consisted by 5 processing cells as shown in the Fig. 3. These 5 processing cells can be stacked up vertically to form a 3-D μP as shown in **Fig. 4**. We could turn this single 3-D μP into multiple or wafer scale 3-D high performance microprocessor easily as shown in **Fig. 5**. This wafer scale 3-D microprocessor could be the ultimate high performance processor or super computer we are searching for.

With multiple identical N x N cells on the same level of wafer and stacked with other 4 different types of cell wafer will lead to an N x N processors. Each processor can process its own instruction or data signal and all N x N different instructions are executed parallel in the same time frame. **Fig. 6** shows a complete 3-D multiple micro-processor system or 3-D super computer. The data signal can go horizontally and vertically in this 5 wafers stack. This 3-D IC design architecture constructed a very powerful N x N processor. For example, let's assume this is a 16 bits word processor. The data on any level of wafer with N x N cells are transmitted to the other level of wafer paralleling through the TSV inter-connected link between levels then executed according to the instruction paralleling. This process is called "words parallel" because the entire N x N data or instructions is processed or executed parallel in the same time. If there is only one TSV for each cell to communicate to the other level, then the 16 bits will be shifted bit by bit through the single TSV and therefore we call it "bit serial" because shifting the word to other level of wafer is bit by bit sequentially.

Typical 2-D processor chip layout

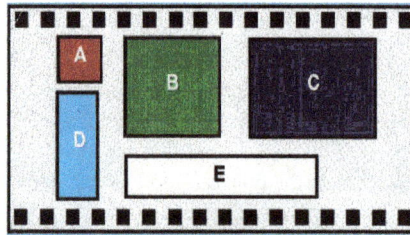

A = BUFFERS
B = MICROCODE ROM
C = ALU
D = I/O INTERFACE
E = CACHE

**Figure 3.** Typical processor chip Implemented with conventional 2-D technology

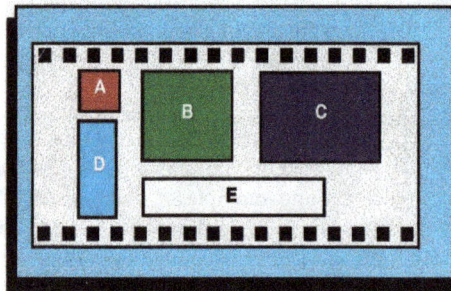

2-D Micro-Processor

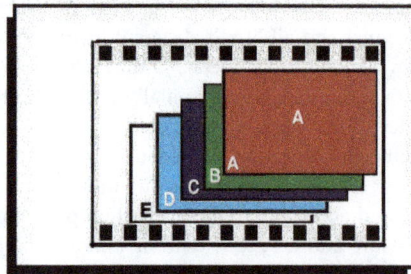

3-D Micro-Processor

**Figure 4.** 2-D vs. 3-D Micro Processor

Impact of Uses of 3-Dimensonal Electronics IC Devices and Computing Systems on the Power
Consumptions and Global Warming Issues

289

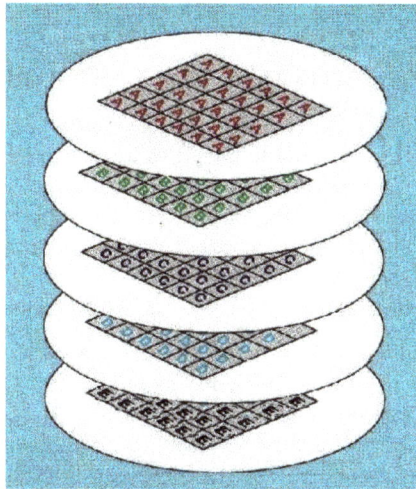

**Figure 5.** 3-D Multi (N x N) Processor in a  5 wafers stack

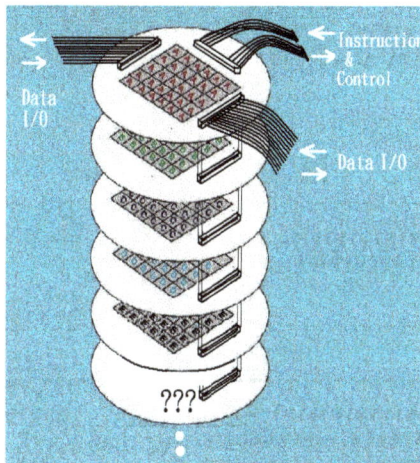

**Figure 6.** 3-D Multi (N x N) Processor System (??? is for adding up new type of wafers)

The command for this BSWP (bit serial word parallel) is instructed by a single instruction
and the operation is performed on all N x N data in the same time. Therefore, we called it
SIMA (single instruction multiple data) operation. For example:

SHIFT A TO C;
ADD DATA;
MOVE DATA TO D;

The above 3 single instructions will do:

i.    shift the data on A level of wafer through TSV to the C level
ii.   do the adding on all N x N data of A with C
iii.  The results of N x N data will be moved to I/O which is D wafer level and accessed to
      outside world such as printer etc…

With the concept of BSWP and SIMD technologies a super high performance computer
could be achieved to enhance the 3-D IC design architecture. Theoretically the N could be a
very large number approaching to the infinite and the system performance could be
tremendously powerful. But in reality the manufacturing encountered the yield and fault
tolerance problem. Redundancy and Interposer technologies could be applied to the circuit
design for enhancing the yield and fault tolerance to make the dream system come true.

Wafers with new function type of cell can be added to the stack randomly to expand the
system because the system can be re-configured by built-in software after stacking. **Fig. 7**
indicated the relative manufacturing cost per system with the size of N x N array related to
the number of stacked wafers in that system will become cheaper as the advancement in
TSV and inter-connection technologies in the next few years to come.

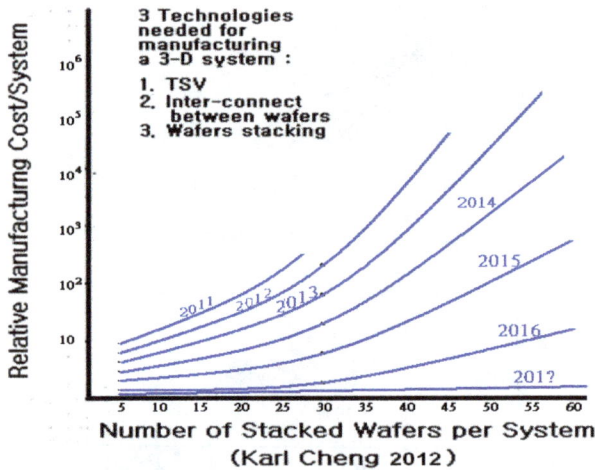

**Figure 7.** Cost of a 3-D system (with N x N array size) versus number of wafers stacked in a 3-D system

### 3.1. 3-D Computer for Image Understanding Application

Image understanding technology has broad application. Especially on Today's LCD display
panel and LED lighting device production, it can be applied to test and evaluate the quality
of the display panel and lighting products. For the industrial applications such as bin
picking, assembly and inspection, image understanding technology can be used to support
those applications at real-time rates. Another area of applicability is that of automatic
interpretation of aerial or satellite imagery for use in GPS or computer aided cartography.

IC design for machine vision and image understanding is one of the most computationally intensive domains of artificial intelligence research. It requires that an interpretation of the changing scene be updated with every new video frame, once every thirtieth of a second. With three quarters of a million color-intensity data values which comprise the picture elements or pixels of the image, performing a single operation on each of these pixels requires an execution rate of about twenty-three million instructions per second for just to keep up with the input. Of course far more than one operation per pixel is necessary. The image understanding is not limited to the data processing in the visible spectrum; it should have more applicable to computer aided systems for image signal processing such as avionics flight control, synthetic array radar and sonar systems etc... The bandwidth requirement of these systems is even higher. In order to achieve these enormous bandwidths a 3-D IC parallel processing design architecture and its algorithm will be discussed. With the advancement of the high density packaging and the level of 3-D IC integration technologies available today, it maybe the best approach to achieve these bandwidths by implementing the SIMD (Single Instruction Multiple Data) and the BSWP (Bit Serial Word Parallel) processing technologies into the 3-D IC circuitry. The SIMD and the BSWP will be discussed and analyzed as approaches to achieving these enormous bandwidths requirement for machine vision and image understanding.

The era of cloud computing is near. The advantage of the cloud computing technology is the tremendous large instant data base available for public. How to handle this large amount of instant data into your local need is one of the most challenges of this new era. In order to quickly manipulating the instant data obtained from the cloud computing to meet your instant need, a local smart terminal computer is required to take the advantage of the cloud computing. 3-D data processing architecture is believed to be the best candidate for this smart terminal computing requirement.

## 4. SIMD and BSWP technologies

With the advancement of the TSV (Through Silicon Via), wafer to wafer interconnection and stacking technologies in 3-D IC manufacturing, it is possible to produce the 3-D smart computer to meet the requirement of the cloud computing at relatively easy and low cost. What will be the different between the 2-D and 3-D computer is that the processing algorithm and the circuit design architecture. With the implementation of SIMD (Single Instruction Multiple Data) and BSWP (Bit Serial Word Parallel) processing technologies, a 3-D smart computer can be designed to execute the data signal paralleling while the conventional 2-D is sequentially. Therefore, the power of the 3-D signal processing could be tremendously better than that of the 2-D version. Although the TSV and the micro-bump technologies developed for stacking the silicon chips or wafers are quite advance now, but the vertical interconnection or wiring density is still significantly lower than in the other two dimensions. Thus, it is still important to partition the system to make best use of the limited vertical dimension in the 3-D IC circuit design.

The MEMS technology applied to form the vertical connection, i.e. TSV and wafer to wafer interconnection, is still in the micron geometrics while the 2-D surface wiring is now in the nanometer technology. There are a million orders of magnitude different between the vertical and horizontal wiring connection. With the limited bandwidth for the vertical interconnection as compared with the other two dimensions how to optimize the 3-D IC circuit design architecture becomes the most popular issue. .

DRIE (Deep Reactive Ion Etcher) or Bosch process is the most popular one used in processing the TSV. The aspect ratio can be reached to the maximum of 13:1 so far. For 3-D stacked chip or wafer the thinnest silicon chip or wafer can be archived now are about 50 um. With the aspect ratio of 13:1 the smallest diameter of through silicon hole is about 3.85um in diameter. It is truly a waste on the wafer surface if this 3.85 um diameter area has to be taken by the TSV. Therefore, the less number of vertical interconnections the more circuit could be deposited on the wafer surface.

*Take for example*; a 2-D conventional microprocessor is formed by 5 processing cells as shown in the **Fig. 3**. The interconnections between the cells are mostly data bus and a few control signals. In order to save the physical silicon space on each chip level after stacking them up to form a 3-D structure, a single wire connection is designed to connect the 5 cells vertically thru each of the 5 chip levels. These 5 processing cells each on 5 different levels are stacked up vertically to form a 3-D microprocessor as shown in **Fig. 4**. The vertical connection is limited to one single wire in order to save the silicon space on each of the silicon chip level. Communication between levels is now can only go thru the single vertical connection wire. In order for the data and a few control signals to go thru the single channel a special shift register circuit must be installed on each cell of the 5 levels. So that signals from a cell can pass thru the single vertical channel to the other cell on the different level sequentially. If the data or the control signal is a 16 bit wide word then the 16 bits will be passed one by one consecutively thru the vertical interconnection wire. By the same token the receiving cell will need to have the similar shift registers to get it back to the original data and signals.

If a system contains more than a single processor, i.e. a multiple N*N microprocessors system, then it can be expanded to an N x N multiple processors system on 5 full size wafers stacked vertically (wafer scale) as shown in the **Fig. 5**. The transfer of the data signals of the system can be performed simultaneously on the all N x N cells on the same level to the other level of N x N cells correspondingly. After the data been transferred to the designated level then all the data signals in the N x N cells will be performed and executed paralleling in the same time on the designated level. This bit by bit way of transferring the data signals sequentially between the wafer levels and executing all the N x N cells data signals on the designated level in a parallel fashion is called BSWP technology. Because an adaptive multi-microprocessors message passing network makes better use of limited bandwidth than any other communication structure we know of, the natural vertical partition in the packaging is likely to match one dimension of the message network.

Based on the above criteria it is suggested that the BSWP (bit serial word parallel) and the SIMD (single instruction multiple data) technologies applied to the 3-D circuit design, together with the circuit partition technique for less interconnection may be a more desired approach for the design of a better circuit performance in regards to limited vertical interconnection bandwidth.

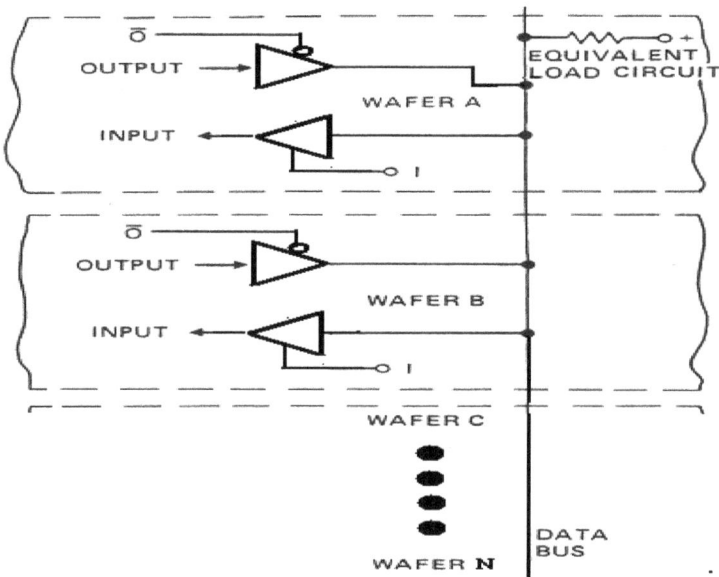

**Figure 8.** Tri-state I/O Buffer

Up to this point questions will be raised for how the cells on a level can transfer their data signals to the designated level or levels; and how the cells on a level or levels can receive data signals from other level without interfering. This is done by the so called wired-AND scheme or tri-state I/O buffer circuit connected to the single vertical interconnection wire as shown in the **Fig. 8**. Supposed data signals on level 1 are going to transfer to level 3 and/or 4. The path between level 1 and level 3 and/or 4 will be open for transfer data signals while other levels will be in tri-state or "Z" state. Sometime the tri-state or "Z" state is referred to electrical potentially "floating" state because it is in the high impedance state condition.

## 5. Example of image understanding by implementing BSWP and SIMD

3 types of cells array are needed for this example and all the cells data are assumed to be the 16-bit shift register with 2 bytes word. Each single cell is a simple logic circuit with only few hundred gates count and all N x N cells on a wafer level has the identical circuit. This will simplify the circuit design tremendously. Type 1 as shown in **Fig. 9** is the N x N cells array data that can be shifted right and left, up and down and wraparound and is named SHIFTER, which in this example is the B wafer level as shown in **Fig. 10**. Type 2 is the N x N cells array data that can do the "XNOR" operation and is named COMPARATOR, which in this example is the C wafer level as shown in Fig. 10. The data of this COMPARATOR can only communicate vertically. Type 3 is the N x N cells array data that can be loaded and unloaded to and from the I/O bus and is named DATA I/O, which in this example is the D wafer level as shown in **Fig. 10**. The rest of the wafers levels are data base for many kinds of

images. Let's assumed DATA-1 is the image of a ball; DATA-2 is the image of a bird and DATA-3 is the image of an airplane etc... All the cells array data can be communicated vertically of course. Fig. 10 shows a simple system for this example which contained all 3 types of wafer levels and the images data base. In order to have better picture of how the image understanding will work, take for an example as below:

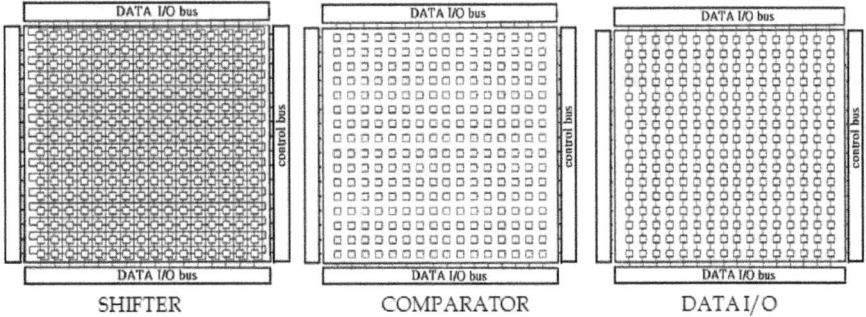

Figure 9. 3 Types of Cells Array

Assuming the pilot of a commercial airline saw a spot far ahead. It could be an airplane heading toward him. Supposed the pilot has an image understanding equipment similar to the one shown in **Fig. 10**. The spot (image) picture was taken by a digital camera that connected to the system. The first thing will be done is to center the spot and magnify (or zoom) it to the suitable size for image processing: Assuming the image is already at the center and zoomed to the right size by the digital camera, and then the following simple instructions program will do the target recognition:

| MOVE D TO B | step 1 |
|---|---|
| SHIFT B RIGHT-LEFT | step 2 |
| SHIFT B UP-DOWN | step 3 |
| MOVE B TO C | step 4 |
| FOR i=1 TO 3 | step 5 |
| MOVE DATA-i TO C | step 6 |
| PERFORM COMPARE | step 7 |
| IF C = "1" THEN | step 8 |
| PRINT "IT'S AN AIRPLANE" | step 9 |
| END ELSE NEXT i | step 10 |

Table 2. 10 single instructions

The above 10 single instructions as shown in **Table 2**, will do:

**Step 1** Image data has been loaded to D level from I/O Bus. Therefore this step is to move the image data in D level to B level (because the B wafer level can do right-left, up-down and wraparound shifting)

**Step 2, 3** will do the fine adjustment for the image to be right on the center with the right size.

**Step 4** then move the image on B to C ready to be compared

**Step 5** indicates the program will do at most 3 times of comparison because system has only 3 images data as exampled.

**Step 6** will move the system image data from DATA-1 to C level where C will perform the "XNOR" operation as shown on top of Fig. 11.

**Step 7** is ordering the C level to perform the "XNOR" manipulation as shown in the Fig. 11 (1), (2) and (3).

**Step 8** is to compare the resultant data whether it has all "1" in the N x N array as shown in the Fig. 11 (1), (2) and (3).

**Step 9** is the end of the program if all the cells data in the N x N array are "1" then the image is understood to be Data-3 which is an airplane as shown in the Fig. 11 (3) results marked with red.

**Step 10** is for continuing the manipulation if the image is still not recognized.

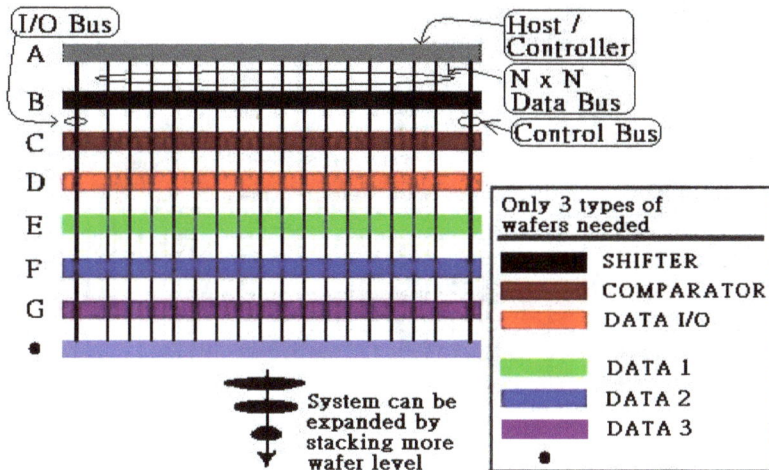

**Figure 10.** Architecture of 3-D Processor System

The example shown above imply the BSWP and SIMD technologies been applied to accomplish the mission at lightening speed. The single instruction contained "MOVE" in step 1, 4 and 6 for moving the N x N data from the wafer level to the destination level are all manipulated by the BSWP technology which the 2-D computer system can hardly do. The single instruction contained "SHIFT" and "PERFORM" are actually manipulating the entire N x N data in the same time which again the 2-D computer system can never do. In the 2-D system, the pixel of the entire N x N data is performed in sequentially while in our 3-D system is paralleling. Therefore, the processing speed is much faster for 3-D system than that of 2-D version.

# COMPARATOR performs the XNOR operation as below:

Rules:

| A | ⊕ | B | = | C |
|---|---|---|---|---|
| ● | ⊕ | ● | = | 1 |
| ● | ⊕ | ● | = | 1 |
| ● | ⊕ | ● | = | 0 |
| ● | ⊕ | ● | = | 0 |

Exclusive NOR ==> XNOR ==> $\overline{A \oplus B}$

**(1) Image of DATA-1 compared with detected mage:**

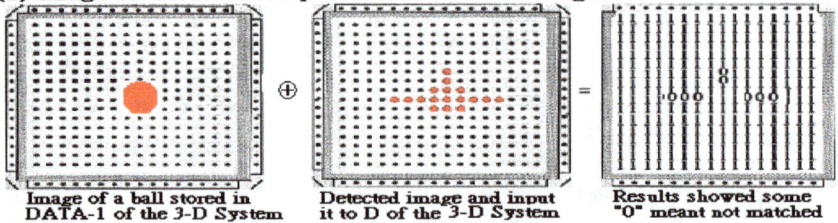

Image of a ball stored in DATA-1 of the 3-D System  |  Detected image and input it to D of the 3-D System  |  Results showed some "0" meant not matched

**(2) Image of DATA-2 compared with detected mage:**

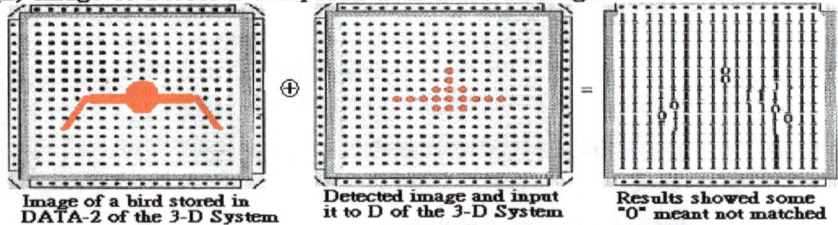

Image of a bird stored in DATA-2 of the 3-D System  |  Detected image and input it to D of the 3-D System  |  Results showed some "0" meant not matched

**(3) Image of DATA-3 compared with detected mage:**

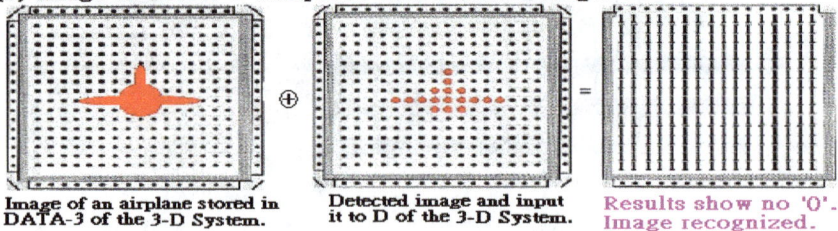

Image of an airplane stored in DATA-3 of the 3-D System.  |  Detected image and input it to D of the 3-D System.  |  Results show no '0'. Image recognized.

**Figure 11.** Performing the comparison on the 3 image data

## 6. Conclusion

From the above study, it is concluded that:

- The 3-D computing system could be custom designed for broad applications with its high performance signal processing power. It will meet the challenges of the coming Cloud Computing.
- The 3-D computing system as described in the above application example can be expanded easily by adding more wafer levels and the size of the N x N array are unlimited. This will make our dream system come true as the advancement of the 3-D computer manufacturing technology progressed.

- The power consumption of the 3-D computer system is about one tenth (1/10) to one seventh (1/7) of that of the 2-D computer system counterpart based on the lab experimental results. The 3-D will be qualified for environmental issue and meet the global warming limits.

It is also assessed that once 2-D IC is swiched to 3-D IC pertaining to all the electronic domestic appliances and or equipments, 3 Dimensional Architecture Integrated Circuits will cut down the residential power requirements from 15% to 2.1% to 1.5% and would help in reduction of power requirement substantially. Apperently 600 megaton's of $CO_2$ emissions per year could be saved on the basis of current data of EIA. It would thus protect our environmental issues to some extent by using less carbon and meet the global warming regulation.

## Author details

Karl Cheng
*Chairman, Innotest Inc., Science Park, Hsinchu, Taiwan*

Bharat Raj Singh
*Director(R&D),School of Management Sciences,Lucknow, India*

Alan Cheng
*Senior Scientist, Innotest Inc., Science Park, Hsinchu, Taiwan*

## Acknowledgement

Authors express their thanks to Ms. Daria Nahtigal, Commissioning Editor, Editor Care and Support Department, for her continued support to get this paper in final shape.

## 7. References

Boguslaw Cyganek, J. Paul Siebert, An Introduction to 3D Computer Vision Techniques and Algorithms, ISBN: 978-0-470-01704-3, pages 504, January 2009,Wiley-eBook
Cheng, K., 2011, A 3-D Super Computer Architecture, 2011 International Symposium, 5-6 Sept. 2011, pp 1 – 31.
Latest industry research & analysis, The power sector energy forecast. Weblink: http://www.confusedaboutenergy.co.uk/index.php/world-energy-issues/energy-consumption-worldwide
Little, M.J., Hughes Res. Lab., Malibu, CA, Etchells, R.D. ; Grinberg, J. ; Laub, S.P. ; Nash, J.G. ; Yung, M.W. 1989, The 3-D Computer, Proceedings, [1st] International Conference, IEEE, 3-5 Jan 1989, Page(s): 55 – 64.
Massoud Pedram, 2009, Green Computing: Reducing the Carbon Footprint of Information Processing Systems in an Energy-Constrained World, Proceeding ICCCN2009, March 08, 2009.

Michaelbluejay, 2012, PC Power Consumption, Article: How much electricity do computers use?; March 2012, Weblink: http://michaelbluejay.com/electricity/computers.html

Noah Horowitz, 2011, Overview of World-Wide Energy Consumption of Consumer Electronics and Energy Savings Opportunities, Natural Resources Defense Council (NRDC), March 2011.

Orion Jones, Moore's Law Maintained by 3D Computer Chip, Article: Shutterstock.com, April 26, 2012.

Prediction of energy consumption world-wide How much energy will we consume in the future? Weblink: http://timeforchange.org/prediction-of-energy-consumption

Sarah Perrin, Jumpstarting computers with 3D chips, source: Mediacom, January 25, 2012.

Singh, A. D., 1985, An area efficient redundancy scheme for wafer scale processor arrays, Proc. of IEEE Int'l Conf. on Computer Design, 1985.

Toborg, S.T., 1990, A 3-D wafer scale architecture for early vision processing, Proceedings of the International Conference, IEEE, Hughes Res. Lab., Malibu, CA,  , 5-7 Sep 1990, Page(s): 247 – 258.

# Chapter 12

# Alternative Resources for Renewable Energy: Piezoelectric and Photovoltaic Smart Structures

D. Vatansever, E. Siores and T. Shah

Additional information is available at the end of the chapter

## 1. Introduction

Energy harvesting is the process of extracting, converting and storing energy from the environment that can also be described as a response of smart materials when they are subjected to an external stimulus such as pressure, vibrations, motion and temperature emanating from wind, rain, waves, tides, light and so on. The efficiency of devices in capturing trace amounts of energy from the environment and transforming it into electrical energy has increased with the development of new materials and techniques. This has sparked interest in the engineering community to establish more and more applications that utilize energy harvesting technologies for power generation.

Some of the energy harvesting systems which use different sources to generate electrical energy and their efficiencies are given below; [1]

- mechanical energy into electricity-generators (20-70% efficiency), piezoelectric systems (0,5-15% efficiency)
- chemical into electricity; fuel cells (25-35% efficiency), primary batteries, rechargeable batteries
- heat/cold into electricity; seebeck-elements (2-5% efficiency)
- electromagnetic radiation into electricity; photovoltaic systems.

Piezoelectric effect is a unique property that allows materials to convert mechanical energy to electrical energy and conversely, electrical energy to mechanical energy. The stimuli for piezoelectric materials can be human walking, wind, rain, tide and wave etc. This effect can be an inherent property of the material or it can be imparted to an existing non-piezoelectric material. However, not every material can be made piezoelectric, only certain ceramics and polymers have the ability to become piezoelectric. Therefore, the chapter will contain fundamentals of piezoelectric effect, a historical review on piezoelectric energy harvesting and recent developments such as flexible piezoelectric fibres which can be integrated or embedded into flexible structures.

Since the sun is the most abundant renewable energy source in the world and the solar energy the earth receives in an hour is greater than the energy consumed in a year. This makes the photovoltaic (solar) materials one of the most significant alternative energy harvesters. This chapter will contain the statistics for solar cell production in EU countries between 2000 and 2010 and also the electricity generated by photovoltaic cells in Europe in 2010 will also be highlighted. The fundamentals of photovoltaic materials and different cell types such as organic, inorganic, dye-sensitized and tandem will be reviewed in this chapter. The chapter will also contain an historical review on the photovoltaic energy harvesters, their efficiencies and the most recent developments are included.

## 2. Piezoelectric energy harvesting

One of the most widely used smart materials is piezoelectric materials because of their wide band width, fast electro mechanical response, relatively low power requirements and high generative forces. Figure 1 presents a market review on piezoelectric materials corresponding to their applications and market share (%) in 2007.

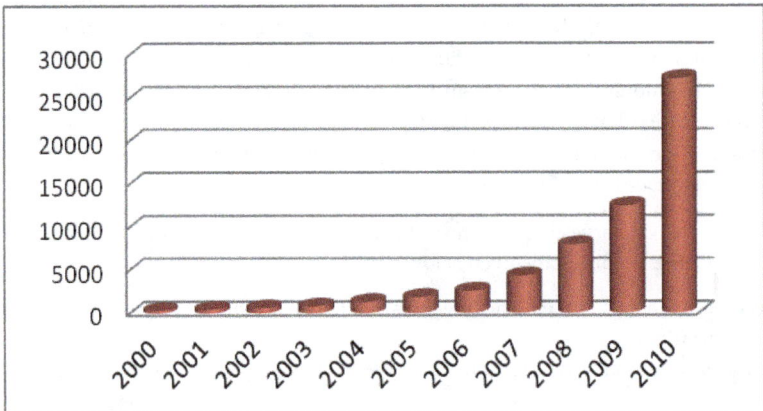

**Figure 1.** Piezoelectric devices market share overview on applications [2]

As it can be seen in Figure 1, information technology/robots is the leader of the market with 31.7% global market share while acoustic devices and resonators have the lowest share in the market with 3.1%. The others in the global market between these two applications can be given from high market share to low; semiconductor manufacturing and precision machines (18.6%), sonar (12.5%), bio/medical (11.1%), ecology and energy harvesting (7%), accelerators and sensors (5.8%), non-destructive testing (5.7%) and miscellaneous which includes gas igniters, piezo printing heads and telecommunication devices (4.5%). It has been reported by Innovative Research and Products (iRAP) Inc. that the global market for piezoelectric devices equals to US$10.6 billion and a high growth is expected over a 5-year period and to reach a value of US$19.5 billion by 2012.

Energy harvesting applications for piezoelectric devices is less than 10% however it can change dramatically if the importance of piezoelectric materials is recognised for alternative energy from nature with zero carbon foot print.

Piezoelectric behaviour was first found in some crystals. According to historical reviews on piezoelectricity [3-4] Charles Coloumb was the first person who theorized in 1817 that electricity may be produced by the application of pressure to certain types of materials. However, it was only a notion until the actual discovery of the "direct-piezoelectric phenomenon" on quartz by Pierre and Jacque Curie [5]. They placed weights on the crystals and detected some charges on the surface and also observed that the magnitude of detected charge was proportional to the applied weight.

Lippmann [6] predicted that if a material could generate electrical charge when a is pressure applied, the reverse effect may be possible so that a mechanical strain could be developed when an electrical charge is applied and this notion was then supported by Curie brothers' experimental results [7]. These two domains had been known as "direct pressure-electric effect" and "converse pressure-electric effect" until Hankel [3] suggested the name "piezoelectricity". Piezoelectricity comes from the Greek words "piezo" and "electricity" that the word "piezo" is a derivative of a Greek word which means "to press" and "electricity" has the same meaning as English word "electricity".

Piezoelectric effect exists in two domains; namely, direct piezoelectric effect and converse piezoelectric effect. Direct piezoelectric effect describes the ability to convert mechanical energy to electrical energy which is also known as generator or transducer effect while the converse piezoelectric effect describes the ability of transforming electrical energy to mechanical energy which is also known as motor/actuator effect. The electrical energy generated by direct piezoelectric effect can be stored to power electronic devices and it is known as "energy/power harvesting".

Piezoelectric materials are member of ferroelectrics so that the molecular structure is oriented such that the material exhibits a local charge separation, known as electric dipole. Electric dipoles in the artificial piezoelectric materials composition are randomly oriented, so the material does not exhibit the piezoelectric effect. However, the electric dipoles reorient themselves when a strong electrical field is applied as shown in Figure 2.

The orientation is dependent on the applied electrical field which is known as poling. Once the electric field is extinguished, the dipoles maintain their orientation and the material then exhibit the piezoelectric effect so that an electrical voltage can be recovered along any surface of the material when the material is subjected to a mechanical stress [8]. However, the alignment of the dipole moments may not be perfectly straight because each domain may have several allowed directions. The piezoelectric property gained is stable unless the material is heated to or above its Curie temperature ($T_c$). However, it can be cancelled by the application of an electric field that is opposite to the direction of the material.

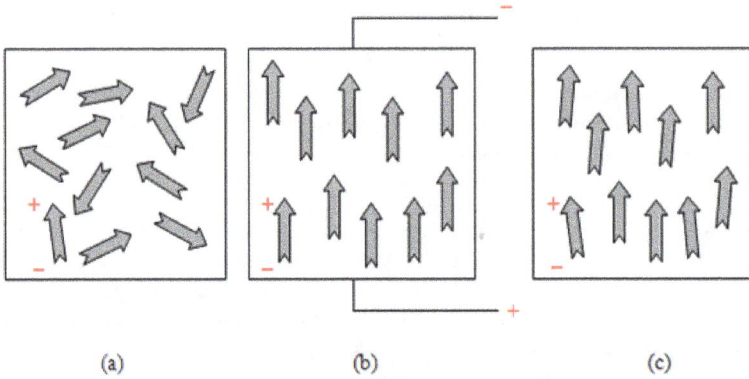

(a)                               (b)                               (c)

**Figure 2.** Orientation of dipoles by polarization, (a) random orientation of polar domains, (b) application of high DC electric field (polarization), (c) remnant polarization after the electric field is extinguished.

According to the definition of "direct piezoelectric effect", when a mechanical strain is applied to crystals by an external stress, an electric charge occurs on the surface(s) of the crystal and the polarity of this observed electric charge on the surface(s) can be reversed by reversing the direction of the mechanical strain applied as shown in Figure 3.

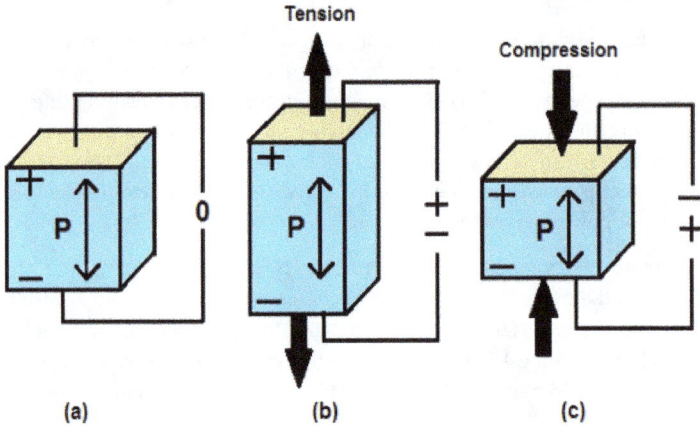

(a)                               (b)                               (c)

**Figure 3.** Schematic of direct piezoelectric effect; (a) piezoelectric material, (b) energy generation under tension, (c) energy generation under compression

On the other hand, according to the definition of "converse piezoelectric effect", when an electric field is applied to a crystal or a crystal is subjected to an electric field, a mechanical deformation on the surface is observed which is generally seen as a change in dimensions of the crystal. The direction of the mechanical strain can also be reversed as shown in Figure 4, by reversing the applied electric field.

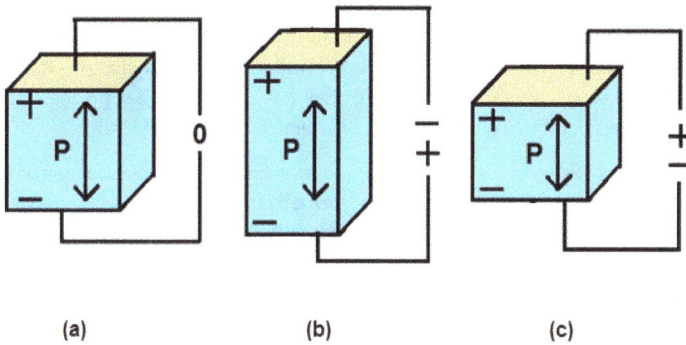

(a)                    (b)                    (c)

**Figure 4.** Schematic of converse piezoelectric effect; (a) piezoelectric material, (b) dimensional change when an electrical charge applied, (c) dimensional change when an opposite electrical charge applied.

Piezoelectricity can be seen in different structures;

- Naturally occurring biological piezoelectric materials
  - Wood [9-10]
  - Bone and tendon [11-12]
  - Keratin, silk [13-14]
  - Enamel [15]
  - Myosin [16]
  - Deoxyribonucleic acid (DNA) [17]
  - Ribonucleic acid (RNA) [18]
- Naturally occurring piezoelectric crystals
  - Quartz [5]
  - Rachell salt
  - Tourmaline
- Man made piezoelectric ceramics
  - Barium titanate – $BaTiO_3$ [19]
  - Lead titanate – $PbTiO_3$ [20]
  - Lead zirconate titanate – $Pb(Zr,Ti)O_3$ – PZT [21-23]
  - Potasium niobate – $KnbO_3$ [24]
  - Lithium niobate – $LiNbO_3$ [25-26]
  - Lithium tantanate – $LiTaO_3$ [26]
- Man made piezoelectric polymers
  - Polyvinylidene fluoride – PVDF [27]
  - Polyparaxylene
  - poly-bischloromethyuloxetane
  - Aromatic polyamides
  - Polysulfone
  - Polyvinyl fluoride
  - Synthetic polypeptide

Polymeric materials can be produced as large thin sheets and then can be cut or stamped into nearly any shape. They also exhibit high mechanical strength and high impact resistance when compared to ceramic materials. Although the piezoelectric charge constant of polymers are lower than that of ceramics, they have much higher piezoelectric voltage constant than that of ceramics which indicates better sensing characteristic.

Polymers consist of two regions; crystalline and amorphous. The percentage of crystalline region in a polymer matrix determines the piezoelectric effect. However, crystallites are dispersed in amorphous region in semi-crystalline polymers as shown in Figure 5.

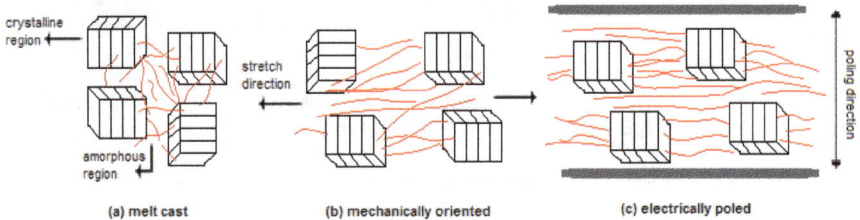

**Figure 5.** Amorphous and crystalline regions in the polymer matrix; from melt cast (a), during mechanical orientation (b) and electrically poling (c) [28]

The melting temperature of a polymer is dependent on the percentage of crystalline region in the polymer while the amorphous region designates the glass transition temperature and mechanical properties of the polymer. As it is seen in Figure 5 crystalline structures and so the molecular dipoles are locked in the amorphous region. Broadhurst et al. [29] studied the molecular and morphological structure of PVDF and its pyroelectric and piezoelectric properties. If a DC voltage is applied across the polymeric piezoelectric material, the material becomes thinner, longer and wider in proportion to the voltage, conversely the film generates a proportional voltage when a mechanical stress is applied either by compression or stretching. The relationship between applied mechanical stress and generated voltage can be defined by stress constants.

## 2.1. Comparing piezoelectric materials

As it can be seen from the Table 1 the piezoelectric constant is lower for polymers as compared to ceramic based piezoelectric materials. Therefore, when the same amount of voltage applied to polymer and ceramic piezoelectric materials, the shape change of ceramic based materials are larger than polymers. Although PVDF has a lower piezoelectric charge coefficient, its piezoelectric voltage coefficient is about 21 times higher than that of PZT and 40 times higher than that of BaTiO$_3$, therefore PVDF is better for sensor applications. Due to being a polymer, PVDF is flexible, light weight, tough, readily manufactured into large areas and can be cut and formed into complex shapes.

The electromechanical coupling constants ($k_{31}$) of PZT is approximately 2.5 times larger than the electromechanical constant of PVDF which means it is able to convert 2.5 times more mechanical stress into electrical energy than that PVDF.

| Property | Units | BaTiO₃ | PZT | PVDF |
|---|---|---|---|---|
| Density | $10^3$ kg/m³ | 5.7 | 7.5 | 1.78 |
| Relative permittivity | $\varepsilon/\varepsilon_0$ | 1,700 | 1,200 | 12 |
| Piezoelectric strain coefficient ($d_{31}$) | $10^{-12}$ C/N | 78 | 110 | 23 |
| Piezoelectric voltage coefficient ($g_{31}$) | $10^{-3}$ Vm/N | 5 | 10 | 216 |
| Pyroelectric voltage coefficient ($P_v$) | V/μm K | 0.05 | 0.03 | 0.47 |
| Electromechanical coupling constant ($k_{31}$) | %@1 kHz | 21 | 30 | 12 |
| Acoustic Impedance | $(10^6)$kg/m²-sec | 30 | 30 | 2.7 |

**Table 1.** Typical properties of commercially available 3 main piezoelectric materials [30]

## 2.2. History and recent developments on piezoelectric energy harvesting

One of the very early studies of energy harvesting by piezoelectric materials was performed in a biological environment by Hausler and Stein [31]. They claimed that a piezoelectric PVDF film and a converter could transform the mechanical energy caused by respiration of a mongrel dog to electrical energy. The piezoelectric material was fixed to the ribs of the dog and a peak voltage of 18V was produced by motions of the ribs during the spontaneous breathing. However, the power generated was about 17μW which was not enough to operate an electronic device.

More than a decade after the study on animals, Starner [32] studied the possibility of energy harvesting from body motions by using piezoelectric materials. He claimed that a human body could be a source for harvestable electric energy. Starner studied different part of the body, such as walking, upper limb motion, finger movements, blood pressure etc., and analysed the possibility of harvestable power from these locations. He claimed that the amount of power lost during walking was about 67W and by mounting a PZT device inside a shoe with an efficiency of 12.5%, up to 8.4W electrical energy could be generated. He also, suggested the possibility of storing the harvested energy by using a capacitor.

Parasitic energy harvesting from walking of a human being to power a radio frequency identification transmitter was studied by Kymissis et al. [33]. They used three different devices which were a thunder actuator consisting of a ceramic based piezoelectric composite material, a rotary magnetic generator and a PVDF stave. Former two structures were integrated into the heel of a shoe to harvest the impact energy while the PVDF stave was integrated into the sole to absorb the bending energy. The researchers constructed a prototype to investigate and compare the energy generation performance of these three different materials. The peak power generated by PZT unimorph structure was 4 times higher than PVDF stave, 80mW and 20mW respectively. However, the peak power generated by the rotary generator was found to be only 0.25mW which was found not to be sufficient to power a radio frequency identification transmitter.

Shenck [34] demonstrated the harvestable power generation from a rigid bimorph piezoceramic transducer, which was integrated into the sole of a shoe. Different regulation systems were evaluated. One of the findings was that the use of a second piezoelectric material leads to more energy generation. Furthermore, it was found that a bimorph transducer was more effective for the application since it was better adapted to various distributions of body weight and footfall velocity. Shenck and Paradiso [35] also studied piezoelectric PVDF and PZT structures embedded in a shoe. A power storage circuit which was designed to power a radio frequency tag was also mounted in a shoe and an offline forward switching DC-DC converter was developed. The experimental results showed that the switching converter harvested energy more efficiently –about twice as much- than the original linear regulator circuit. The whole set-up was successful to power low energy electronic devices since the switching circuit provided continuous power during walking.

Churchill et al. [36] investigated the power harvesting capability of a piezoelectric fibre composite structure consisting of unidirectionally aligned PZT fibres of 250μm diameter embedded in a resin matrix. It was found that 7.5mW of power could be harvested from a piezoelectric fibre composite material - with a length of 130 mm, a width of 13 mm and a thickness of 0.38 mm – when a vibration of 180 Hz was applied. In another work the possibility of power harvesting was performed by Renaud et al. [37].They studied the wrist and arm motions during walking. They found that a spring mass resonant system was not appropriate for energy harvesting from arm since motions caused by arm movements were low in frequency. An analytic model for a non-resonant system was developed and it showed that a maximum power of 40 μW could be generated from the wrist movements during walking.

Granstrom et al. [38] developed a theoretical model of an energy harvesting backpack that can generate electrical energy from flexible piezoelectric PVDF films integrated into the straps. It was found that 45.6mW of power could be generated from a complete backpack with two piezoelectric straps with an efficiency of more than 13%. Swallow et al. [39] developed a micropower generator using micro composite based piezoelectric materials for energy reclamation in glove structures. They developed fibre composite structures by using different fibre diameters embedded between two copper electrodes and both the effect of fibre diameter and the materials thickness were investigated. Their results showed that the composite structure was able to produce a voltage up to 6 volts. Siores and Swallow [40] developed an apparatus for detection and suspension of muscle tremors.

A multi-material piezoelectric fibre production has been reported by Egusa et al. [41] however it was produced by a multi-process method where a copolymer of PVDF, P(VDF-TrFE) and polycarbonates were used, which makes the fibre expensive and difficult to scale up for production. The first flexible piezoelectric fibre has been produced successfully by Siores et al. [42] via a continuous process on a customised melt extruder. This is a cost effective process since the polarisation of the fibre is carried out during the fibre production and the process is easy to scale up for production.

## 3. Photovoltaic effect and photovoltaic energy harvesting

The sun is the most abundant renewable energy source in the World. The solar energy which the Earth receives in an hour is greater than the energy consumed in a year. If we need to present the situation by numbers, the received solar power is about 120,000 Terawatts while the global energy consumption is about 13 Terawatts [43].

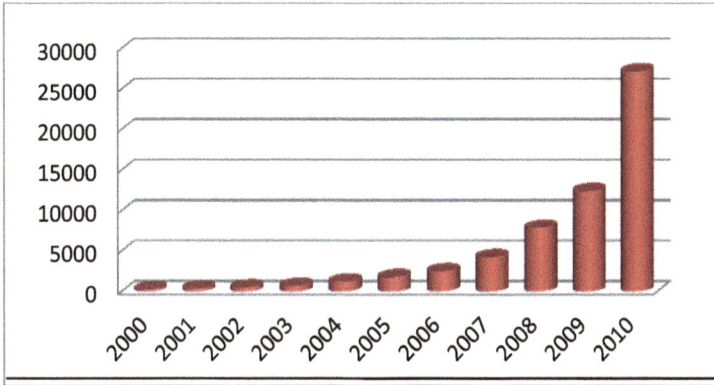

**Figure 6.** Number of solar cell production between 2000 and 2010 in European Countries [44]

The importance of renewable energy generation increases significantly with an increase in global warming, air and water pollution etc. The most of the European countries have started using PV cells for their electrical energy need. Figure 6 clearly shows the dramatic increase in the PV cell production in Europe over a 10-year period. Increasing demand on the solar cell production has shown a steady increase since 2000. This may be a result of the increased awareness of global warming and the need for using environmentally friendly materials and techniques. The number of solar cell production in EU countries was more than doubled in a year between 2009 and 2010.

Figure 7 shows the projected solar power generation in European countries. It is clear that the largest solar power generator is Germany followed by Italy, Czech Republic and France. Although Spain is known as a sunny country, the production of power from solar cells is almost 20 times less than that of Germany. Mostly dull and cloudy countries like Latvia and Estonia pointed as "Rest of the EU" in the figure and United Kingdom have much lower solar power generation and their portion is under 1%.

Photovoltaic effect was first observed by Alexandre-Edmond Becquerel in 1839 when he subjected an AgCl electrode in an electrolyte solution to the light. The word "photo" is a Greek word used for light and "voltaic" named after Alessandro Volta. The beam of sunlight contains photons which may contain different amount of energy related to the different wavelengths of the solar spectrum. When a photovoltaic material is exposed to sunlight, photons may be reflected, absorbed or transmitted. Only the absorbed photons with energy greater than the bandgap energy can generate electricity by causing the breakage of covalent bonds and dislodging of the electrons from the atoms of the cell. The free electrons start moving through the cell and during this movement they create and fill in the cell's vacancies to generate electricity. The ability of materials to absorb photons and convert into electricity is known as *photovoltaic effect* [46-47]. Two fundamental processes of PV effect, light absorption and charge separation, are the basis of all inorganic PV cells.

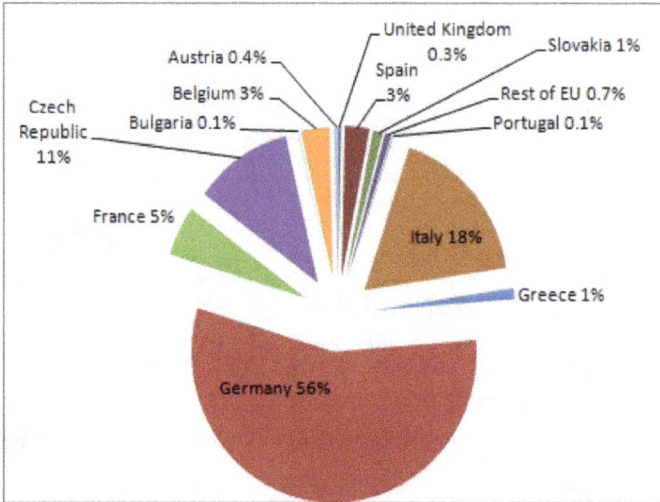

**Figure 7.** PV market share (MW, %) in EU in 2010 and evaluation until 2015 [45]

The proportion of sunlight energy is significant for the conversion efficiency of a PV cell which converts sunlight energy to electrical energy. The efficiency of PV energy is important to make PV energy competitive with more traditional sources of energy, such as fossil fuels. For comparison, the earliest PV devices converted about 1%-2% of sunlight energy into electric energy. Today, it is likely to produce photovoltaic structures made of pure silicon with 24.7% efficiency [48-49] however, due to the rigidity of silicon based solar cells and pursuit of light weight and flexible photovoltaic materials for curved structures, applications are limited. Photovoltaic materials based on conjugated polymers, due to ease of processing, low-cost fabrication, being light weight and flexible, are evolving into a promising alternative to silicon based solar cells [50-51].

## 3.1. Inorganic photovoltaic materials

The best example for inorganic photovoltaic material is silicon. It is the most commonly used material which absorbs light and creates electron-hole pairs. The individual inorganic solar cells are designed with a positive (p-junction) and a negative (n-junction) layer to create an electric field. When n-type layer is doped, the element with an extra electron, generally phosphorous, is used to give a negative charge to the layer. On the other hand, when p-type layer is doped, the element with a less electron, generally boron, is used to give a positive charge to the layer. The place in between these two layers is called p-n cell junction.

Electrons in n-type layer are free and travel through the material to lower energy levels while holes travel to higher energy levels when the photovoltaic cell is exposed to the sunlight. Free electrons jump across the p-n cell junction. These electrons then return to the n-type layer when the two sides of the cell are connected with a wire and this electron flow is known as "the electric current". The Figure 8 clearly presents the layers of an inorganic PV cell and the generation of electric current by flowing electrons.

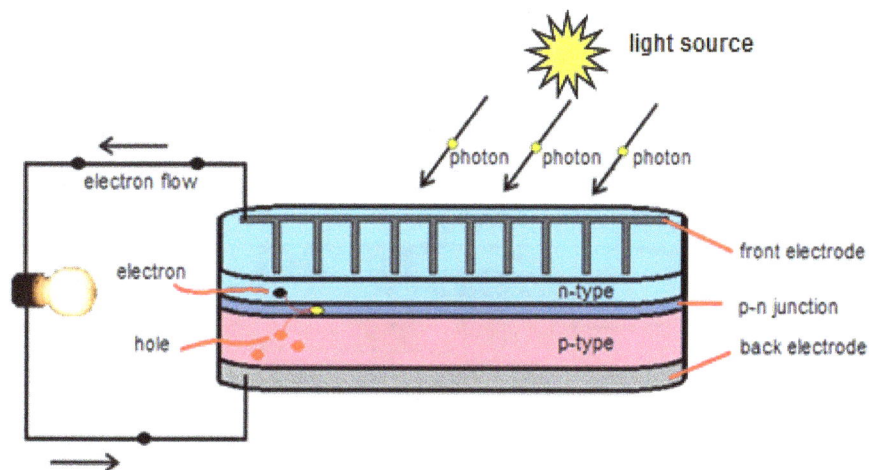

**Figure 8.** Layers and working principle of a silicon solar cell

Crystalline, multi-crystalline, amorphous and microcrystalline silicon, copper indium gallium diselenide (CIGS), the III-V compounds and alloys, CdTe, InP, Cu2Se, WSe2, GaAs etc. are mostly used as inorganic semiconductor materials for PV cells [52-53]. These semiconductor materials, used for inorganic PV cell fabrication, have energy bandgaps within the range of 1.1-1.7 eV which make them desirable due to being near to the optimum energy bandgap of 1.5 eV for PV energy conversion by a single junction solar cell [54]. Many researchers have concentrated on increasing the efficiency and achieving maximum power. Recorded efficiency for a free-standing 50µm thin film monocrystalline silicon solar cell is 17% [55], for 47µm thin film silicon cell is 21.5% [56] and maximum recorded efficiency for inorganic solar cells is 24.7% [57].

## 3.2. Organic photovoltaic materials

Semiconducting polymers with suitable bandgaps, absorption characteristics and physical properties can be used for the fabrication of organic photovoltaic materials. They are cheaper raw materials as compared to silicon based inorganic solar cells and they can also be fabricated by using cheap processing techniques. Photovoltaic effect of organic PV cells is based on electron transfer from donor-type semiconducting conjugated polymers to acceptor-type conjugated polymers or acceptor molecules, such as fullerenes [58]. These materials have donor-acceptor heterojunctions to achieve separation of the electron-hole pairs. Most of the semiconducting polymers are hole-conductors and known as electron donor polymers.

There are six basic operational principles for a polymer solar cell [59-60] as listed below:

1. Coupling of the photons
2. Photon absorption by active layer, $\eta_{abs}$,
3. Electron-hole pair creation (excited state) and diffusion, $\eta_{diff}$,
4. Charge separation, $\eta_{tc}$,
5. Charge transportation within the respective polymer to the respective electrodes, $\eta_{tr}$,
6. Charge collection, $\eta_{cc}$

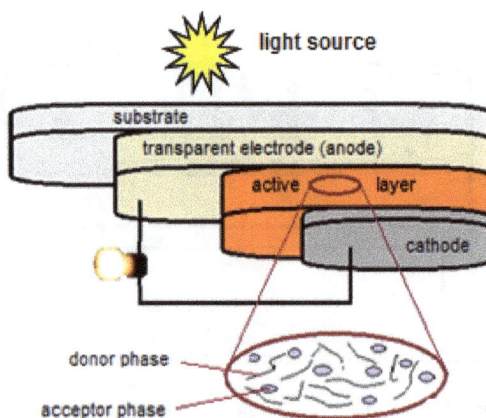

**Figure 9.** Donor-acceptor heterojunction configurations in a typical organic solar cell

When the photoconductive properties of organic polymers was first observed, the most widely studied polymer was poly(vinyl carbazole), PVK [61]. Other suitable electron donor polymers for organic photovoltaics include;

- poly(3-hexylthiophene), P3HT,
- po5ly(3-octylthiophene), (P3OT)
- polyphenylenevinylene, (PPV)
- polyfluorene, (PFO)
- poly[2,7-(9,9-dioctyl-fluorene)-alt-5,5-(4,7'-di-2-thienyl-2',1',3',-benzothiadiazole), (PFO-DBT)
- poly[2-methoxy-5-(2'-ethyl-hexyloxy)-1,4-phenylene vinylene], (MEH-PPV)
- poly[2-methoxy-5-(3,7-dimethyloxy)]-1,4-phenylenevinylene), (MDMO-PPV)
- poly[N-9'-hepta-decanyl-2,7-carbazole-alt-5,5-(4',7'-di-thienyl-2',1',3'-benzothiadiazole, (PCDTBT)

Semiconducting polymers have lower dielectric constant but higher extinction constant than that of inorganic PV materials. To absorb the most incident light about 300nm thickness is enough for a film material [62]. However, the optimized thickness for most polymer solar cells is less than 100nm [63] due to the low carrier mobility.

Electron acceptors with high electron mobility are the most suitable materials for polymer solar cells. Due to exhibiting $1cm^2V^{-1}s^{-1}$ electron mobility [64], ultrafast photo induced charge transfer and derivatives of $C_{60}$ and $C_{70}$ are the best electron acceptors so far. Suitable electron acceptor polymers for organic photovoltaics include;

- 6,6-phenyl-C61-butric acid methyl ester, ($PC_{60}BM$)
- 6,6-phenyl-C71-butric acid methyl ester, ($PC_{70}BM$)
- poly(9,9'-dioctylfluorene-co-bis-N,N'-(4-butylphenyl)-bis-N,N'-phenyl-1,4-phenylenediamine, (F8TB)
- poly-[2-methoxy-5,2'-ethylhexyloxy]-1,4-(1-cyanovinylene)-phenylene,(CN-MEH-PPV)

### 3.3. Dye-sensitized photovoltaic materials

The dye-sensitized solar cells (DSSC or DSC) are thin film photovoltaic materials. They are also known as "the 3rd generation solar cells". The first DSSCs were studied by Gerischer et al in late 1960s who illustrated that organic dyes can generate electricity at oxide electrodes in electrochemical cells [65]. The first actual work on DSSCs was carried out with a chlorophyll sensitized zinc oxide (ZnO) electrode. In this work, photons were converted into an electric current by charge injection of excited dye molecules into a wide bandgap semiconductor for the first time [66]. DSSCs have slightly different working principle than traditional silicon solar cells. Light absorption and charge carrier transport processes are separated in DSSCs. Light is absorbed by a sensitizer, which is affixed to the surface of a wide band semiconductor. Charge separation takes place at the surface via photo-induced electron injection between dye, semiconductor and electrolyte [67-68].

When a DSSC is exposed to the sun light, photons pass through the transparent electrode into the dye (active) layer and excite electrons. Excited electrons move toward the transparent electrode where they are collected. Once an electron completes its travel through the external circuit, it is re-induced into the DSSC on the back electrode and flows into the electrolyte and then it is transported back to the dye molecules [69].

After the discovery of DSSCs in late 1960s and early 1970s, DSSCs have attracted many researchers' attention and a significant number of works have been carried out on suitable transparent electrodes [70-71], and electrolytes [72] but mostly on increasing the efficiency of DSSCs [73-78].

DSSCs have some advantages over inorganic solar cells as given below:

- low cost materials,
- the electron is injected from a dye into $TiO_2$, there is no electron-hole pair,
- DSSCs can work even in low density light conditions which make them possible to be used for some indoor applications,

• DSSCs can operate at lower internal temperatures even in a hot environment.

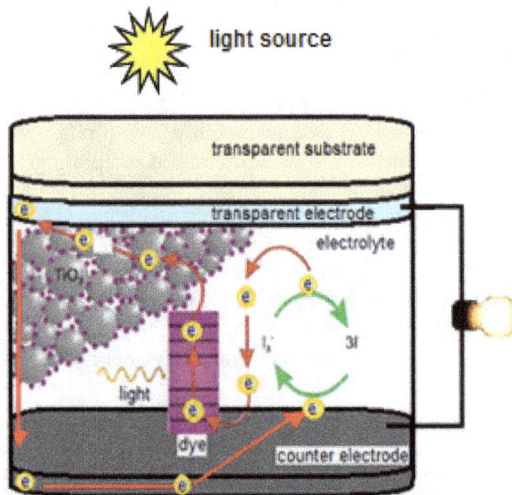

**Figure 10.** Dye-sensitized solar sell structure; transparent electrode coated transparent substrate and over it a TiO$_2$ layer sensitized by a monolayer of adsorbed dye (photo-electrode), electrolyte and counter electrode.

On the other hand, the power conversion efficiency of DSSCs is lower than silicon based inorganic solar cells. There is also a possibility of breakdown of the dye material and leakage of liquid electrolyte.

## 3.4. Tandem cell photovoltaic materials

Tandem solar cells (TSCs) are developed to overcome some drawbacks of conventional solar cells. Each active material used to fabricate a solar cell can only convert certain wavelength of the light to electricity. To achieve better photon absorption efficiency, two or more active materials with different bandgaps are linked to built-up a TSC. Two or more heterojunction solar cells are deposited on top of each other to create a TSC. One of the photo-active materials with a higher bandgap collects photons with higher energy while the other with a lower bandgap absorbs photons with lower energy (Figure 11).

Since solar cells with different bandgaps are used, when the structure is built-up, semiconductor material with a wide band gap is used as the first active layer and semiconductor material with a smaller band gap is used as the second active layer. When the individual cells are connected in series to create a TSC, the open-circuit voltage (V$_{oc}$) of tandem cell is increased to the sum of the V$_{oc}$ of individual cells [79]. The maximum efficiency calculated for a tandem solar cell consisting of 2 sub-cells is 42% with band gaps of 1.9 and 1.0 eV and calculated maximum efficiency for a tandem solar cell consisting of 3 sub-cells is 49% with band gaps of 2.3, 1.4 and 0.8 eV [80]. However, experimental studies on tandem solar cells consisting of GaInP/GaInAs/GaInAs showed only an efficiency of 33.8% [81] and 38.9% [82]. The maximum efficiency calculated for organic tandem solar cells consisting of 2 sub cells is close to 14% [83].

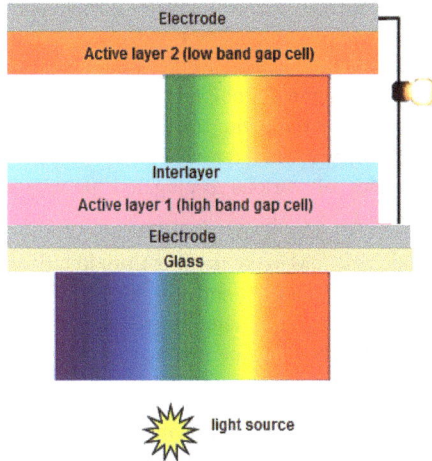

**Figure 11.** Tandem solar cell structure consisting of two photovoltaic cells having different band gaps

## 3.5. Hybrid photovoltaic materials

To combine the unique properties of inorganic semiconductor nanoparticles with organic polymeric materials, both organic and inorganic nanostructures are combined and named as "hybrid solar cell" (HSC). Organic materials absorb light as a donor and transport holes while inorganic materials act as an acceptor to transport electrons. The combination of organic and inorganic photoactive materials provides some advantages over individual organic and inorganic solar cells. The overall cost of the solar material is reduced by using organic thin film technology which is low cost, easy to manufacture and versatile while inorganic nanoparticles add high absorption coefficient and bandgap tenability [84].

The idea of making hybrid solar cells has attracted many researchers who then worked on different concepts of HSC manufacturing by using bulk heterojunction concept with different nanoparticles such as $TiO_2$ [85], PbS [86-87], ZnO [88-89], CdS [90], CdSe [91-92], CdTe [93] and $CuInS_2$ [94]. Although HSCs provide some advantages over inorganic solar cells, such as low cost, reduced thickness (being thin film), easy manufacturing, versatility, tuneable nanoparticle size thus tuneable bandgap etc., the power conversion efficiency of HSCs is still lower than that of silicon based inorganic solar cells.

## 3.6. Brief history and recent developments on photovoltaic energy harvesting

Since the discovery of photovoltaic effect by Becquerel, researchers have studied and worked on various photoactive materials and methods of making photovoltaic cells. The first solar cell was developed at Bell Laboratories [95], which was silicon based inorganic solar cell with power conversion efficiency of 6%. The highest reported power conversion efficiency for inorganic solar cells today is 24.7% [96].

Polymers including poly(sulphur nitride) and polyacetylene were investigated for their photoelectric property in the 1980s. Using a donor and an acceptor material in a cell was a real breakthrough for organic photovoltaics. A donor - acceptor cell may consist of dye - dye, polymer - dye, polymer - polymer or polymer - fullerene blends [97]. Due to having high electron affinity, fullerenes have become the most widely used acceptor materials in organic solar cells and thus polymer - fullerene blends have received a particular interest from researchers. Photophysics of various conjugated polymer/$C_{60}$ blends have been extensively studied and reported [98-104].

MEH-PPV:$C_{60}$ and MDMO-PPV:PCBM were the most predominant active layer materials. However, due to exhibiting large bandgap and low mobility of the PPV type polymers, efficiencies are limited to 3% [105-108]. Therefore, researchers have started to work on different polymers and P3HT has become the most predominant active layer material for OPVs and also its blends with PCBM.

Probably, the starting point of the rapid developments on P3HT:PCBM based OPVs was the work published in 2002 [109]. These researchers investigated the short-circuit current density of P3HT:PCBM based organic solar cells with a weight ratio of 1:3 in active layer. They also recorded that it was the largest short-circuit current density (8.7mAcm$^{-2}$) observed in OPVs at that time.

A number of studies have been carried out to increase the efficiency of P3HT:PCBM cells by thermal annealing [109-116]. It was found that the $V_{oc}$ was usually slightly decreased after annealing process while both the $I_{sc}$ and FF increased significantly [117] and provides optimum charge carrier creation and extraction.

The morphology and the optimization of the weight ratios for donor and acceptor are also important for a desirable performance. Studies showed that morphology of P3HT and PCBM can be modified upon [118-120]. Padinger et al. [121] applied a post-treatment to P3HT:PCBM based solar cell by annealing and applying an external voltage greater than the open-circuit voltage, simultaneously. They reported that the post treatment increased all the parameters, such as $I_{sc}$, $V_{oc}$ and FF, thus the overall efficiency reached 3.5% from 0.4% (without any post treatment).

There have been other approaches to control the morphology of P3HT:PCBM blends. It has been reported by Li et al. [122] that controlling the morphology of P3HT and PCBM in the blend is possible by slow drying. It has also been reported that additives, such as n-hexylthiol, n-octylthiol, or n-dodecylthiol [123], can also contribute to the hole mobility enhancement slightly and charge-carrier lifetime significantly. Another approach to control the morphology was addition of nitrobenzene to P3HT:PCBM solution (in chlorobenzene) that increased the efficiency as high as 4% without thermal annealing [124-125].

The effect of weight ratio of P3HT and PCBM on the power conversion efficiency of OPVs has been extensively studied. Reports from various researchers confirmed each other's work and the optimum weight ratio is considered as 1:1 [122, 126-129]. Table 2-4 shows the improvements in the efficiencies of P3HT:PCBM based organic photovoltaic materials.

One of the most recent approaches is based on the growth of fibres by slow cooling of P3HT solutions [124]. The crystalline fibres are isolated from the amorphous material by centrifugation and filtration and then reformulated in dispersions with PCBM.

The highest PCE reported recently is just than 6% for OPV [130]. Researchers used a co-polymer,poly[$N$-9''-hepta-decanyl-2,7-carbazole-alt-5,5-(4',7'-di-2-thienyl-2',1',3'-benzothiad iazole) (PCDTBT) as an electron donor material with the fullerene derivative [6,6]-phenyl $C_{70}$-butyric acid methyl ester (PC$_{70}$BM) as an acceptor to fabricate a bulk heterojunction solar cell. They also investigated that the internal quantum efficiency was close to 100% so that essentially every absorbed photon resulted in a separated pair of charge carriers which were collected at the electrodes. Another group of researchers have also reported development of a simple solar cell based on a mixture of fluorinated PTB4 and PC$_{61}$BM with higher than 6% power conversion efficiency [131]. The best life time recorded by Konarka is more than a year for polymer based PV [132].

### 3.7. Photovoltaic fibre attempts

There are also a significant number of approaches to produce solar cells in fibre form. However, Konarka Technologies, Inc. was the first one who announced and patented the idea of producing a flexible photovoltaic fibre via a continuous process in 2005 [133]. They have used an electrically conductive fibre core which passes through a titania (TiO$_2$) suspension and thus coated with the interconnected nanoparticles. The interconnected nanoparticle coated fibre is dried and passed through a dye solution and dried again. The dried fibre is then passed through a polymeric electrolyte and thus coated with the transparent electrode.

Kuraseko et al [134] reported flexible fibre-type poly-Si solar cell. Glass fibre was used in the core of the fibre like photovoltaic structure and p-type poly-Si and n-type poly-Si was deposited onto the core. They studied two different methods; atmospheric thermal CVC and microwave PECVD and the top (TCO) and bottom (metal) electrodes were deposited by thermal evaporation technique. There are also more recent works on the design of OPV based fibres [135-137] and DSSC based fibres [138-140].

# 4. A new approach to energy harvesting by Hybrid Piezoelectric-Photovoltaic (HPP) materials

Renewable energy sources are endless but not available at all times at a given location. For instance, the electrical energy generation by a photovoltaic material is dependent on the light density and the number of photons absorbed by the photoactive layer. If the solar radiation is scarce in a region, for example on a cloudy day, the electrical energy generation will be affected. If flexible solar cells are coupled with flexible piezoelectric materials in a combined structure, then the hybrid structure can generate energy from solar radiation as well as mechanical energy, such as wind, rainfall, waves etc.

A novel technology has been developed by Siores et al. [141] that integrates piezoelectric polymer substrate and photovoltaic coating system to create a film or a fibre structure (Figure 12) which is able to transform both mechanical energy (by using the piezoelectric part) and light energy (by using organic photovoltaic part). Since the organic photovoltaic material system is made in a normal atmospheric environment and the usage of ITO is eliminated, the cost associated with the whole structure is manifold less than silicon based photovoltaic. The resultant material system is flexible and can be incorporated in textiles for a wide variety of applications, under different environments on earth, underwater and possibly space.

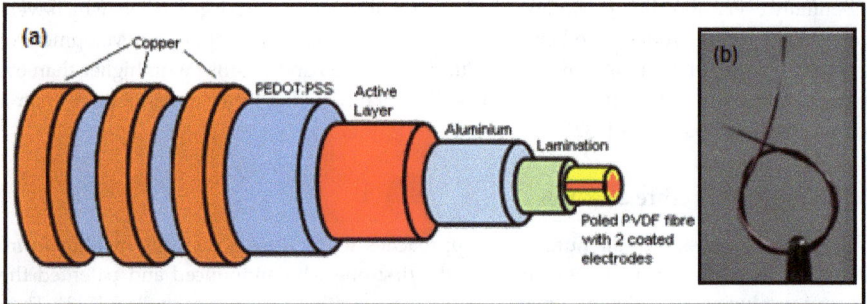

**Figure 12.** Sketch (a) and photograph (b) of hybrid fibre: OPV cell layers developed onto Al evaporated piezoelectric fibre

The HPP materials are able to produce electrical energy from the environment and provide almost uninterrupted energy generation to power small electronic devices. The flexible HPP structure can be part of any material such as sail, window curtain, tent etc. to generate renewable energy even in the absence of sunlight. One possible configuration for land-based applications of hybrid fibre is a pine tree like structure where the needles are made of HPP fibres. Such a structure may replace the conventional photovoltaic parks that require large panels and sun tracking devices to operate. The surface area that fibres provide is substantially more compared to the solar panels, thus they may be able to generate more energy in a confined area. The tree structure also costs less to manufacture and can harvest energy not only through the photovoltaic but also through the piezoelectric material. Furthermore, the aesthetic aspects of parks incorporating them cannot be overstated. Once flexible fibres are incorporated in textile structures, a plethora of opportunities exist, limited only by the imagination.

Since the HPP structures produce combined piezoelectric technology which converts mechanical energy to fluctuating electrical energy (AC) and organic photovoltaic technology which converts solar energy to constant electrical energy (DC), an associated rectifying circuit consisting of 4 diodes and a capacitor can be used to rectify the fluctuating voltage of various frequencies to a constant DC voltage. The constant voltage generated and rectified can then be either stored in an electrical storage device such as batteries and super capacitors or can be utilised on-line directly.

## 5. Conclusions

The term "global warming" has been highlighted more and more every day since it is considered as one of the biggest dangers to life on earth. It is a fact that one of the factors which cause global warming is high carbon emission. Growing population and the increasing technology consumerism contribute to the enhanced usage of energy from coal, oil, electricity etc. However, sooner or later the mankind is anticipated to run out of the coal and oil reserves since they are finite and are not renewable. Energy harvesting properties of both piezoelectric and photovoltaic materials have been known for a long period of time however recently more attention has been paid to produce usable materials for energy generation in the form of electricity to decrease carbon foot print.

Piezoelectric materials can convert almost any kind of mechanical energy to electrical energy. The most suitable piezoelectric material is chosen for a particular application depending on the properties needed. Thus, the maximum energy output, with minimum carbon emission, can be provided to power an electronic device on-line or to be stored. Photovoltaic materials use the biggest energy source to generate green energy and many countries, including Germany and Italy, are well aware of the advantages of using green energy. Furthermore, the increasing solar cell production in general is considerably promising for a cleaner world. Hybrid photovoltaic and piezoelectric structures are capable of converting photons to electrical energy by using photovoltaic part and mechanical energy to electrical energy by using piezoelectric part, in the presence of rain, wind etc, where there is not enough sunlight for photo-conversion. The advantages of the hybrid photovoltaic/piezoelectric materials are their flexibility, light weight, low production cost and the possibility of almost undisturbed energy generation from nature, such as sunlight, wind, rain and other mechanical resources.

The smart materials discussed in this chapter are responsive to many natural resources for green energy generation. The increase in the use of alternative resources for renewable energy can substantially decrease carbon foot print and consequently the effects of global warming.

## Author details

D. Vatansever, E. Siores and T. Shah
*University of Bolton, Institute for Materials Research and Innovation, United Kingdom*

## 6. References

[1] Flipsen, S.P.J. 2005, "Alternative Power Sources for Portables & Wearables, Part 1: Power Generation & Part 2: Energy Storage", *Delft University of Technology*, ISBN: 90 5155-021-9 / 90-5155-022-7.

[2]  IRAP (Innovative Research and Product) 2008, Piezoelectric Ceramic, Polymer and Ceramic/Polymer Composite Devices – Types, materials, Applications, New Developments, Industry Structure and Global Markets.

[3]  Cady, W.G. 1946, "Piezoelectricity", *New York: McGraw-Hill*, pp. 699.

[4]  Ballato, A. 1996, "Piezoelectricity: history and new thrusts", *IEEE Proceeding of Ultrasonic Symposium*, vol. 13, pp. 575-583.

[5]  Curie, J. and Curie, P. 1880, "Development par compression de l'electricite polaire dans les cristaux hemiedres a faces inclinees", *Compt. Rend*, vol. 91, pp. 383-386.

[6]  Lippmann, G. 1881, "Sur le principe de la conversation de l'electricite", *Compt. Rend.*, vol. 92, pp. 1149-1152.

[7]  Curie, J. and Curie, P. 1881, "Contractions et dilations produites par des tensions electriques dans les cristaux hemiedres a faces inclinees", *Compt. Rend.*, vol. 93, pp. 1137-1140.

[8]  Minazara, E. and Vasic, D. and Costa, F. 2007 "Piezoelectric Generator Harvesting Bike Vibrations Energy to Supply Portable Devices" *Applied Mathematics and Mechanics.*

[9]  Fukada, E. 1955, "Piezoelectricity of Wood", *Journal of the Physical Society of Japan*, vol. 108, no. 2, pp. 149.

[10] Bazhenov, V.A. 1961, *Piezoelectric properties of wood*, Consultant Bureau Enterprises, Inc., New York.

[11] Fukada, E. and Yasuda, I. 1957, "On the Piezoelectric Effect of Bone", *Journal of the Physical Society of Japan*, vol. 12, no. 10, pp. 1158.

[12] Fukada, E. and Yasuda, I. 1964, "Piezoelectric Effects in Collagen", *Japanese Journal of Applied Physics*, vol. 3, no. 2, pp. 117.

[13] Fukada, E. 1956, "On the Piezoelectric Effect of Silk Fibers", *Journal of the Physical Society of Japan*, vol. 11, no. 12, pp. 1301

[14] Yucel, T., Cebe, P. and Kaplan, D.L. 2011, "Structural Origins of Silk Piezoelectricity", *Advanced Functional Materials*, vol. 21, no. 4, pp. 779-785.

[15] Wang, T., Feng, Z., Song, Y. and Chen, X. 2007, "Piezoelectric properties of human dentin and some influencing factors", *Dental materials official publication of the Academy of Dental Materials*, vol. 23, no. 4, pp.450-453.

[16] Ueda, H. and Fukada, E. 1971, "Piezoelectricity in Myosin and Actin", *Japanese Journal of Applied Physics*, vol. 10, no. 11, pp. 1650.

[17] Fukada, E. 1972, "Piezoelectricity in Oriented DNA Films", *Journal of Polymer Science Part A-2: Polymer Physics*, vol. 10, no. 3.

[18] Duchesne, J., Depireux, J., Bertinchamps, A. , Cornet, N.and van der Kaa, J. M. 1960, "Thermal and electric properties of nucleic acids and proteins", *Nature*, vol. 188, pp. 405.

[19] Robert, S. 1947, "Dielectric and Piezoelectric properties of Barium Titanate", Phys. Rev. vol. 71, pp. 890-895.

[20] Shirane, G. and Suzuki, K. 1952, "Crystal Structure of $Pb(Zr-Ti)O_3$", *J. Phys. Soc. Jpn.*, vol. 7, pp. 333.

[21] Shirane, G., Hoshino, S. and Suzuki, K. 1950, "X-ray study of the phase transition in lead titanate", *Physical Review*, vol. 80, pp. 1105-1106.

[22] Sawaguchi, E. 1953, "Ferroelectricity versus antiferroelectricity in the solid solutions of PbZrO₃ and PbTiO₃", *J. Phys. Soc. Jpn.*, vol. 8, pp. 615-629.

[23] Jaffe, B., Roth, R.S. and Marzullo, S. 1954, "Piezoelectric Properties of lead zirconate-lead titanate solid-solution ceramic ware", *J. Appl. Phys.*, vol. 25, pp. 809-810.

[24] Egerton, L. and Dillon, D. M. 1959, "Piezoelectric and Dielectric Properties of Ceramics in the System Potassium-Sodium Niobate", *Journal of the American Ceramic Society*, vol. 42, no. 9, pp. 438-442.

[25] Weis, R. S. and Gaylord, T. K. 1985, "Lithium niobate: Summary of physical properties and crystal structure", *Applied Physics A: Materials Science & Processing*, vol. 37, no. 4, pp. 191-203.

[26] Smith, R. T. and Welsh, F. S. 1971, "Temperature Dependence of the Elastic, Piezoelectric, and DIelectric Constants of Lithium Tantanate and Lithium Niobate", *Journal of Applied Physics*, vol. 42, no. 6, pp. 2219-2230.

[27] Kawai, H. 1969, "The piezoelectricity of poly(vinylidene fluoride)", *Jpn. J. Appl. Phys.*, vol. 8, pp. 975-976.

[28] Harrison, J.S. 2001, "Piezoelectric Polymers", NASA/CR-2001-211422, ICASE Report No. 2001-43, pp. 1-26.

[29] Broadhurst, M.G., Davis, G.T., McKinney, J. E. and Collins, R. E. 1978, "Piezoelectricty and pyroelectricity in polyvinylidene flupride-A model", *J. Appl. Phys.*, vol. 49, no. 10, pp. 4992-4997.

[30] MEAS Piezo 1999, "MEAS Piezo Sensors Technical Manual", *Measurement Specialties, Inc.* (MEAS, 1999)

[31] Hausler, E. and Stein, L. 1984, "Implantable Physiological Power Supply with PVDF Film", *Ferroelectrics*, vol. 60, pp. 277-282.

[32] Starner, T. 1996, "Human-powered wearable computing", *IBM Systems Journal*, vol. 35, no. 3&4, pp. 618-629.

[33] Kymissis, J., Kendall, C., Paradiso, J. and Gershenfeld, N. 1998, "Parasitic power harvesting in shoes", *Digest of Papers. Second International Symposium on Wearable Computers*, pp. 132-139.

[34] Shenck, N.S. 1999, "A demonstration of useful electric energy generation from piezoceramics in a shoe", *MSc Thesis @ Department of Electrical Engineering and Computer Science, Massachusetts Institute of Technology*.

[35] Shenck, N. S. and Paradiso, J. A. 2001, "Energy scavenging with shoe-mounted piezoelectrics", *Micro, IEEE*, vol. 21, no. 3, pp. 30-42.

[36] Churchill, D.L., Hamel, M.J., Townsend, C. P. and Arms, S. W. 2003, "Strain energy harvesting for wireless sensor networks", *Proc. Smart Struct. and Mater. Conf.; Proc. SPIE*, vol. 5055, pp. 319-327.

[37] Renaud, M., Sterken, T., Fiorini, P., Puers, R., Baert, K. and van Hoof, C. 2005, "Scavenging energy from human body: design of a piezoelectric transducer", *Proc. 13th Int. Conf. on Solid-State Sensors and Actuators and Microsystems*, vol. 1, pp. 784-787.

[38] Granstrom, J., Feenstra, J., Sodano, H.A. and Farinholt, K. 2007, "Energy harvesting from a backpack instrumented with piezoelectric shoulder straps", *Smart Materials and Structures*, vol. 16, no. 5, pp. 1810-1820.

[39] Swallow, L.M., Luo, J.K., Siores, E., Patel, I. and Dodds, D. 2008, *A piezoelectric fibre composite based energy harvesting device for potential wearable applications*.

[40] Siores, E. and Swallow, L. 2008, "Apparatus for detection and suppression of muscle tremors", *GB Patent No 2 444 393*.

[41] Egusa, S., Wang, Z., Chocat, N., Ruff, Z.M., Stolyarov, A.M., Shemuly, D., Sorin, F., Rakich, P.T. , Joannopoulos, J. D. and Fink, Y. 2010, "Multimaterial piezoelectric fibres", *Nat Mater*, vol. 9, no. 8, pp. 643-648.

[42] E. Siores, R.L. Hadimani, D. Vatansever, Piezoelectric Polymer Element and Production Method and Apparatus Thereof, U.K. Patent 1015399.7.

[43] Anonymous 2005 "U.S. Department of Energy, Basis Research Needs for Solar Energy Utilization".

[44] *Photon International Magazine*, March2011

[45] *EPIA Global Market Outlook*, March2011, visited on 30th May 2011 http://www.epia.org/publications/photovoltaic-publications-global-market-outlook/gl obal -market-outlook-for-photovoltaics-until-2015.html

[46] Anonymous 1998 "Fundamentals of Photovoltaic Materials" *National Solar Power Research Institute, Inc.*,pp. 1-10.

[47] Nogueira, M. and Black, A. 2003, "Basics of Solar Electricity: Photovoltaics (PV)", *Northern California Solar Energy Resource Guide,* , pp. 1-4.

[48] Green, M. 2001, "Third generation photovoltaics: Ultra-high conversion efficiency at low cost", *Prog.Photovolt: Res.Appl.*, vol. 9, no. 2, pp. 123-135.

[49] Chopra, K.L., Paulson, P.D. and Dutta, V. 2004, "Thin-Film Solar Cells: An Overview", *Prog. Photovolt: Res. Appl.*, vol. 12, pp. 69-92.

[50] Gunes, S., Neugebauer, H. and Sariciftci, N. S. 2007, "Conjugated Polymer-Based Organic Solar Cells", *Chemical Reviews*, vol. 107, pp. 1324-1338.

[51] Li, G., Shrotriya, V., Huang, J., Yao, Y., Moiarty, T., Emery, K. and Yang, Y. 2005, "High-efficiency solution processable polymer photovoltaic cells by self-organization of polymer blends", *Nature Materials*, vol. 4, pp. 864-868.

[52] Miles, R.W., Zoppi, G. and Forbes, I. 2007, "Inorganic photovoltaic cells", *Materials Today*, vol. 10, no. 11, pp. 20-27.

[53] Pagliaro, M., Palmisano, G. and Ciriminna, R. 2008, *Flexible Solar Cells*, WILEY-VCH Verlag GmbH, Weinheim.

[54] Partain, L.D. 1995, *Solar Cells and Their Applications*, John Wiley & Sons, New York.

[55] Reuter, M., Brendle, W., Tobail, O. and Werner, J. H. 2009, "50 μm thin solar cells with 17.0% efficiency", *Solar Energy Materials and Solar Cells*, vol. 93, no. 6-7, pp. 704-706.

[56] Wang, A., Zhao, J., Wenham, S.R. and Green, M.A. 1996, "21.5% Efficient thin silicon solar cell", *Progress in Photovoltaics: Research and Applications*, vol. 4, no. 1, pp. 55-58.

[57] Chopra, K.L., Paulson, P.D. and Dutta, V. 2004, "Thin-Film Solar Cells: An Overview", *Prog. Photovolt: Res. Appl.*, vol. 12, pp. 69-92.

[58] Sariciftci, N.S. 1999, "Polymeric photovoltaic materials", *Current Opinion in Solid State and Materials Science*, vol. 4, no. 4, pp. 373-378. [59]

[59] Cai, W. , Gong, X. and Cao, Y. 2010, "Polymer solar cells: Recent development and possible routes for improvement in the performance", *Solar Energy Materials and Solar Cells*, vol. 94, no. 2, pp. 114-127.

[60] Li, G., Shrotriya, V., Yao, Y., Huang, J. and Yang, Y. 2007, "Manipulating regioregular poly(3-hexylthiophene) : [6,6]-phenyl-C61-butyric acid methyl ester blends-route towards high efficiency polymer solar cells", *Journal of Materials Chemistry*, vol. 17, no. 30, pp. 3126-3140.

[61] Cozzens, R.F. 1982, in *Electrical Properties of Polymers*, ed. D.A. Seanor, Academic Press, New York, pp. 93.

[62] Ameri, T., Dennler, G., Waldauf, C., Denk, P., Forberich, K., Scharber, M.C., Brabec, C.J. and Hingerl, K. 2008, "Realization, characterization, and optical modeling of inverted bulk-heterojunction organic solar cells", *Journal of Applied Physics*, vol. 103, no. 8, pp. 084506.

[63] Liang, Y., Wu, Y., Feng, D., Tsai, S.-., Son, H.-. and Li, G. and Yu, L. 2008; 2009, "Development of New Semiconducting Polymers for High Performance Solar Cells", *Journal of the American Chemical Society*, vol. 131, no. 1, pp. 56-57.

[64] Singh, T.B., Marjanović, N., Matt, G.J., Günes, S., Sariciftci, N.S., Montaigne Ramil, A., Andreev, A., Sitter, H., Schwödiauer, R. and Bauer, S. 2005, "High-mobility n-channel organic field-effect transistors based on epitaxially grown C60 films", *Organic Electronics*, vol. 6, no. 3, pp. 105-110.

[65] Gerischer, H., Michel-Beyerle, M.E., Rebentrost, F. and Tributsch, H. 1968, "Sensitization of charge injection into semiconductors with large band gap", *Electrochimica Acta*, vol. 13, no. 6, pp. 1509-1515.

[66] Tributsch, H. 1972, "Reaction of Excited Chorophyll Molecules at Electrodes and in Photosynthesis", *Photochem.Photobiol*, vol. 16, pp. 261-269.

[67] Hagfeldt, A., Boschloo, G., Sun, L., Kloo, L. and Pettersson, H. 2010, "Dye-Sensitized Solar Cells", *Chemical reviews*, vol. 110, no. 11, pp. 6595-6663.

[68] Gratzel, M. 2005, "Solar Energy Conversion by Dye-Sensitized Photovoltaic Cells", *Inorganic chemistry*, vol. 44, no. 20, pp. 6841-6851.

[69] Cahen, D., Hodes, G., Gratzel, M., Guillemoles, J.F. and Riess, I. 2000, "Nature of Photovoltaic Action in Dye-Sensitized Solar Cells", *The Journal of Physical Chemistry B*, vol. 104, no. 9, pp. 2053-2059.

[70] Wang, X., Zhi, L. and Mullen, K. 2008, "Transparent, Conductive Graphene Electrodes for Dye-Sensitized Solar Cells", *Nano Letters*, vol. 8, no. 1, pp. 323-327.

[71] Calogero, G., Calandra, P., Irrera, A., Sinopoli, A., Citro, I. and Di Marco, G. 2011, "A new type of transparent and low cost counter-electrode based on platinum nanoparticles for dye-sensitized solar cells", *Energy & Environmental Science*, vol. 4, no. 5, pp. 1838-1844.

[72] Wang, P., Zakeeruddin, S.M., Comte, P., Exnar, I. and Gratzel, M. 2003, "Gelation of Ionic Liquid-Based Electrolytes with Silica Nanoparticles for Quasi-Solid-State Dye-Sensitized Solar Cells", *Journal of the American Chemical Society*, vol. 125, no. 5, pp. 1166-1167.

[73] Gratzel, M. 2005, "Solar Energy Conversion by Dye-Sensitized Photovoltaic Cells", *Inorganic chemistry,* vol. 44, no. 20, pp. 6841-6851.

[74] Bach, U., Lupo, D., Comte, P., Moser, J.E., Weissortel, F., Salbeck, J., Spreitzer, H. and Gratzel, M. 1998, "Solid-state dye-sensitized mesoporous TiO2 solar cells with high photon-to-electron conversion efficiencies", *Nature,* vol. 395, no. 6702, pp. 583-585.

[75] Han, L., Fukui, A., Chiba, Y., Islam, A., Komiya, R., Fuke, N., Koide, N., Yamanaka, R. and Shimizu, M. 2009, "Integrated dye-sensitized solar cell module with conversion efficiency of 8.2%", *Applied Physics Letters,* vol. 94, no. 1, pp. 013305.

[76] Law, M., Greene, L.E., Johnson, J.C., Saykally, R. and Yang, P. 2005, "Nanowire dye-sensitized solar cells", *Natural Materials,* vol. 4, pp. 455-459.

[77] Horiuchi, T., Miura, H., Sumioka, K. and Uchida, S. 2004, "High Efficiency of Dye-Sensitized Solar Cells Based on Metal-Free Indoline Dyes", *Journal of the American Chemical Society,* vol. 126, no. 39, pp. 12218-12219.

[78] Chiba, Y., Islam, A., Watanabe, Y., Komiya, R., Koide, N. and Han, L. 2006 "Dye-Sensitized Solar Cells with Conversion Efficiency of 11.1%", *Japanese Journal of Applied Physics,* vol. 45, no. 25, pp. L638.

[79] Wanlass, M.W., Coutts, T.J., Ward, J.S., Emery, K.A., Gessert, T.A. and Osterwald, C.R. 1991, "Advanced high-efficiency concentrator tandem solar cells", IEEE *Preceeding on Photovoltaic Specialists Conference,* 07th-11th October, pp. 38.

[80] De Vos, A. 1980, "Detailed balance limit of the efficiency of tandem solar cells", *J. Phys. D: Appl. Phys.,* vol. 13, pp. 839-846.

[81] Green, M.A., Emery, K., Hisikawa, Y. and Warta, W. 2007, "Solar cell efficiency tables (version 30)", *Progress in Photovoltaics: Research and Applications,* vol. 15, no. 5, pp. 425-430.

[82] Geisz, J.F., Kurtz, S., Wanlass, M.W., Ward, J.S., Duda, A., Friedman, D.J., Olson, J.M., McMahon, W.E., Moriarty, T.E. and Kiehl, J.T. 2007, "High-efficiency GaInP/GaAs/InGaAs triple-junction solar cells grown inverted with a metamorphic bottom junction", *Applied Physics Letters,* vol. 91, no. 2, pp. 023502-3.

[83] Dennler, G., Scharber, M.C., Ameri, T., Denk, P., Forberich, K., Waldauf, C. and Brabec, C.J. 2008, "Design Rules for Donors in Bulk-Heterojunction Tandem Solar Cells: Towards 15 % Energy-Conversion Efficiency", *Advanced Materials,* vol. 20, no. 3, pp. 579-583.

[84] Gunes, S. and Sariciftci, N. S. 2008, "Hybrid solar cells", *Inorganica Chimica Acta,* vol. 361, no. 3, pp. 581-588.

[85] van Hal, P.A., Wienk, M.M., Kroon, J.M., Verhees, W.J.H., Slooff, L.H., van Gennip, W. J. H., van Gennip, W. J. H. and Janssen, R.A.J. 2003, "Photoinduced Electron Transfer and Photovoltaic Response of a MDMO-PPV:TiO2 Bulk-Heterojunction", *Advanced Materials,* vol. 15, no. 2, pp. 118-121.

[86] McDonald, S.A., Konstantatos, G., Zhang, S., Cyr, P.W., Klem, E.J.D., Levina, L. and Sargent, E.H. 2005, "Solution-processed PbS quantum dot infrared photodetectors and photovoltaics", *Nat Mater,* vol. 4, no. 2, pp. 138-142.

[87] Zhang, S., Cyr, P.W., McDonald, S.A., Konstantatos, G. and Sargent, E.H. 2005, "Enhanced infrared photovoltaic efficiency in PbS nanocrystal/semiconducting polymer

composites: 600-fold increase in maximum power output via control of the ligand barrier", *Applied Physics Letters*, vol. 87, no. 23, pp. 233101.

[88] Beek, W.J.E., Wienk, M.M. and Janssen, R.A.J. 2006, "Hybrid Solar Cells from Regioregular Polythiophene and ZnO Nanoparticles", *Adv. Funct. Mater.*, vol. 16, no. 8, pp. 1112-1116.

[89] Olson, D.C., Piris, J., Collins, R.T., Shaheen, S.E. and Ginley, D.S. 2006, "Hybrid photovoltaic devices of polymer and ZnO nanofiber composites", *Thin Solid Films*, vol. 496, no. 1, pp. 26-29.

[90] Greenham, N.C., Peng, X. and Alivisatos, A.P. 1996, "Charge separation and transport in conjugated-polymer/semiconductor-nanocrystal composites studied by photoluminescence quenching and photoconductivity", *Physical Review B*, vol. 54, no. 24, pp. 17628-17637.

[91] Ginger, D.S. and Greenham, N.C. 1999, "Photoinduced electron transfer from conjugated polymers to CdSe nanocrystals", *Physical Review B*, vol. 59, no. 16, pp. 10622-10629.

[92] Huynh, W.U., Dittmer, J.J. and Alivisatos, A.P. 2002, "Hybrid Nanorod-Polymer Solar Cells", *Science*, vol. 295, no. 5564, pp. 2425-2427.

[93] Gur, I., Fromer, N.A., Geier, M.L. and Alivisatos, A.P. 2005, "Air-Stable All-Inorganic Nanocrystal Solar Cells Processed from Solution", *Science*, vol. 310, no. 5747, pp. 462-465.

[94] Arici, E., Sariciftci, N.S. and Meissner, D. 2003, "Hybrid Solar Cells Based on Nanoparticles of CuInS2 in Organic Matrices", *Advanced Functional Materials*, vol. 13, no. 2, pp. 165-171.

[95] Chapin, D.M., Fuller, C.S. and Pearson, G.L. 1954, "A New Silicon pin Junction Photocell for Converting Solar Radiation into Electrical Power", *Journal of Applied Physics*, vol. 25, no. 5, pp. 676.

[96] Chopra, K.L., Paulson, P.D. and Dutta, V. 2004, "Thin-Film Solar Cells: An Overview", *Prog. Photovolt: Res. Appl.*, vol. 12, pp. 69-92.

[97] Spanggaard, H. and Krebs, F. C. 2004, "A brief history of the development of organic and polymeric photovoltaics", *Solar Energy Materials and Solar Cells*, vol. 83, no. 2-3, pp. 125-146.

[98] Sariciftci, N., Smilowitz, L.B., Zhang, C., Srdanov, V.I., Heeger, A.J. and Wudl, F. 1993a, "Photoinduced electron transfer from conducting polymers onto Buckminsterfullerene", *Proceedings of SPIE*, vol. 1852, no. 1, pp. 297.

[99] Sariciftci, N.S., Smilowitz, L., Heeger, A.J. and Wudl, F. 1993b, "Semiconducting polymers (as donors) and buckminsterfullerene (as acceptor): photoinduced electron transfer and heterojunction devices", *Synthetic Metals*, vol. 59, no. 3, pp. 333-352.

[100] Sariciftci, N.S. and Heeger, A.J. 1994, *Conjugated Polymer-Acceptor Heterojunctions; Diodes, Photodiodes, and Photovoltaic Cells*, 5,331,183, California, USA.

[101] Sariciftci, N. and Heeger, A.J. 1995, "Role of buckminsterfullerene, C60, in organic polymeric photoelectric devices", *Proceedings of SPIE*, vol. 2530, no. 1, pp. 76.

[102] Yu, G., Gao, J., Hummelen, J.C., Wudl, F. and Heeger, A.J. 1995, "Polymer Photovoltaic Cells: Enhanced Efficiencies via a Network of Internal Donor-Acceptor Heterojunctions", *Science*, vol. 270, no. 5243, pp. 1789-1791.

[103] Morita, S., Kiyomatsu, S., Yin, X.H., Zakhidov, A.A., Noguchi, T., Ohnishi, T. and Yoshino, K. 1993, "Doping effect of buckminsterfullerene in poly(2,5-dialkoxy-p-phenylene vinylene)", *Journal of Applied Physics*, vol. 74, no. 4, pp. 2860.

[104] Brabec, C., Schilinsky, P. and Waldauf, C. 2007, *Method for Treating a Photovoltaic Active Layer and Organic Photovoltaic Element*, 7,306, edn, Lowel, MA, USA.

[105] Brabec, C.J., Shaheen, S.E., Winder, C. and Sariciftci, N. S. and Denk, P. 2002, "Effect of LiF/metal electrodes on the performance of plastic solar cells", *Applied Physics Letters*, vol. 80, no. 7, pp. 1288-1290.

[106] Blom, P.W.M., de Jong, M. J. M. and Breedijk, S. 1997, "Temperature dependent electron-hole recombination in polymer light-emitting diodes", *Applied Physics Letters*, vol. 71, no. 7, pp. 930-932.

[107] Melzer, C., Koop, E.J., Mihailetchi, V.D. and Blom, P.W.M. 2004, "Hole Transport in Poly(phenylene vinylene)/Methanofullerene Bulk-Heterojunction Solar Cells", *Advanced Functional Materials*, vol. 14, no. 9, pp. 865-870.

[108] Mihailetchi, V.D., Koster, L.J.A., Blom, P.W.M., Melzer, C., Boer, B., Duren, J.K.J. and Janssen, R.A.J. 2005, "Compositional Dependence of the Performance of Poly(p-phenylene vinylene):Methanofullerene Bulk-Heterojunction Solar Cells", *Advanced Functional Materials*, vol. 15, no. 5, pp. 795-801.

[109] Schilinsky, P., Waldauf, C. and Brabec, C.J. 2002, "Recombination and loss analysis in polythiophene based bulk heterojunction photodetectors", *Applied Physics Letters*, vol. 81, no. 20, pp. 3885-3887.

[110] Li, G., Shrotriya, V., Huang, J., Yao, Y., Moiarty, T., Emery, K. and Yang, Y. 2005, "High-efficiency solution processable polymer photovoltaic cells by self-organization of polymer blends", *Nature Materials*, vol. 4, pp. 864-868.

[111] Padinger, F., Rittberger, R.S. and Sariciftci, N.S. 2003, "Effects of Postproduction Treatment on Plastic Solar Cells", *Advanced Functional Materials*, vol. 13, no. 1, pp. 85-88.

[112] Kim, Y., Choulis, S.A., Nelson, J., Bradley, D.D.C., Cook, S. and Durrant, J.R. 2005, "Device annealing effect in organic solar cells with blends of regioregular poly(3-hexylthiophene) and soluble fullerene", *Applied Physics Letters*, vol. 86, no. 6, pp. 063502-3.

[113] Shrotriya, V., Ouyang, J., Tseng, R.J. & Li, G. and Yang, Y. 2005, "Absorption spectra modification in poly(3-hexylthiophene):methanofullerene blend thin films", *Chemical Physics Letters*, vol. 411, no. 1-3, pp. 138-143.

[114] Reyes-Reyes, M., Kim, K., Dewald, J., LÃ³pez-Sandoval, R., Avadhanula, A., Curran, S. and Carroll, D.L. 2005, "Meso-Structure Formation for Enhanced Organic Photovoltaic Cells", *Organic letters*, vol. 7, no. 26, pp. 5749-5752.

[115] Ma, W., Yang, C., Gong, X., Lee, K. and Heeger, A.J. 2005, "Thermally Stable, Efficient Polymer Solar Cells with Nanoscale Control of the Interpenetrating Network Morphology", *Advanced Functional Materials*, vol. 15, no. 10, pp. 1617-1622.

[116] Kim, Y., Cook, S., Tuladhar, S.M., Choulis, S.A., Nelson, J., Durrant, J.R., Bradley, D.D.C., Giles, M., McCulloch, I., Ha, C.-. and Ree, M. 2006, "A strong regioregularity effect in self-organizing conjugated polymer films and high-efficiency polythiophene:fullerene solar cells", *Nat Mater*, vol. 5, no. 3, pp. 197-203.

[117] Yang, X., Loos, J., Veenstra, S.C., Verhees, W.J.H., Wienk, M.M., Kroon, J.M., Michels, M.A.J. and Janssen, R.A.J. 2005, "Nanoscale Morphology of High-Performance Polymer Solar Cells", *Nano Letters*, vol. 5, no. 4, pp. 579-583.

[118] Ma, W., Yang, C., Gong, X., Lee, K. and Heeger, A.J. 2005, "Thermally Stable, Efficient Polymer Solar Cells with Nanoscale Control of the Interpenetrating Network Morphology", *Advanced Functional Materials*, vol. 15, no. 10, pp. 1617-1622.

[119] Yang, X., Loos, J., Veenstra, S.C., Verhees, W.J.H., Wienk, M.M., Kroon, J.M., Michels, M.A.J. and Janssen, R.A.J. 2005, "Nanoscale Morphology of High-Performance Polymer Solar Cells", *Nano Letters*, vol. 5, no. 4, pp. 579-583.

[120] Savenije, T.J., Kroeze, J.E., Yang, X. and Loos, J. 2005, "The Effect of Thermal Treatment on the Morphology and Charge Carrier Dynamics in a Polythiophene-Fullerene Bulk Heterojunction", *Advanced Functional Materials*, vol. 15, no. 8, pp. 1260-1266.

[121] Padinger, F., Rittberger, R.S. and Sariciftci, N.S. 2003, "Effects of Postproduction Treatment on Plastic Solar Cells", *Advanced Functional Materials*, vol. 13, no. 1, pp. 85-88.

[122] Li, G., Shrotriya, V., Huang, J., Yao, Y., Moiarty, T., Emery, K. and Yang, Y. 2005, "High-efficiency solution processable polymer photovoltaic cells by self-organization of polymer blends", *Nature Materials*, vol. 4, pp. 864-868.

[123] Wang, W., Wu, H., Yang, Y., Luo, C., Zhang, Y., Chen, J. and Cao, Y. 2007, "High-efficiency polymer photovoltaic devices from regioregular-poly(3-hexylthiophene-2,5-diyl) and [6,6]-phenyl-C61-butyric acid methyl ester processed with oleic acid surfactant", *Applied Physics Letters*, vol. 90, no. 18, pp. 183512.

[124] Berson, S., De Bettignies, R., Bailly, S. and Guillerez, S. 2007, "Poly(3-hexylthiophene) Fibers for Photovoltaic Applications", *Advanced Functional Materials*, vol. 17, no. 8, pp. 1377-1384.

[125] Moule, A. J. and Meerholz, K. 2008, "Controlling Morphology in Polymer-Fullerene Mixtures", *Advanced Materials*, vol. 20, no. 2, pp. 240-245.

[126] Li, G., Shrotriya, V., Yao, Y., Huang, J. and Yang, Y. 2007, "Manipulating regioregular poly(3-hexylthiophene) : [6,6]-phenyl-C61-butyric acid methyl ester blends-route towards high efficiency polymer solar cells", *Journal of Materials Chemistry*, vol. 17, no. 30, pp. 3126-3140.

[127] Dennler, G., Scharber, M. C. and Brabec, C. J. 2009, "Polymer-Fullerene Bulk-Heterojunction Solar Cells", *Adv Mater*, vol. 21, pp. 1323-1338.

[128] Shrotriya, V., Ouyang, J., Tseng, R.J. & Li, G. and Yang, Y. 2005, "Absorption spectra modification in poly(3-hexylthiophene):methanofullerene blend thin films", *Chemical Physics Letters*, vol. 411, no. 1-3, pp. 138-143.

[129] Chirvase, D., Parisi, J., Hummelen, J.C. and Dyakonov, V. 2004, *Influence of nanomorphology on the photovoltaic action of polymer - fullerene composites*.

[130] Park, S.H., Roy, A., Beaupre, S., Cho, S., Coates, N., Moon, J.S., Moses, D., Leclerc, M., Lee, K. and Heeger, A.J. 2009, "Bulk heterojunction solar cells with internal quantum efficiency approaching 100%", *Nat Photon*, vol. 3, no. 5, pp. 297-302.

[131] Liang, Y., Feng, D., Wu, Y., Tsai, S.-., Li, G., Ray, C. and Yu, L. 2009, "Highly Efficient Solar Cell Polymers Developed via Fine-Tuning of Structural and Electronic Properties", *Journal of the American Chemical Society*, vol. 131, no. 22, pp. 7792-7799.

[132] Hauch, J.A., Schilinsky, P., Choulis, S.A., Childers, R., Biele, M. and Brabec, C.J. 2008, Flexible organic P3HT:PCBM bulk-heterojunction modules with more than 1 year outdoor lifetime", *Solar Energy Materials and Solar Cells*, vol. 92, no. 7, pp. 727-731.

[133] Chittibabu, K., Eckert, R., Gaudiana, R., Li, L., Montello, A., Montello, E. and Wormser, P. 2005, *Photovoltaic Fibers*, 6,913,713 edn, USA.

[134] Kuraseko, H., Nakamura, T., Toda, S., Koaizawa, H., Jia, H. and Kondo, M. 2006, Development of flexible fiber-type poly-Si solar cell", *Conference Record of the 2006 IEEE 4th World Conference on Photovoltaic Energy Conversion*, pp. 1380.

[135] Li, Y., Zhou, W., Xue, D., Liu, J., Peterson, E.D., Nie, W. and Carroll, D. L. 2009, Origins of performance in fiber-based organic photovoltaics", *Applied Physics Letters*, vol. 95, no. 20, pp. 203503-3.

[136] O'Connor, B., Pipe, K. P. and Shtein, M. 2008, "Fiber based organic photovoltaic devices", *Applied Physics Letters*, vol. 92, no. 19, pp. 193306-3.

[137] Bedeloglu, A., Demir, A. and Bozkurt, Y. and Sariciftci, N. S. 2010, "A Photovoltaic Fiber Design for Smart Textiles", *Textile Research Journal*, vol. 80, no. 11, pp. 1065-1074.

[138] Toivola, M., Ferenets, M., Lund, P. and Harlin, A. 2009, "Photovoltaic fiber", *Thin Solid Films*, vol. 517, no. 8, pp. 2799-2802.

[139] Baps, B., Eber-Koyuncu, M. and Koyuncu, M. 2002, "Ceramic Based Solar Cells in Fiber Form", *Key Engineering Materials*, vol. 206-213, pp. 937-940.

[140] Ramier, J., Plummer, C.J.G., Leterrier, Y., Månson, J.-.E., Eckert, B. and Gaudiana, R. 2008, "Mechanical integrity of dye-sensitized photovoltaic fibers", *Renewable Energy*, vol. 33, no. 2, pp. 314-319.

[141] Siores, E., Hadimani, R.L. and Vatansever, D. 2010, "Hybrid Energy Conversion Device", *GB Patent No.* 1016193.3.

# Chapter 13

# Study of the Consequences of Global Warming in Water Dynamics During Dormancy Phase in Temperate Zone Fruit Crops

Robson Ryu Yamamoto, Paulo Celso de Mello-Farias, Fabiano Simões and Flavio Gilberto Herter

Additional information is available at the end of the chapter

## 1. Introduction

For a normal development of temperate zone fruit trees, a very complex phase of their life-cycle called dormancy is regulated by environmental conditions during autumn-winter season. This phase allows those species to survive under adverse environmental conditions, and one of the most important factors which induce and release dormancy stage is temperature. It was adopted a terminology suggested by Lang et al. (1987) for the different stages of dormancy: para-, endo-, and ecodormancy.

From the end of the 19th century, cultivation of temperate zone fruit crops were set up in areas warmer than those traditionally cultivated. As the cultivars grown were those used traditionally, the chilling requirements were not adequately fulfilled. These difficulties led to identification and study problems of growing temperate zone fruit crops in warm areas, which was approached in two different ways. On the one hand, traditional cultivars were selected and bred to obtain new cultivars with commercial quality and low chilling requirement, which is a heritable character. On the other hand, new cultural practices, such as applications of rest-breaking agents, were developed to avoid or reduce the negative consequences of an insufficient chilling accumulation. The most useful option for temperate zone fruit crops in warm areas has frequently been a combination of low-chill cultivars and the adoption of cultural practices to break or avoid dormancy. Although different species of the *Rosaceae* family have been bred for low-chill cultivars, the furthest advances have been made in peaches and apples.

During dormancy, several events occur simultaneously, which modifies water and carbohydrate dynamics, hormonal balance, among others. Water is important for bud and plant development, such in solute translocation, enzymatic reactions, and osmotic regulated events, reason why many studies are focused on its dynamics. Recent studies in water dynamics during dormancy phase, such as the embolism (loss of hydraulic conductivity) in xylem of plants and water status (free or bound water) in dormant buds determined by magnetic resonance (MR) techniques, showed the importance of water on dormancy release process. And under stressed conditions this importance seems to be strongly accentuated.

## 1.1. Physiology of dormancy

Dormancy is an adaptive behavior of temperate zone tree species that allows the plant to survive under unfavorable conditions during winter. From the appearance of the bud only, it is difficult to ascertain what kind of biochemical and physiological changes are happening during this period (Faust et al., 1997; Yooyongwech et al., 2008a).

Winter dormancy is an important adaptive mechanism for plant survival in temperate and cold climates. It is essential that the dormant condition is established within the plant well in advance of the cold season. This requires the timely sensing and physiological processing of a regular and reliable environmental signal (Heide & Prestrud, 2005).

Lang et al. (1987) classified the different phases of dormancy as ecodormancy, which is found in late winter and spring and imposed by unfavorable conditions to growth; paradormancy, which is equivalent to correlative inhibition or apical dominance; and endodormancy, also called deep or winter dormancy. The last is the genuine dormancy that characterizes woody plants in temperate zones and has been subject of many studies that have shown the enormous complexity of this phenomenon.

Lack of plant environmental synchrony is considered to be the primary cause of abiotic stress injury. Plant synchrony requires timely responses to environmental cues to minimize risk from abiotic stresses. The timing of growth cessation and dormancy, and subsequent cold acclimation, deacclimation and the depth of cold hardiness are all critical components of winter survival in temperate climates. The degree to which temperature mediates this response is important in order to determine the impact of future temperature change in a global warming context on timing of growth cessation and cold acclimation in woody plants (Tanino et al., 2010).

In temperate zone deciduous fruit trees, the most important factor in dormancy release process is the accumulation of a certain amount of chilling (Lang, 1996). During dormancy, chilling temperature are associated with changes in carbohydrate contents and other substances, such as nucleic acids, proteins, polyamines, amino acids, organic acids and in the respiration rate, that might be related with bud break and the time of bloom (Wang et al., 1987).

Most of the temperature effects on plants are mediated by their effects on plant biochemistry. Carbohydrates are the main source of energy for the metabolic changes that occurred during the period of dormancy release. Carbohydrate availability is presumably of major relevance to the control of bud growth and development during dormancy induction and its release (Sherson et al., 2003), and might be related to the bud abortion (Cottignies, 1986). Starch, which is accumulated in reserve tissues during the preceding summer, is converted to sucrose and other soluble sugars during winter. Effects of chilling accumulation on changes in both starch and sugar concentrations may be explained because amylase activity is induced by cold temperature, which result in increasing of starch hydrolysis and, consequently, sugar concentration (Elle & Sauter, 2000).

During winter, dynamics of carbohydrates in the different tissues reflects the inter-conversion between starch and soluble sugars as described previously by Améglio et al. (2001), Lacointe et al. (1993), and Sauter (1980). Just before bud break, a strong increase in hexoses in apical buds seems closely linked to the decrease in sucrose and starch. The sucrose decrease revealed high activity of enzymes involved in sucrose catabolism (Bonhomme et al., 2009). Starch, which is degraded by amylases, and is used to the sucrose synthesis by the sucrose-6-phosphate synthase (SPS) in response to decreasing temperature. The sucrose produced in reserve tissue is transported by the xylem pathway to the bud and hydrolyzed to glucose and fructose to supply energy and carbonic precursors (Marafon et al., 2011; Yoshioka et al., 1988).

Researchers have investigated the relationships among carbohydrates (Fahmi, 1958; Monselise & Goldschmidt, 1981; Sarmiento et al., 1976; Stutte & Martin, 1986), hormones (Chen, 1987; Lavee et al., 1983; Mullins & Rajasekaran, 1981; Ramirez & Hoad, 1981; Stephan et al., 1999), mineral nutrients (Golomp & Goldschmidt, 1981; Hartmann et al., 1966; Priestly, 1977), and flower bud formation in different fruit species.

Ulger et al. (2004) observed that the abscisic acid (ABA), indole-3-acetic acid (IAA), and gibberellic acid (GA₄) levels in leaves, nodes and fruits of olive (*Olea europea* L.) during the induction, initiation, and differentiation periods in the on year were lower than those in the off year. Similarly, Pal & Ram (1978) in mango (*Mangifera indica* L.), Ulger et al. (1999) in olive, and Chen (1990) in litchi (*Litchi chinensis* Sonn.), demonstrated higher GA₃ level in the on year. The fact that GA₃ decreased and GA₄ levels increased during the induction and initiation periods in the off year suggest that they affect flower bud formation. Looney et al. (1985) determined that exogenous GA₄ application on apple (*Malus domestica* Borkh.) trees promoted flowering and yield, and Stephan et al. (1999) also noted that lack of GA₄ induced a biennial bearing habit.

An increase in zeatin levels during the induction period in the off year suggests that had possibly a positive effect on floral formation (Chen, 1987, 1990; Mullins & Rajasekaran, 1981; Ramirez & Hoad, 1981; Ulger et al., 2004). Many experiments confirmed the direct involvement of different growth regulators in promoting or inhibiting flower bud induction and differentiation. However, all these studies were related to the effect of a single regulator or its quantitative changes before, during or after flower bud induction (Lavee, 1989). On contrary to plant growth regulators, Ulger et al. (2004) suggested that carbohydrates and mineral nutrients might not have a direct effect to induce flower initiation.

## 1.2. Global warming and dormancy

Global climate change can alter significantly plant phenology because temperature influences the timing of development, both alone and through interactions with other cues, such as photoperiod (Bernier, 1988; Partanen et al., 1998). Temperature records showed that, over the past 30 years, global average surface temperatures increased by 0.28°C per decade (Cleland et al., 2007).

Fruit and vegetable growth and development are influenced by different environmental factors. Studies conducted by Moretti et al. (2012) showed that warmer temperatures affect photosynthesis directly, causing alterations in carbohydrates, organic acids, and flavonoid contents, firmness and antioxidant activities; carbon dioxide ($CO_2$) accumulation in the atmosphere had direct effects on postharvest quality causing tuber malformation, occurrence of common scab, and changes in reducing sugars contents on potatoes; high concentrations of atmospheric ozone can potentially caused reduction in the photosynthetic process, growth, and biomass accumulation. Understanding how climate changes will impact fruit crop production in the next decades is extremely important for survival.

All economically important fruit and nut tree species that originated from temperate and cold subtropical regions have chilling requirements that need to be fulfilled each winter to ensure homogeneous flowering and fruit set, and generate economically sufficient yields (Luedeling et al., 2009). Elevated concentration of atmospheric $CO_2$, warmer temperatures in general, and changing precipitation regimes will affect the exchange of energy, carbon, water and nutrients between forests and the environment, leading to changes in forest growth, survival and structure. Interactions with biotic and abiotic disturbance agents will also shape future forests (Chmura et al., 2011).

Lack of chilling, associated with mild winter conditions, results in abnormal patterns of bud break and development in temperate zone fruit trees (Mauget & Rageau, 1988) and is known as the main factor of pear flower bud abortion in Brazil (Petri et al., 2002; Petri & Herter, 2002) and New Zealand (Do Oh & Klinac, 2003; Klinac & Geddes, 1995). Rakngan et al. (1996) observed that the Japanese pear trees under enough chilling released from dormancy earlier than plants with insufficient chilling accumulation. The occurrence of intermittent warm days during dormancy period, with temperatures higher than 27 C, delayed the dormancy release more than mild temperature fluctuations (2-3 C) (Marafon et al., 2011).

## 1.3. Apple and pear production under warm winter conditions of Brazil

Apple orchards in Brazil has been highlighted by growth in harvested area of 34%, 30% in production, 948% in exported quantity and 1328% in exportation value in the last 10 years. These results reinforce the trend that has been happening since the beginning of pomiculture in the country and reveal the great exportation potential of Brazil. The responsible factors for this great development of the apple orchards is due to the development of technologies used in crops, the logistics in place, the definition of cultivars (Gala 58% and Fuji 36%) and clones capable to meet consumer demands (Fachinello et al., 2011).

In addition to government incentives to apple crop on 1970's decade, which resulted in improvement of plant breeding, nutrition and phytotechnical management, post-harvest technology, made by research centers and extension programs, enabled to change from an importer country to export in a few years. One decade after starting the apple crop production on a commercial scale, Brazilian apple importations began to decline, and actually is considered self-sufficient (Boneti et al., 2002; Fachinello et al., 2011).

On contrary, pear crop production is not considered a fruit of great expression despite the large domestic market for pear fruit, moreover, it has the lowest expression in terms of production, crop area and production value (Fioravanço, 2007). In the beginning of 1960's decade, the pear crop in Brazil was economically more important than apple. According to FAOSTAT (2012), in 1961, the harvested area, production, and yield of pear crop were 70%, 22%, and 34% higher than apple, but in 2010 it was 25 times, 78 times, and 3 times lower, respectively. Such situation was changed at the end of 1970's, and nowadays Brazil imports around 90% of fresh pear fruit (140,000 tones), corresponding to USD 130 million in 2009.

It is therefore possible to note that pear crop represents an important market opportunity. Isolated initiatives of pear (European and Japanese cultivars) crop in highlands of Southern Brazil have confirmed the viability of its commercial production. A great demand of this fruit in local market occurs in Brazil, but it is observed a technological lack available for growers, such as development of cultivars of both rootstock and scion adapted to mild winter conditions, crop management adapted to such environmental conditions, appropriate logistic systems, among other factors (Faoro, 2001). Low rates of fruit set (Petri, 2008) and high incidence of physiological disorder called locally "flower bud abortion" (Marafon et al., 2011; Petri & Herter, 2002; Petri et al., 2001; Trevisan et al., 2005) are cited as factors of low yield in pear crops. In mild winter conditions of Southern Brazil, over 60% of "flower bud abortion" was found in 2001 (Petri et al., 2002) and more than 90% in 1999, resulting in low numbers of opened flowers at the bud break and consequently low production (Veríssimo et al., 2002). Another consequence of low chilling accumulation during dormancy stage is a prolonged and poorly synchronized flowering, which results in poor uniformity of fruit size at harvest (Yamamoto, 2010).

Nishimoto et al. (1995) cited the value of 750 chilling hours as the requirement in 'Housui' buds, but the amount of 600 hours below $7.2°C$ (80% of requirement) brought about the release from endodormancy stage (Yamamoto et al., 2010a). According to Yamamoto (2010), however, abnormalities on floral primordia (partial or total necrosis) and eventual development of new inflorescences, during dormancy progression (Fig. 1) and at flowering, were observed in all treatments in lateral buds of Japanese pear shoots (Fig. 2). These symptoms became severe with prolonged cold deprivation before chilling accumulation (simulation of delayed mild winter, treatment 3) and after consecutive seasons (simulation of permanent global warming situation) (Yamamoto et al., 2010a, b).

## 2. Past works carried out in water dynamics by using magnetic resonance techniques

Water is one of the basic determining factors of bud development because it is quantitatively the major component of plant tissues. Besides, it is the essential medium for many metabolic processes, such as transpiration, $CO_2$ uptake, and photosynthesis, which regulates plant production, yield, and reproduction. Dormancy, an important phase which allow temperate zone fruit crops to overwinter, is considered to be closely related to changes in water movement (Welling & Palva, 2006).

Winter temperatures can impair the hydraulic functions of trees because water uptake from soil water reservoirs is very limited when upper soil layers are cool or frozen during winter months or early spring (Mellander et al., 2006; Peguero-Pina et al., 2011).

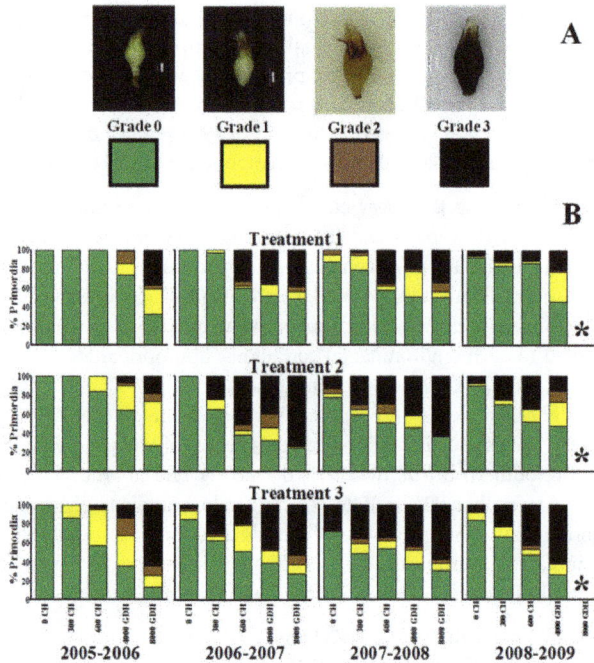

**Figure 1.** Scale adopted to evaluate the severity of floral primordia necrosis: grade 0, normal primordia (green); grade 1, yellow primordia; grade 2, primordia with partial necrosis; grade 3, completely necrosed primordia (A). Incidence and severity of floral primordia necrosis in buds of 'Housui' Japanese pear grown under mild (600 CH) winter conditions during four consecutive seasons (from 2005-2006 to 2008-2009), expressed as percentages of primordia (B). (*n* = 10). CH, chilling hours; GDH, growing degree hours; Treatment 1, normal chilling treatment; Treatment 2, one month of cold deprivation before exposure to chill; Treatment 3, two months of cold deprivation; Asterisks, not analyzed (Yamamoto, 2010).

## 2.1. Hydraulic conductivity in plants

In temperate zone woody plants, hydraulic conductivity describes resistance of xylem against embolism formation (Mayr et al., 2003) and can be impaired by xylem embolism, which is mainly caused by water stress or frost (Cochard et al., 2001; Cochard & Tyree, 1990; Sperry & Sullivan, 1992). Frost-induced embolism can occur as a consequence of alternating frost–thaw events (Améglio et al., 1995; Cochard & Tyree, 1990; Just & Sauter, 1991; Pockman & Sperry, 1997). As the sap freezes, previously dissolved gases escape because of their low solubility in ice (Sperry & Sullivan, 1992). On subsequent thawing, these bubbles can either dissolve back into the xylem sap or grow to obstruct the entire xylem conduit (Cruiziat et al., 2002, 2003; Yang & Tyree, 1992), resulting in an embolized conduit that is unable to transport water (Améglio et al., 2002).

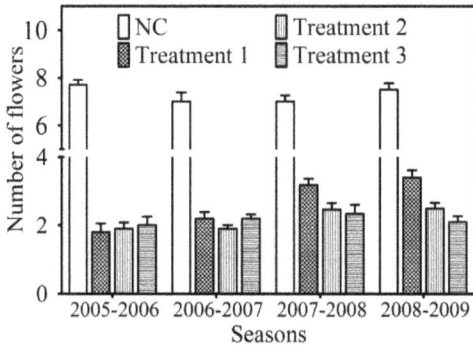

**Figure 2.** Average number of opened flowers per bud, from 2005-2006 to 2008-2009 seasons in treatments 1 (normal chilling treatment), 2 (one month of cold deprivation before exposure to chill), 3 (two months of cold deprivation), and natural conditions (NC). The vertical bars are mean ± SD (n = 10) (Yamamoto, 2010).

It was demonstrated loss of hydraulic conductivity in xylem vessels during winter on peach trees conditioned to temperatures between 4-19 C after several cycles of freeze-thaw (Améglio et al., 2002). In deciduous fruit species, an increase in sap tension caused by dehydration of the tissue can affect the resumption of growth and production, moreover, in the xylem pathway between buds and twigs, affecting translocation of nutrients to the developing bud (Lauri & Cochard, 2008).

## 2.2. Water status determined by MRI techniques in buds during dormancy

Nuclear magnetic resonance (NMR) possibly the determination of water properties in biological systems by measuring proton spin density and relaxation time by spectroscopy (Chudek & Hunter, 1997). Magnetic resonance imaging (MRI) technique, a spatially resolved NMR in essence, is cited as an important tool for providing detailed and quantitative information on both water transport and status in intact plants (Van As et al., 2009). In addition, it is a non-destructive technique, allowing a continuous developmental analysis that provides morphological and molecular structure information, measurements of biophysical parameters, including diffusion, viscosity, and solute status (Van As et al., 2009; Van der Toorn et al., 2000).

The long-distance transport of water (sap) plays an important and crucial role in the exchange of nutrients and plant hormones between different organs. MR techniques have been used to study the sap flow in xylem (Johnson et al., 1987; Köckenberger et al., 1997; Scheenen et al., 2007; Van As, 2007; Wistuba et al., 2000) as well as in its embolism occurrence (Clearwater & Clark, 2003; Fukuda et al., 2007; Holbrook et al., 2001; Umebayashi et al., 2011; Utsuzawa et al., 2005). In contrast to MRI, images obtained by synchrotron X-ray has higher spatial resolution, with possibility to obtain in real time. Such technique was used to study the mechanism of water-refilling of embolized xylem vessels, and provided a dynamic data on water transport (Lee & Kim, 2008). Brodersen et al. (2010) presented the in vivo 3D visualization and quantification of the vessel refilling process in grapevine (*Vitis vinifera*) by using a high-resolution X-ray computed tomography.

Portable MRI systems were developed recently (Rokitta et al., 2000; Van As et al., 1994), but limitations on magnetic field or MRI probes, equipments weight, external electromagnetic noise, temperature drift, among others, were cited. Umebayashi et al. (2011) monitored the developmental process of xylem embolism in Pine wilt disease by using a compact magnetic resonance imaging system with a U-shaped probe coil. Kimura et al. (2011) adapted a probe with a local electromagnetic shielding, and a flexible rotation and translation mechanism in an electric wagon system. These authors did a whole-day outdoor MRI measurements in not detached Japanese pear shoots, and a correlation between water flow and solar radiation were observed. Fukuda et al. (2007) found that cavitation and embolism events in water-stressed Japanese black pine (*Pinus thunbergii* Parl.) seedlings were detected by acoustic emission coupled with magnetic resonance microscopy, respectively.

NMR and MRI has been used to provide details during cold acclimatization and/or dormancy stage control in organs and tissues of several species, such as in dogwoods (Burke et al., 1974), Norway spruces (de Fay et al., 2000), tulip bulbs (Kamenetsky et al., 2003; Van der Toorn et al., 2000b), hybrid poplars (Kalcsits et al., 2009), blueberries (Parmentier et al., 1998; Rowland et al., 1992), grapes (Fennell & Line, 2001; Gardea et al., 1994), peaches (Erez et al., 1998; Sugiura et al., 1995; Yooyongwech et al., 2008b), Japanese pears (Yamamoto et al., 2010a), apples (Faust et al., 1991; Liu et al., 1993; Millard et al., 1993; Snaar & Van As, 1992), among others.

Differences in MRI signal strength between endo- and ecodormant buds is attributed to water content and its mobility (Faust et al., 1991). MRI primarily detects differences in proton relaxation time of water, spin-lattice or longitudinal $T_1$, and spin-spin or transverse $T_2$. A short relaxation time is interpreted as an indicator of water associated or bounded to macromolecules and a longer relaxation time is indicative of freer water (Parmentier et al., 1998).

A mobility restricted form or bounded water become freer with progression of endodormancy or chilling accumulation in apple (Faust et al., 1991) and blueberry buds (Rowland et al., 1992), among other fruit species. At endodormancy stage, freer water appeared but as a slower response to forcing conditions, indicating that changes on water status is more correlated to growth resumption in peach buds (Erez et al., 1998). Apple buds in terminal portions of shoot had both higher water content and proportionally more free water that did buds in the lower portions, which demonstrate a gradient in dormancy (or resumption of growth ability) along the shoot (Liu et al., 1993). The relaxation time measured in flower buds is an indicator of the physiological development of ecodormancy because of water absorption (Sugiura et al., 1995). Peach cultivars with lower chilling requirement increased both $T_2$ and PD after release of dormancy, which occurred earlier compared to high-chill cultivars (Yooyongwech et al., 2008b). Water transport in twigs of *Picea* was firstly slow during winter season and then increased relatively fast in early spring (de Fay et al., 2000).

In addition to these two parameters, apparent coefficient of self-diffusion (ADC) behavior of molecules can also contribute to the image contrast (Chen et al., 1978), by providing information on water diffusion at the cellular or tissue level (de Fay et al., 2000). Proton density (PD) predominantly represents water molecular mobility at cellular level and water membrane permeability (Van der Toorn et al., 2000).

Under water deficit conditions, membrane permeability seems to have two opposite actions: increase permeability to facilitate water transport to maintain cell expansion and rehydration of tissues, or induces a decrease permeability to preserve cellular water within the tissues (Barrieu et al., 1999; Smart et al., 2001; Van der Weerd et al., 2002). The discovery of aquaporins in the beginning of 1990's popularized the concept of membrane water channels (Connolly et al., 1998; Engel et al., 2000; Preston & Agre, 1991), and their roles were studied on membrane permeability (Johansson et al., 2000; Maurel, 2007; Sakr et al., 2003; Sarda et al., 1999; Van der Weerd et al., 2002). The most immediate decrease in ADC under dormancy inducing conditions occurred in the axillary buds of poplar followed by the vascular bud trace region (de Fay et al., 2000). Yooyongwech et al. (2008), by comparing two cultivars of peaches, found increase of such proteins in bud portions after the endodormancy release irrespective of their chilling requirements. From the same experiment, they suggested that an increase in ADC can be used as a indicator of ecodormancy release in peach buds.

de Fay et al. (2000) used diffusion-weighted images to demonstrate water movement and activity during budburst of spruce (*Picea abies* (L.) Karst.) after fulfillment of chilling requirement. Similar methodology was used by Van der Toorn et al. (2000) to assess water mobility, which is proportional to ADC value. Past investigations had been limited by low resolution of the images, but recently the ADC measurements correlated better with dormancy development than did the vascular tissue measurements (Kalcsits et al., 2009).

## 3. Present work: Determination of water status by MRI in flower buds under mild winter conditions

The following experiment was conducted in order to verify the effect of delayed mild winter conditions and consecutive seasons on water dynamics by using the MRI techniques, and to analyze a possible correlation with floral primordia necrosis.

Plants of Japanese pear (*Pyrus pyrifolia* (Burm. f.) Nakai) var. culta 'Housui' grafted on *P. pyrifolia* var. *pyrifolia* ('Yamanashi') rootstock were obtained from a commercial nursery, transplanted into a 30 cm diameter plastic pots in April 2005 and grown at the Agricultural and Forestry Research Centre, University of Tsukuba, Japan (36°2' N, 140°4' E, 25 m above sea level). One-year old shoots were reoriented to horizontal position in order to promote flower bud formation (Ito et al., 1999).

Plants were divided into three groups, in order to simulate different conditions of chilling accumulation: plants were kept under natural conditions, with chilling accumulation superior that than required for this cultivar (NC); two groups of plants were submitted to two months (November and December) of cold deprivation in a heated greenhouse before exposure to chilling amount of 600 hours below 7.2°C (CH), in order to simulate a delayed mild winter condition of subtropical climates. These two groups of plants were exposure to such conditions for either a single season of 2008-2009 (treatment SS) and during four consecutive seasons, from 2005-2006 until 2008-2009 (treatment CS). The later condition was set as a tentative to simulate conditions of permanent warm winter. After chilling accumulation, plants were again moved to a heated greenhouse, with minimum temperature set at 16°C, to force bud break (Fig. 3).

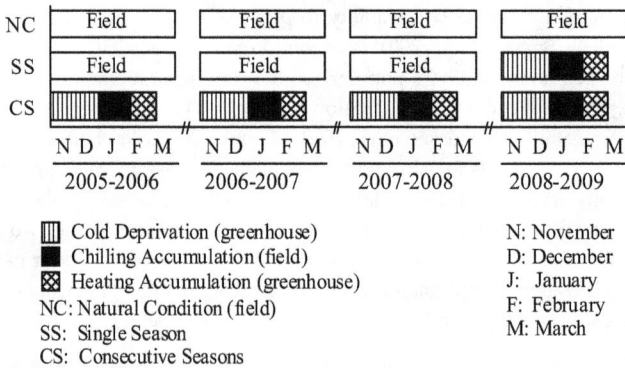

**Figure 3.** Experimental design for MRI measurements, from November (N) to March (M). Plants were kept under natural conditions (NC) and cold deprivation during two months before exposure to 600 hours of chilling accumulation below 7.2°C under two regimes: a single season of 2008-2009 (SS), and four consecutive seasons from 2005-2006 to 2008-2009 (CS) (Yamamoto et al., 2010a).

Both field and greenhouse air temperature was recorded by a data logger (TR-51A, T and D Co., Matsumoto, Japan). Chilling calculation started when two consecutive hours below 7.2°C were detected in field. In the heated greenhouse, maximum temperature was not controlled and plants were submitted do natural photoperiod.

All MRI measurements were made in 2008-2009 season. Three lateral buds were collected randomly from one-year old shoots. Samples were taken from SS and CS after one month of cold deprivation (CD1), at the end of cold deprivation and just before the chilling treatment (0 CH, SS only), at the end of chilling treatment (600 CH), and before bud break under forced conditions (4000 GDH). Under NC, samples were taken in the middle of endodormancy (December), during transition from endo- to ecodormancy (January), middle (February) and end of ecodormancy stage (March).

MRI measurements were performed using an NMR spectrometer (DRX 300WB, Bruker, Karlsruhe, Germany) equipped with a microimaging accessory at a magnetic field of 7.1 Tesla at ≈21°C (Fig. 4A). Magnetic resonance images were acquired and reconstructed with ParaVision imaging software (ver. 3.0.2 Bruker). The sample was placed on a homemade plastic holder and putted in a 15 mm NMR coil (Fig. 4B). Morphological images and 32 sequential echo images in longitudinal sections of flower buds were obtained by a multi-slice multi-echo MRI pulse program. For morphological images the repetition time was set to 1 s with the echo time of 5.524 ms, the matrix size of $256 \times 256$, the field of view of $15 \times 15$ mm$^2$ or $18 \times 15$ mm$^2$, and the slice thickness of 0.5 mm. The sequential echo images were obtained with the repletion time of 5 s, the echo time of 3.069 to 115.5 ms with a constant interval of 3.069 ms, the matrix size of $128 \times 128$, and same field of view and slice thickness with morphological images. $T_2$ (spin-spin relaxation time) and relative proton density (PD) maps were calculated from 32 sequential images using the image sequential tool in ParaVision. Three regions of interest (ROIs) of grouped floral primordia, bud base, and whole bud were determined manually in a longitudinal section at the highest bud base portion of flower bud (Fig. 4C) and $T_2$ values of each ROI was also calculated. There were three replications of each analysis.

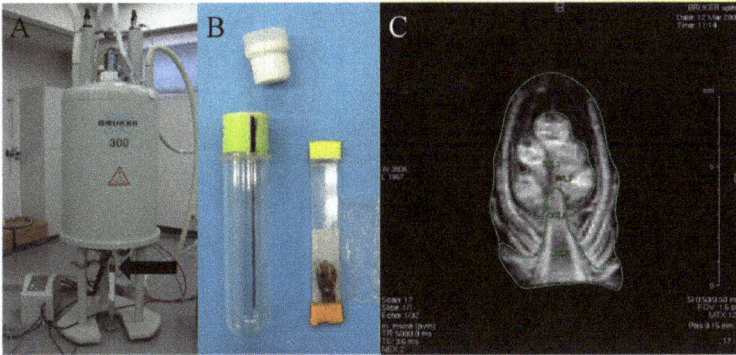

**Figure 4.** NMR spectrometer (Bruker DRX 300WB) equipped with a 7.1 Tesla magnetic field tank (A). The arrow indicates the NMR coil with the glass tube. A 15 mm glass tube and a homemade plastic holder with sampled bud of Japanese pear (B). MR morphological image of Japanese pear bud with regions of interest (ROI) determined manually (C) The scale bars represent 8mm (adapted from Yamamoto, 2010).

## 3.1. Results

$T_2$ values of floral primordia increased considerably at the transition from endo- to ecodormancy stage (January) in buds grown under normal winter conditions (NC) (Fig. 5A). Under warm and delayed winter conditions (SS and CS), however, $T_2$ values of floral primordia increased only after the end of chilling accumulation (600 CH to 4000 GDH). The bud base of plants under NC showed a gradual increase in $T_2$ values until the end of chilling accumulation, but under SS were lower than those of NC until after submitting plants to heating accumulation (Fig. 5B). After several season under warm winter conditions (CS) buds showed constant $T_2$ values until submitting plants to heating accumulation, when increased abruptly. The average $T_2$ values in whole bud increased gradually in plants kept under NC, whereas treated plants (SS and CS) had low values at the end of the chilling treatment (600 CH) (Fig. 5C).

$T_2$ and proton density (PD) maps (Fig. 6) were almost entirely calculated in detected parts of the morphological images (Fig. 7). Water mobility, as determined by $T_2$ (Fig. 6A), showed intermediate values (16 to 24 ms) in the floral primordia, bud base, and bud scales of plants kept under fulfillment of chilling requirement (NC) on March. In contrast, at the end of chilling treatment (600 CH), buds of plants kept just a single season under warm winter conditions (SS) showed high $T_2$ values (over than 24 ms) in the lower portions of the bud scales, while values were similar to NC in the bud base. Compared to SS and NC, $T_2$ values increased (to higher than 16 ms) after several seasons of mild and delayed winter conditions (CS) before flowering only in specific portions of bud base and scales near the bud base.

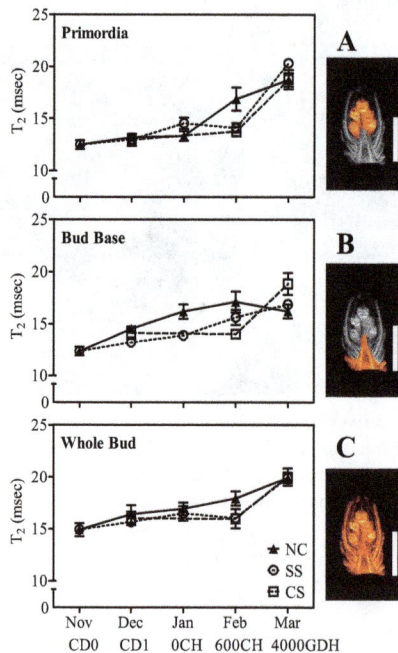

**Figure 5.** Averages of $T_2$ values (ms) of floral primordia (A), bud base (B), and whole bud (C), measured in mixed buds of 'Housui' Japanese pear grown under natural conditions (NC), one (SS) and four consecutive seasons of mild and delayed winter conditions. Means ± SE ($n$ = 3). CD0: start of cold deprivation; CD1: one month of cold deprivation. Scale bars in each ROI represent 5 mm (adapted from Yamamoto, 2010).

Relative amount of water, represented by PD maps (Fig. 6B), gradually increased in the floral primordia of buds grown under normal winter conditions (NC) during the transition from endo- to ecodormancy (January). Similar increase in PD was observed in primordia under SS. At the end of chilling accumulation (600 CH) in CS, the PD value was medium to

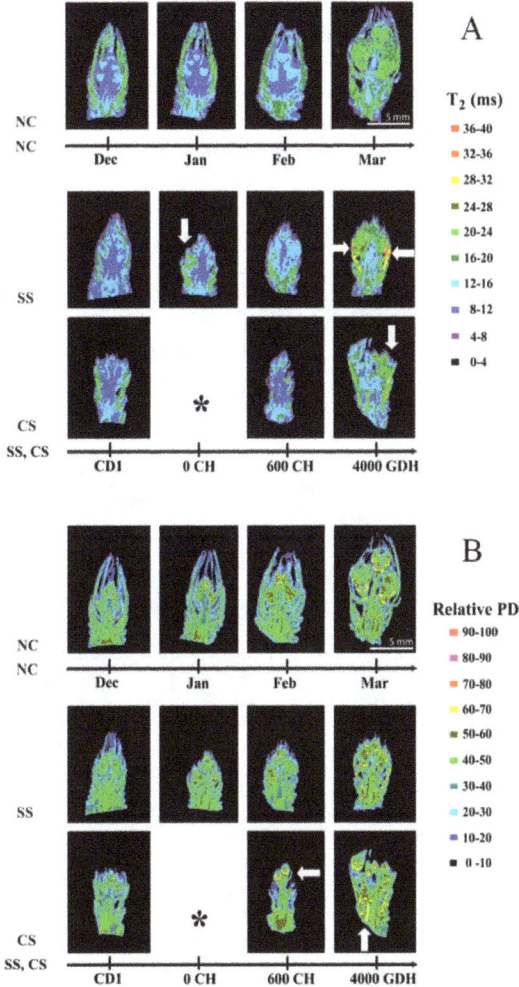

**Figure 6.** Transverse relaxation ($T_2$) time maps (A) and Proton density (PD) maps in longitudinal sections of flower buds of Japanese pear 'Housui', obtained by magnetic resonance imaging, determined after a single season of 2008-2009 (SS) and four consecutive seasons from 2005-2006 to 2008-2009 under partial chilling accumulation (CS), and from plants kept under natural conditions (NC). The arrows indicate significant changes on water dynamics. Asterisk: not analyzed. CD1: one month of cold deprivation (adapted from Yamamoto, 2010).

high (40-70 %) in the bud base, and high (until 80%) in the floral primordia. During the heating accumulation (4000 GDH), PD was higher in primordia and some specific portions of bud base after consecutive seasons of mild winter conditions (CS).

Structural changes during dormancy until flowering were observed in morphological MR images of longitudinal sections at the central portion of 'Housui' flower buds (Fig. 7). Plants grown under normal conditions of winter (NC) had a single inflorescence, with a high and uniform signal intensity of bud scales and floral primordia during all dormancy stage. However, bud scales and floral primordia configurations were remarkably different under mild winter conditions. Signals from upper portion of bud scales could not be detected at the end of cold deprivation (0 CH). Signals in floral primordia were not uniform from the end of chilling accumulation, and numbers of detected primordia and inflorescences were different. After four consecutive seasons of mild winters (CS), buds showed also variations on primordia size especially during heating accumulation (600 CH and 4000 GDH).

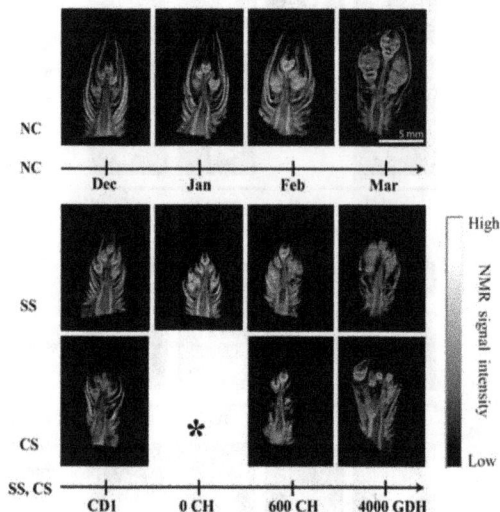

**Figure 7.** Morphological images in longitudinal sections of Japanese pear 'Housui' buds, obtained by magnetic resonance imaging, determined after a single season of 2008-2009 (SS) and four consecutive seasons from 2005-2006 to 2008-2009 under partial chilling accumulation, and from plants kept under natural conditions (NC). Asterisk: not analyzed. CD1: one month of cold deprivation (Yamamoto et al., 2010a).

Morphological comparison between buds of plants growing under natural (NC) and warm winter conditions for several seasons (CS) is showed in Fig. 8. Externally there was no morphological differences between then in photographs obtained from digital microscope (Figs. 8A, D), but under NC both floral primordia and bud scales remained green (Fig. 8B), whereas necrosed primordia was observed under CS (Fig. 8E). MR images showed such differences (Figs. 8C, F). Differences on coloration of bud scales between bud of plants grown under natural and warm winters could be observed also in photographs obtained from digital microscope (Fig. 9).

During November, there were no morphological differences observed in SEM images, but during cold deprivation (December), buds of plants grown under mild winter conditions (SS and CS) developed a second inflorescence (Fig. 10). Digital photographs obtained in February (after chilling treatment) showed a progression of floral primordia necrosis and development of new inflorescences in different levels. Reduced number of opened flowers, variations on length of pedicels, and differences on flowering stage were observed in buds of plants grown under mild winter conditions.

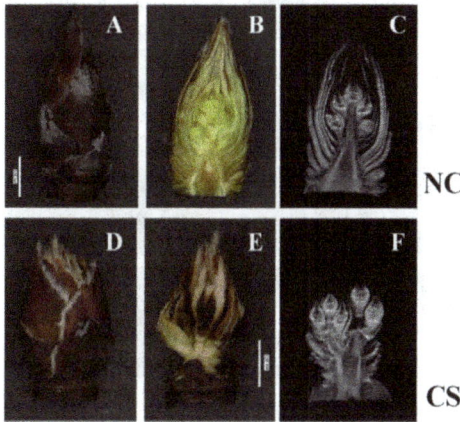

**Figure 8.** Differences between flower buds of 'Housui' Japanese pear under natural conditions (NC) and four consecutive seasons of plants submitted to warm and delayed mild winter conditions. Photographs obtained from digital microscope (A, D), longitudinal section (B), and after removing all twelve scales (E), and morphological images of longitudinal sections of same flower bud obtained by MRI (C, F). Scale bars represent 3 mm.

**Figure 9.** Photographs obtained from digital microscope of flower buds sampled on February from plants grown under natural conditions (A) and a single season of 2008-09 under artificial mild winter conditions (B). From left to right: normal bud; after removed 4, 8, all 12 scales, and inner scales. Scale bars represent 3 mm (Yamamoto, 2010).

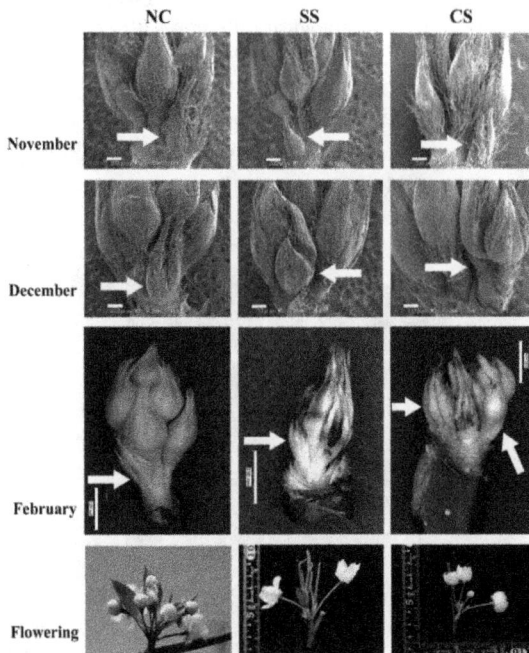

**Figure 10.** Micro-images of flower buds obtained from scanning electron microscope (SEM) before any chilling accumulation (November) and after one month of cold deprivation conditions (December). Photographs of buds during the transition from chilling to heating treatment (February) obtained from digital microscope. NC: natural conditions; a single (SS) and four consecutive seasons (CS) under mild and delayed winter conditions. The arrows indicate the leaf primordia or inflorescence duplication. The scale bars represent 333 μm in SEM and 3 mm in digital microscope images (Yamamoto et al., 2010a).

## 3.2. Discussion

The data presented suggest that MRI techniques can be an useful non-destructive and rapid analysis tool of internal anatomical details of tissues or organs, the spatial distribution of water, dynamics and mobility of water (Yamamoto et al., 2010a). Japanese pear plants grown under conditions of sufficient chilling accumulation showed a gradual increase in $T_2$ values in the whole bud, indicating a conversion from bound to free water (Fig. 6A). In other words, the endodormancy stage takes place to the ecodormant phase, which is correlated to free water (Faust et al., 1991; Liu et al., 1993; Sugiura et al., 1995). However, plants grown under mild and delayed winter conditions kept high levels of bound water (low $T_2$ values) in the floral primordia, bud base, and whole bud until the end of chilling treatment, and an abrupt change to free-state water during the afterward heating accumulation period was observed. Previous studies in peaches suggested a correlation between the level of bound water and cold hardiness (Erez et al., 1998; Faust, 1991; Parmentier et al., 1998) through the activation of dehydrin, a hydrophilic protein (Kalberer et al., 2006; Rowland & Arora, 1997; Yamane et al., 2006). Differences in $T_2$ values at the end of chilling accumulation (600 CH) is probably resulted from reduction on percentage of normal primordia occurred during cold deprivation in buds grown under mild winter conditions (data not shown). Differences in $T_2$ signal intensity with time, as a consequence of environmental temperature variation, were detected in buds of plants grown for only one season under mild winter conditions, whereas response under several seasons tended to occur slowly and was specific to some portion of the bud, like in the duplicated inflorescence.

The proton density (PD), which is also called spin density, indicates the concentration of MRI-visible protons, which were associated with water protons (Brown et al., 1986). PD maps (Fig. 6B) showed a low water content in the bud scales under mild winter conditions, and a high water content in specific portions of the floral primordia (600 CH) and bud base (4000 GDH) in bud after continuous mild winter seasons. A possible loss of functionality in vascular connections or dysfunction of water channel proteins between bud base and primordia under mild winter conditions might be resulted in progression of floral primordia necrosis (Yamamoto et al. 2010a). This irregular distribution of water resulted in a high water mobility in morphologically modified portions of the bud. However, it was not possible to determine the cause and effect essay.

The highest incidence of flower bud abortion in Japanese pear under mild winter conditions was observed in lateral buds of young shoots (Klinac & Geddes, 1995), reason why this bud type was used in this experiment. It was demonstrated that lateral (axillary) buds have the greatest sensitivity to temperature-induced dormancy (Kalcsits et al., 2009). Bud scales, which are modified leaves responsible for enclosing and protecting buds of perennial plants, had high $T_2$ and PD values in the buds of plants grown under sufficient chilling conditions, during all dormancy stages of our experiment. In contrast, the low water mobility ($T_2$) and low water content (PD) detected by MRI and in the scales of buds exposed to mild and delayed winter conditions might have been related to an increase on sensibility of the floral primordia to external temperature oscillations and/or reduced protection from freezing during winter (Yamamoto et al., 2010a).

Changes of amplitude on NMR signal intensity is related to the number of protons, and is proportional to the water content per volume element (voxel) (Donker et al., 1997; Van der Weerd et al., 2002). Morphological images of longitudinal sections showed different signal intensities in buds of plants kept under delayed and warm winter conditions. Photographs and morphological images of buds showed a clear difference in conditions of these bud scales (Figs. 7, 8, 9). Yooyongwech et al. (2008b) reported that oscillation temperature conditions accelerate water movement in peach buds, but promoted an irregular bud growth. The effect of consecutive seasons of mild winter (CS) on reduction of water mobility and content in bud scales was also observed by a more accentuated reduction on signal intensity (Yamamoto et al., 2010a). An important detail in MR studies is that only detectable portions are used. Dehydrated bud scales or necrosed floral primordia, and also structures located in different plans from selected section is not considered.

Buds under normal winter conditions had all of floral primordia developed normally, only a single inflorescence was observed during the dormant period, and all primordia sprouted simultaneously (Figs. 9, 10). However, plants grown under mild winter conditions developed a second (sometimes a third) inflorescence in buds of Japanese pear, resulting in more than 8 floral primordia per bud in average (data not shown). In such artificial conditions of 80% of chilling fulfillment, a prolonged flowering period, with low number of opened flowers was observed, regarding of necrosis incidence in floral primordia. It was possible to reproduce, in this experiment, similar phenology observed in a warm winter conditions of Southern Brazil and described by several authors as "flower bud abortion" (Marafon et al., 2011; Petri & Herter, 2002; Petri et al., 2002; Petri et al., 2001; Trevisan et al., 2005; Veríssimo et al., 2002).

## 4. Conclusions and future prospects

A cumulative effect of several seasons of mild winter conditions could be observed, where flower primordia necrosis occurrence were anticipated with delaying chilling accumulation. Moreover, a dehydration of bud scales, development of new inflorescences, different degrees of necrosis, and development of abnormal flowers were observed. MR images indicated that low levels of mobile water ($T_2$) and relative water content (PD) in bud scales might affect the sensibility of flower primordia to external temperature oscillations under mild winter conditions. Another possible consequence of such changes on water dynamics is a reduction on protection from freezing temperatures (hardening level) during winter. Floral primordia necrosis might reallocate water, by affecting its absorption potential, and other functional elements (carbohydrates or plant hormones) for development of new inflorescences.

Studies on hydraulic conductivity are focused on situations of severe winters, generally observed in the Northern hemisphere, where sap freezing occurs in xylem. But, how is the functioning of winter embolism in deciduous woody plants in regions where winter is more affected by global warming?

The advanced measurement techniques above mentioned, which can facilitate investigations of water movement in intact plants, seems to be a promising methodology to study in long-distance water transport and water dynamics in specific portions of plant tissues grown under conditions of global warming context. Unfortunately, major part of these studies were done in a typical temperate zones (Germany, Japan, Netherlands, USA, among others), where just recently faced the effect of global climate changes. In subtropical climates of Southern Brazil, for example, temperate zone fruit crops were produced under marginal conditions of mild winters. Actually, in such locations basically phenotypical observations and superficial determinations of water or carbohydrate contents were done experimentally. So, a lack of more specific studies, by using modern techniques described in this chapter under "natural" marginal conditions during dormancy phase, is nowadays observed. A multidisciplinary approach, including multi-institutional collaborations and international cooperation, is needed to study the consequences of global warming on water dynamics during dormancy phase in temperate zone fruit crops with economic importance.

## Author details

Robson Ryu Yamamoto[1], Paulo Celso de Mello-Farias[1],
Fabiano Simões[2] and Flavio Gilberto Herter[1]
[1]Federal University of Pelotas. 'Eliseu Maciel' Faculty of Agronomy, Brazil,
[2]Rio Grande do Sul State University, Vacaria Unidity, Brazil

## Acknowledgement

We thank the Nippon Foudation and the Coordenação de Aperfeiçoamento de Pessoal de Nível Superior (CAPES) for providing the scholarships, Dr. A.K. Horigane and Dr. M. Yoshida for helping the MRI experiment, Dr. Y. Sekozawa, Dr. S. Sugaya, and Dr. H. Gemma for skilful help, Ms. N.M. Nishihata for assist the edition of MR images.

## 5. References

Améglio, T.; Bodet, C.; Lacointe, A. & Cochard, H. (2002). Winter embolism, mechanisms of xylem hydraulic conductivity recovery and springtime growth patterns in walnut and peach trees. *Tree Physiology*, Vol.22, No.17, pp. 1211-1220.

Améglio, T.; Cruiziat, P. & Béraud, S. (1995). Alternance tension/pression de la sève dans le xylème chez le noyer pendant l'hiver: conséquences sur la conductance hydraulique des rameaux. *Comptes Rendus de l'Académie des Sciences Paris Série III*, No.318, pp. 351–357 (in French).

Améglio, T.; Ewers, F.W.; Cochard, H.; Martignac, M.; Vandame, M.; Bodet, C. & Cruiziat, P. (2001). Winter stem pressures in walnut trees: effects of carbohydrates, cooling and freezing. *Tree Physiology*, Vol.21, No.1, pp. 387-394.

Barrieu, F.; Marty-Mazars, D.; Thomas, D.; Chaumont, F.; Charbonnier, M. & Marty, F. (1999). Desiccation and osmotic stress increase the abundance of mRNA of the tonoplast aquaporin BobTIP26-1 in cauliflower cells. *Planta*, Vol.209, pp.77-86.

Bernier, G. (1988). The control of floral evocation and morphogenesis. *Annual Review of Plant Physiology and Plant Molecular Biology*, Vol.39, pp. 175-219.

Boneti, J.I.S.; Cesa, J.D.; Petri, J.L. & Bleicher, J. (2002). Evolução da cultura da macieira, In: *A cultura da macieira*, EPAGRI (ed.), pp. 37-58, ISBN 85-85014-45-8, Florianópolis, Brazil (in Portuguese).

Bonhomme, M.; Peuch, M.; Améglio, T.; Rageau, R.; Guilliot, A.; Decourteix, M.; Alves, G.; Sakr, S. & Lacointe, A. (2009). Carbohydrate uptake from xylem vessels and its distribution among stem tissues and buds in walnut (*Juglans regia* L.). *Tree Physiology*, Vol.30, No.1, pp. 89-102.

Brodersen, C.R.; McElrone, A.J.; Choat, B.; Matthews, M.A. & Shackel, K.A. (2010). The dynamics of embolism repair in xylem: In vivo visualizations using high-resolution computed tomography. *Plant Physiology*, Vol.154, No.3, pp. 1088-1095.

Brown, J.M.; Johnson, G.A. & Kramer, P.J. (1986). In vivo magnetic resonance microscopy of changing water content in *Pelargonium hortorum* roots. *Plant Physiology*, Vol.82, No.4, pp. 1158-1160.

Burke, M.J.; Bryant, R.G. & Weiser, C.J. (1974). Nuclear magnetic resonance of water in cold acclimating red osier dogwood stem. *Plant Physiology*, Vol.54, pp. 392-398.

Chen, P.M.; Gusta, L.V. & Stout, D.G. (1978). Changes in membrane permeability of winter wheat cells following freeze-thaw injury as determined by nuclear magnetic resonance. *Plant Physiology*, Vol.61, No.6, pp. 878-882.

Chen, W.S. (1987). Endogenous growth substances in relation to shoot growth and flower bud development of mango. *Journal of American Society for Horticultural Science,* Vol.112, pp. 360-363.

Chen, W.S. (1990). Endogenous growth substance in xylem and shoot tip diffusate of lychee in relation to flowering. *Horticultural Science,* Vol.25, No.3, pp. 314-315.

Chmura, D.J.; Anderson, P.D.; Howe, G.T.; Harrington, C.A.; Halofsky, J.E.; Peterson, D.L.; Shaw, D.C. & Clair, B.S. (2011). Forest responses to climate change in the northwestern United States: Ecophysiological foundations for adaptive management. *Forest Ecology and Management,* Vol.261, No.1, pp. 1121-1142.

Chudek, J.A. & Hunter, G. (1997). Magnetic resonance imaging of plants. *Progress in Nuclear Magnetic Resonance Spectroscopy,* Vol.31, No.1, pp. 43-62.

Clearwater, M.J. & Clark, C.J. (2003). In vivo magnetic resonance imaging of xylem vessel contents in woody lianas. *Plant Cell and Environment,* Vol.26, No.8, pp. 1205-1214.

Cleland, E.E.; Chuine, I.; Menzel, A.; Mooney, H.A. & Schwartz, M.D. (2007). Shifting plant phenology in response to global change. *Trends in Ecology and Evolution,* Vol.22, No.7, pp. 357-365.

Cochard, H. & Tyree, M.T. (1990). Xylem dysfunction in *Quercus*: vessel sizes, tyloses, cavitation and seasonal changes in embolism. *Tree Physiology,* Vol.6, pp. 393-407.

Cochard, H.; Lemoine, D.; Améglio, T. & Granier, A. (2001). Mechanisms of xylem recovery from winter embolism in *Fagus sylvatica. Tree Physiology,* Vol.1, pp. 27-33.

Connolly, D.L.; Shanahan, C.M. & Weissberg, P.L. (1998). The aquaporins. A family of water channel proteins. *The International Journal of Biochemistry & Cell Biology,* Vol.30, No.2, pp. 169-172.

Cottignies, A. (1986). The hydrolisis of starch as related to the interruption of dormancy in the ash bud. *Journal of Plant Physiology,* Vol.123, No.4, pp. 381-388.

Cruiziat, P.; Cochard, H. & Améglio, T. (2002). Hydraulic architecture of trees: main concepts and results. *Annals of Forest Science.* Vol.59, pp. 723-752.

Cruiziat, P.; Cochard, H. & Améglio, T. (2003). L'embolie des arbres. *Pour La Science,* Vol.305, pp. 50-56.

de Fay, E.; Vacher, V. & Humbert, F. (2000). Water-related phenomena in winter buds and twigs of *Picea abies* L. (Karst.) until bud-burst: A biological, histological and NMR study. *Annals of Botany,* Vol.86, No.6, pp. 1097-1107.

Donker, H.C.W.; Van As, H.; Snijder, H.J. & Edzes, H.T. (1997). Quantitative 1H-NMR imaging of water in white button mushrooms (*Agaricus bisporus*). *Magnetic Resonance Imaging,* Vol.15, No.1, pp. 113-121.

Do Oh, S. & Klinac, D. (2003). Relationship between incidence of floral bud death and temperature fluctuation during winter in Japanese Pear (*Pyrus pyrifolia*) cv. Housui under New Zealand climate conditions. *Journal of Korean Society for the Horticultural Science,* Vol.44, No.1, pp. 162-166.

Elle, D. & Sauter, J.J. (2000). Seasonal changes of activity of a starch granule bound endoamylase and of a starch phosphorylase in poplar wood (*Populus × canadensis* Moench ‹robusta›) and their possible regulation by temperature and phytohormones. *Journal of Plant Physiology,* Vol.156, No.5-6, pp. 731-740.

Engel, A.; Fujiyoshi, Y. & Agre, P. (2000). The importance of aquaporin water channel protein structures. *EMBO Journal*, Vol.19, No.5, pp. 800-806.

Erez, A.; Faust, M. & Line, M.J. (1998). Changes in water status in peach buds on induction, development and release from dormancy. *Scientia Horticulturae*, Vol.73, No.2-3, pp. 111-123.

Fachinello, J.C.; Pasa, M.S.; Schmitz, J.D. & Betemps, D.L. (2011). Situation and perspectives of temperate fruit crops in Brazil. *Revista Brasileira de Fruticultura*, Vol.33, No.1 (supplement),pp. 109-120 (in Portuguese with English abstract).

Fahmi, I. (1958). Changes in carbohydrate and nitrogen content in 'Souri' olive leaves in relation to alternate bearing. *Proceedings of the American Society for Horticultural Science*, Vol.78, No.1, pp. 252-256.

Faoro, I. (2001). História e produção, In: *Nashi, a pêra japonesa*, EPAGRI (ed.), pp. 15-66, EPAGRI/JICA, ISBN 85-85014-42-3, Florianópolis, Brazil (in Portuguese).

FAOSTAT. (2012). *Food and Agriculture Organization of the United Nations Statistical Databases*. Available from <http://faostat.fao.org/default.aspx>

Faust, M. (1991). Magnetic resonance imaging: A nondestructive analytical tool for developmental physiology. *HortScience*, Vol.26, No.7, pp. 818-935.

Faust, M.; Erez, A.; Rowland, L.J.; Wang, S.Y. & Norman, H.A. (1997). Bud dormancy in perennial fruit trees: Physiological basis for dormancy induction, maintenance, and release. *HortScience*, Vol.32, No.4, pp. 623-629.

Faust, M.; Liu, D.; Millard, M.M. & Stutte, G.W. (1991). Bound versus free water in dormant apple buds A theory for endodormancy. *HortScience*, Vol.26, No.7, pp. 887-890.

Fennell, A. & Line, M.J. (2001). Identifying differential tissue response in grape (*Vitis riparia*) during induction of endodormancy using nuclear magnetic resonance imaging. *Journal of the American Society for Horticultural Science*, Vol.126, No.6, pp. 681-688.

Fioravanço, J.C. (2007). A Cultura da Pereira no Brasil: situação econômica e entraves para o seu crescimento. *Informações Econômicas*, Vol.37, No.3, pp. 52-60 (in Portuguese).

Fukuda, K.; Utsuzawa, S. & Sakaue, D. (2007). Correlation between acoustic emission, water status and xylem embolism in pine wilt disease. *Tree Physiology*, Vol.27, No.7, pp. 969-976.

Gardea, A.A.; Daley, L.S.; Kohnert, R.L.; Soeldner, A.H.; Ning, L.; Lombard, P.B. & Azarenko, A.N. (1994). Proton NMR signals associated with eco- and endodormancy in winegrape buds. *Scientia Horticulturae*, Vol.56, No.4, pp. 339-358.

Golomp, A. & Goldschmidt, F.F. (1981). Mineral balance of alternate bearing "Wilking" mandarins. *Alon Hanotea*, Vol.35, pp. 639-647.

Hartmann, H.T.; Uriu, K. & Lilleland, O. (1966). Olive nutrition, In: *Fruit nutrition*, Childers, N.F. (ed.), pp. 252–268, Horticultural Publications, Rutgers University, New Jersey, USA.

Heide, O.M. & Prestrud, A.K. (2005). Low temperature, but not photoperiod, controls growth cessation and dormancy induction and release in apple and pear. *Tree Physiology*, Vol.25, pp. 109-114.

Holbrook, N.M.; Ahrens, E.T.; Burns, M.J. & Zwieniecki, M.A. (2001). In vivo observation of cavitation and embolism repair using magnetic resonance imaging. *Plant Physiology*, Vol.126, No.1, pp. 27-31.

Ito, A.; Yaegaki, H.; Hayama, H.; Kusaba, S.; Yamaguchi, I. & Yoshioka, H. (1999). Bending shoots stimulates flowering and influences hormone levels in lateral buds of Japanese pear. Vol. 34, pp. 1224-1228).

Johansson, I.; Karlsson, M.; Johanson, U.; Larsson, C. & Kjellbom, P. (2000). The role of aquaporins in cellular and whole plant water balance. *Biochimica et Biophysica Acta (BBA) - Biomembranes*, Vol.1465, No.1-2, pp. 324-342.

Johnson, G.A., Brown, J. & Kramer, P.J. (1987). Magnetic resonance microscopy of changes in water content in stems of transpiring plants. *Proceedings of the National Academy of Sciences*, Vol. 84, No.9, pp. 2752-2755.

Just, J. & Sauter, J.J. (1991). Changes in hydraulic conductivity upon freezing of the xylem of *Populus × canadensis* Moench 'robusta'. *Trees*, Vol.5, pp. 117–121.

Kalberer, S.R., Wisniewski, M. & Arora, R. (2006). Deacclimation and reacclimation of cold-hardy plants: Current understanding and emerging concepts. *Plant Science*, Vol.171, No.1, pp. 3-16.

Kalcsits, L.; Kendall, E.; Silim, S. & Tanino, K. (2009). Magnetic resonance microimaging indicates water diffusion correlates with dormancy induction in cultured hybrid poplar (*Populus* spp.) buds. *Tree Physiology*, Vol.29, No.10, pp. 1269-1277.

Kamenetsky, R.; Barzilay, A.; Erez, A. & Halevy, A.H. (2003). Temperature requirements for floral development of herbaceous peony cv. 'Sarah Bernhardt'. *Scientia Horticulturae*, Vol.97, No.3-4, pp. 309-320.

Kimura, T.; Geya, Y.; Terada, Y.; Kose, K.; Haishi, T.; Gemma, H. & Sekozawa, Y. (2011). Development of a mobile magnetic resonance imaging system for outdoor tree measurements. *Review of Scientific Instruments*, Vol.82, No.5, pp. 53704-53709.

Klinac, D.J. & Geddes, B. (1995). Incidence and severity of the floral bud disorder budjump on Nashi (*Pyrus serotina*) grown in the Waikato region of New Zealand. *New Zealand Journal of Crop and Horticultural Science*, Vol.23, No.2, pp. 185-190.

Köckenberger, W.; Pope, J.M.; Xia, Y.; Jeffrey, K.R.; Komor, E. & Callaghan, P.T. (1997). A non-invasive measurement of phloem and xylem water flow in castor bean seedlings by nuclear magnetic resonance microimaging. *Planta*, Vol.201, No.1, pp. 53-63.

Lacointe, A.; Kajji, A.; Daudet, F.A.; Archer, P. & Frossard, J.S. (1993). Mobilization of carbon reserves in young walnut trees. *Acta Botanica Gallica*, Vol.140, No.4, pp. 435-441.

Lang, G.A.; Early, J.D.; Martin, G.C. & Darnell, R.L. (1987). Endo-, Para-, and Ecodormancy - Physiological terminology and classification for dormancy research. *HortScience*, Vol.22, No.3, pp. 371-377.

Lang, G.A. (1996). *Plant Dormancy:* Physiology, biochemistry and molecular biology, CAB International,408 pp., ISBN 085-198-97-80, Wallingford, UK.

Lauri, P-E. & Cochard, H. (2008). Bud development and hydraulics: An innovative way to forecast shoot architecture. *Communicative & Integrative Biology*, Vol.1, pp. 1-2.

Lavee, S.; Haskal, A. & Ben-Tal, Y. (1983). Girdling olive trees, a partial solution to biennial bearing. I. Methods, timing and direct tree response. *Journal for Horticultural Science*, Vol.58, pp. 209-218.

Lavee, S. (1989). Involvement of plant growth regulators and endogenous growth substances in the control of alternate bearing. *Acta Horticulturae*, Vol.239, pp. 311-322.

Lee, S.-J. & Kim, Y. (2008). In vivo visualization of the water-refilling process in xylem vessels using X-ray micro-imaging. *Annals of Botany*, Vol.101, No.4, pp. 595-602.

Liu, D.; Faust, M.; Millard, M.M.; Line, M.J. & Stutte, G.W. (1993). States of water in summer-dormant apple buds determined by proton magnetic resonance imaging. *Journal of the American Society for Horticultural Science*, Vol.118, No.5, pp. 632-637.

Looney, N.E.; Pharis, R.P. & Noma, M. (1985). Promotion of flowering in apple-trees with gibberellin-A4 and C3 epi-gebberellin-A4. *Planta,* Vol.165, No.2, pp. 292-294.

Luedeling, E.; Gebauer, J. & Buerkert, A. (2009). Climate change effects on winter chill for tree crops with chilling requirements on the Arabian Peninsula. *Climatic Change,* Vol.96, No.1, pp. 219-237.

Marafon, A.C.; Citadin, I.; Amarante, L.d.; Herter, F.G. & Hawerroth, F.J. (2011). Chilling privation during dormancy period and carbohydrate mobilization in Japanese pear trees. *Scientia Agricola,* Vol.68, pp. 462-468.

Mauget, J.C. & Rageau, R. (1988). Bud dormancy and adaptation of apple tree to mild winter climates. *Acta Horticulturae,* Vol.232, No.1, pp. 101-108.

Maurel, C. (2007). Plant aquaporins: Novel functions and regulation properties. *FEBS Letters,* Vol.581, No.12, pp. 2227-2236.

Mayr, S.; Rothart, B. & Dämon, B. (2003). Hydraulic efficiency and safety of leader shoots and twigs in Norway spruce growing at the alpine timberline. *Journal of Experimental Botany.* Vol.54, No.392, pp. 2563-2568.

Mellander, P.E.; Stahli, M.; Gustafsson, D. & Bishop, K. (2006). Modelling the effect of low soil temperatures on transpiration by Scots pine. *Hydrological Processes,* Vol.20, No.9, pp. 1929-1944.

Millard, M.M.; Liu, D.; Line, M.J. & Faust, M. (1993). Method for imaging the states of water by nuclear magnetic resonance in low-water-containing apple bud and stem tissues. *Journal of the American Society for Horticultural Science,* Vol.118, No.5, pp. 628-631.

Monselise, S.P. & Goldschmidt, E.E. (1981). Alternate bearing in citrus and ways of control. *Proceedings of the International Society of Citriculture,* Vol.1, pp. 239-242.

Moretti, C.L.; Mattos, L.M.; Calbo, A.G. & Sargent, S.A. (2010). Climate changes and potential impacts on postharvest quality of fruit and vegetable crops: A review. *Food Research International,* Vol.43, No.7, pp. 1824-1832.

Mullins, M.G. & Rajasekaran, K. (1981). Fruiting cuttings: A revised method for producing test plants of grapevine cuttings. *American Journal of Enology and Viticulture,* Vol. 32, pp. 35-40.

Nishimoto, N.; Kisaki, K. & Fujisaki, M. (1995). Estimation of chilling requirement of using a grafting method in Japanese pear. *Journal of the Japanese Society for Horticultural Science,* Vol.64, No.2, pp. 140-141 (in Japanese).

Pal, S. & Ram, S. (1978). Endogenous gibberellins of mango shoot-tips and their significance in flowering. *Scientia Horticulturae,* Vol.9, pp. 369-379.

Parmentier, C.M., Rowland, L.J., & Linc, M.J. (1998). Water status in relation to maintenance and release from dormancy in blueberry flower buds. *Journal of the American Society for Horticultural Science,* Vol.123, No.5, pp. 762-769.

Partanen, J.; Koski, V. & Hänninen, H. (1998) Effects of photoperiod and temperature on the timing of bud burst in Norway spruce (*Picea abies*). *Tree Physiology,* Vol.18, pp. 811-816.

Peguero-Pina, J.J.; Alquézar-Alquézar, J.M.; Mayr, S.; Cochard, H. & Gil-Pelegrín, E. (2011). Embolism induced by winter drought may be critical for the survival of *Pinus sylvestris* L. near its southern distribution limit. *Annals of Forest Science,* Vol.68, No.3, pp. 565-574.

Petri, J.L. & Herter, F.G. (2002). Nashi pear (*Pyrus pyrifolia*) dormancy under mild temperature climate conditions. *Acta Horticulturae,* Vol. 587, pp. 353-361.

Petri, J.L.; Leite, G.B. & Yasunobu, Y. (2002). Studies on the causes of floral bud abortion of Japanese pear (*Pyrus pyrifolia*) in Southern Brazil. *Acta Horticulturae,* Vol.587, pp. 375-380.

Petri, J.L.; Schuck, E. & Leite, G.B. (2001). Effects of thidiazuron (TDZ) on fruiting of temperate tree fruits. *Revista Brasileira de Fruticultura*, Vol.23, No.3, pp. 513-517 (in Portuguese with English abstract).

Pockman, W.T. & Sperry, J.S. (1997). Freezing-induced xylem cavitation and the northern limit of *Larrea tridentata*. *Oecologia*, Vol.109, pp. 19-27.

Preston, G.M. & Agre, P. (1991). Isolation of the cDNA for erythrocyte integral membrane protein of 28 kilodaltons: member of an ancient channel family. *Proceedings of the National Academy of Sciences*, Vol.88, No.24, pp. 11110-11114.

Priestly, G.A. (1977). The annual turnover resources in young olive trees. *Journal of Horticultural Science*, Vol.52, pp. 105-112.

Rakngan, J.; Gemma, H. & Iwahori, S. (1996). Phenology and carbohydrate metabolism of Japanese pear trees grown under continuously high temperatures. *Journal of the Japanese Society for Horticultural Science*, Vol.65, No.1, pp. 55-65.

Ramirez, H. & Hoad, G.V. (1981). Effects of growth substances on fruit-bud initiation in apple. *Acta Horticulturae*, Vol.120, pp. 131–136.

Rokitta, M.; Rommel, E.; Zimmermann, U. & Haase, A. (2000). Portable nuclear magnetic resonance imaging system. *Review of Scientific Instruments*, Vol.71, No.11, pp. 4257-4262.

Rowland, L.J. & Arora, R. (1997). Proteins related to endodormancy (rest) in woody perennials. *Plant Science*, Vol.126, No.2, pp. 119-144.

Rowland, L.J.; Liu, D.; Millard, M.M. & Line, M.J. (1992). Magnetic resonance imaging of water in flower buds of blueberry. *HortScience*, Vol.27, No.4, pp. 339-341.

Sakr, S.; Alves, G.; Morillon, R.; Maurel, K.; Decourteix, M.; Guilliot, A.; Fleurat-Lessard, P.; Julien, J.L. & Chrispeels, M.J. (2003). Plasma membrane aquaporins are involved in winter embolism recovery in walnut tree. *Plant Physiology*, Vol.133, No.2, pp. 630-641.

Sarda, X.; Tousch, D.; Ferrare, K.; Cellier, F.; Alcon, C.; Dupuis, J. M.; Casse, F. & Lamaze, T. (1999). Characterization of closely related δ-TIP genes encoding aquaporins which are differentially expressed in sunflower roots upon water deprivation through exposure to air. *Plant Molecular Biology*, Vol.40, No.1, pp. 179-191.

Sarmiento, R.; Valpuestra, V.; Catalina, L. & Gonzales-Garcia, F. (1976). Variation of the contents of starch and soluble carbohydrate of leaves and buds of plants of *Olea europaea* var. Manzanillo in relation to their vegetative or reproductive process. *Aneles de Edafologia y Agrobiologia*, Vol.35, No.1, pp. 683-695.

Sauter, J.J. (1980). Seasonal variation of sucrose content in the xylem sap of *Salix*. *Zeitschrift fur Pflanzenphysiologie*, Vol.98, No.1, pp. 377-391.

Scheenen, T.W.J.; Vergeldt, F.J.; Heemskerk, A.M. & Van As, H. (2007). Intact plant magnetic resonance imaging to study dynamics in long-distance sap flow and flow-conducting surface area. *Plant Physiology*, Vol.144, No.2, pp. 1157-1165.

Sherson, S.M.; Alford, H.L.; Forbes, S.M., Wallace, G. & Smith, S.M. (2003). Roles of cell-wall invertases and monosaccharide transporters in the growth and development of Arabidopsis. *Journal of Experimental Botany*, Vol.54, No.382, pp. 525-531.

Smart, L.B.; Moskal, W.A.; Cameron, K.D. & Bennett, A.B. (2001). MIP Genes are down-regulated under drought stress in *Nicotiana glauca*. *Plant and Cell Physiology*, Vol.42, No.7, pp. 686-693.

Snaar, J.E.M. & Van As, H. (1992). Probing water compartments and membrane permeability in plant cells by 1H NMR relaxation measurements. *Biophysical Journal*, Vol.63, No.6, pp. 1654-1658.

Sperry, J.S. & Sullivan, J.E.M. (1992). Xylem embolism in response to freeze–thaw cycles and water stress in ring-porous, diffuse-porous, and conifer species. *Plant Physiology*, Vol.100, pp. 605-613.

Stephan, M.; Bangerth, F. & Schneider, G. (1999). Quantification of endogenous gibberellins in exudates from fruits of *Malus domestica*. *Plant Growth Regulators*, Vol.28, pp. 55-58.

Stutte, G.W. & Martin, G.C. (1986). Effect of light intensity and carbohydrate reserves on flowering in olive. *Journal of the American Society for Horticultural Science*, Vol.111, No.1, pp. 27-31.

Sugiura, T.; Yoshida, M.; Magoshi, J. & Ono, S. (1995). Changes in water status of peach flower buds during endodormancy and ecodormancy measured by differential scanning calorimetry and nuclear magnetic resonance spectroscopy. *Journal of the American Society for Horticultural Science*, Vol.120, No.2, pp. 134-138.

Tanino, K.; Kalcsits, L.; Silim, S.; Kendall, E. & Gray, G. (2010). Temperature-driven plasticity in growth cessation and dormancy development in deciduous woody plants: a working hypothesis suggesting how molecular and cellular function is affected by temperature during dormancy induction. *Plant Molecular Biology*, Vol.73, No.1, pp. 49-65.

Trevisan, R.; Chavarria, G.; Herter, F.G.; Gonçalves, E.D.; Rodrigues, A.C.; Veríssimo, V. & Pereira, I.S. (2005). Bud flower thinning on the reduction of abortion in pear (*Pyrus pyrifolia*) in Pelotas region. *Revista Brasileira de Fruticultura*, Vol.27, No.3, pp. 504-506 (in Portuguese with English abstract).

Ulger, S.; Baktir, I. & ve Kaynak, L. (1999). Determination of the effects of endogenous plant hormones on alternate bearing and flower bud formation. *Turkish Journal of Agriculture and Forestry*, Vol.23, No.3, pp. 519-523.

Ulger, S.; Sonmez, S.; Karkacier, M.; Ertoy, N.; Akdesir, O. & Aksu, M. (2004). Determination of endogenous hormones, sugars and mineral nutrition levels during the induction, initiation and differentiation stage and their effects on flower formation in olive. *Plant Growth Regulation*, Vol.42, No.1, pp. 89-95.

Umebayashi, T.; Fukuda, K.; Haishi, T.; Sotooka, R.; Zuhair, S. & Otsuki, K. (2011). The developmental process of xylem embolisms in pine wilt disease monitored by multipoint imaging using compact magnetic resonance imaging. *Plant Physiology*, Vol.156, No.2, pp. 943-951.

Utsuzawa, S.; Fukuda, K. & Sakaue, D. (2005). Use of magnetic resonance microscopy for the nondestructive observation of xylem cavitation caused by pine wilt disease. *Phytopathology*, Vol.95, No.7, pp. 737-743.

Van As, H. (2007). Intact plant MRI for the study of cell water relations, membrane permeability, cell-to-cell and long distance water transport. *Journal of Experimental Botany*, Vol.58, No.4, pp. 743-756.

Van As, H., Reinders, J.E.A., de Jager, P.A., van de Sanden, P.A.C. M., & Schaafsma, T.J. (1994). In situ plant water balance studies using a portable NMR spectrometer. *Journal of Experimental Botany*, Vol.45, No.1, pp. 61-67.

Van As, H., Scheenen, T., & Vergeldt, F. (2009). MRI of intact plants. *Photosynthesis Research*, Vol.102, No.2, pp. 213-222.

Van der Toorn, A.; Zemah, H.; Van As, H.; Bendel, P. & Kamenetsky, R. (2000). Developmental changes and water status in tulip bulbs during storage: visualization by NMR imaging. *Journal of Experimental Botany*, Vol.51, No.348, pp. 1277-1287.

Van der Weerd, L.; Claessens, M.M.A. E.; Efdé, C. & Van As, H. (2002). Nuclear magnetic resonance imaging of membrane permeability changes in plants during osmoticstress. *Plant, Cell & Environment,* Vol.25, No.11, pp. 1539-1549.

Veríssimo, V.; Gardin, J.P.; Trevisan, R.; da Silva, J.B. & Herter, F.G. (2002). Morphological and physical parameters of flower buds of trees of two Japanese pear cultivars grown at three different areas of Southern Brazil, and their relationship with flower bud abortion intensity. *Acta Horticulturae,* Vol. 587, pp. 381-387.

Wang, S.Y.; Ji, Z.L. & Faust, M. (1987). Metabolic changes associated with bud break induced by thidiazuron. *Journal of Plant Growth Regulation,* Vol.6, No.2, pp. 85-95.

Welling, A. & Palva, E.T. (2006). Molecular control of cold acclimation in trees. *Physiologia Plantarum,* Vol.127, No.2, pp. 167-181.

Wistuba, N.; Reich, R.; Wagner, H.J.; Zhu, J.J.; Schneider, H.; Bentrup, F.W.; Haase, A. & Zimmermann, U. (2000). Xylem flow and its driving forces in a tropical Liana: Concomitant flow-sensitive NMR imaging and pressure probe measurements. *Plant Biology,* Vol.2, No.6, pp. 579-582.

Yamamoto, R.R. (2010) Study on dormancy progression and floral primordia abortion in 'Housui' Japanese pear grown under mild winter conditions. PhD thesis. University of Tsukuba, Japan, 177pp.

Yamamoto, R.R.; Katsumi-Horigane, A.; Yoshida, M.; Sekozawa, Y.; Sugaya, S. & Gemma, H. (2010a). "Floral primordia necrosis" incidence in mixed buds of Japanese pear (*Pyrus pyrifolia* (Burm.) Nakai var. culta) 'Housui' grown under mild winter conditions and the possible relation with water dynamics. *Journal of the Japanese Society for Horticultural Science,* Vol.79, No.3, pp. 246-257.

Yamamoto, R.R.; Sekozawa, Y.; Sugaya, S. & Gemma, H. (2010b). Influence of chilling accumulation time on "flower bud abortion" occurrence in Japanese pear grown under mild winter conditions. *Acta Horticulturae,* Vol.872, No.8, pp. 69-76.

Yamane, H.; Kashiwa, Y.; Kakehi, E.; Yonemori, K.; Mori, H.; Hayashi, K.; Iwamoto, K.; Tao, R. & Kataoka, I. (2006). Differential expression of dehydrin in flower buds of two Japanese apricot cultivars requiring different chilling requirements for bud break. *Tree Physiology,* Vol.26, No.12, pp. 1559-1563.

Yang, S. & Tyree, M.T. (1992). A theoretical model of hydraulic conductivity recovery from embolism with comparison to experimental data on *Acer saccharum. Plant Cell Environmental,* Vol.15, pp. 633-643.

Yooyongwech, S.; Horigane, A.K.; Yoshida, M.; Sekozawa, Y.; Sugaya, S. & Gemma, H. (2008a). Effect of oscillating temperature on the expression of two aquaporin genes (Pp-delta TIP1, Pp-PIP2) involved in regulating intercellular water status in flower buds of peach. *Journal of Horticultural Science & Biotechnology,* Vol.83, pp. 784-790.

Yooyongwech, S.; Horigane, A.K.; Yoshida, M.; Yamaguchi, M.; Sekozawa, Y.; Sugaya, S. & Gemma, H. (2008b). Changes in aquaporin gene expression and magnetic resonance imaging of water status in peach tree flower buds during dormancy. *Physiologia Plantarum,* Vol.134, No.3, pp. 522-533.

Yoshioka, H.; Nagai, K.; Aoba, K. & Fukumoto, M. (1988). Seasonal changes of carbohydrates metabolism in apple trees. *Scientia Horticulturae,* Vol.36, No.3-4, pp. 219-227.

# Permissions

The contributors of this book come from diverse backgrounds, making this book a truly international effort. This book will bring forth new frontiers with its revolutionizing research information and detailed analysis of the nascent developments around the world.

We would like to thank Prof. (Dr.) Bharat Raj Singh, for lending his expertise to make the book truly unique. He has played a crucial role in the development of this book. Without his invaluable contribution this book wouldn't have been possible. He has made vital efforts to compile up to date information on the varied aspects of this subject to make this book a valuable addition to the collection of many professionals and students.

This book was conceptualized with the vision of imparting up-to-date information and advanced data in this field. To ensure the same, a matchless editorial board was set up. Every individual on the board went through rigorous rounds of assessment to prove their worth. After which they invested a large part of their time researching and compiling the most relevant data for our readers. Conferences and sessions were held from time to time between the editorial board and the contributing authors to present the data in the most comprehensible form. The editorial team has worked tirelessly to provide valuable and valid information to help people across the globe.

Every chapter published in this book has been scrutinized by our experts. Their significance has been extensively debated. The topics covered herein carry significant findings which will fuel the growth of the discipline. They may even be implemented as practical applications or may be referred to as a beginning point for another development. Chapters in this book were first published by InTech; hereby published with permission under the Creative Commons Attribution License or equivalent.

The editorial board has been involved in producing this book since its inception. They have spent rigorous hours researching and exploring the diverse topics which have resulted in the successful publishing of this book. They have passed on their knowledge of decades through this book. To expedite this challenging task, the publisher supported the team at every step. A small team of assistant editors was also appointed to further simplify the editing procedure and attain best results for the readers.

Our editorial team has been hand-picked from every corner of the world. Their multi-ethnicity adds dynamic inputs to the discussions which result in innovative outcomes. These outcomes are then further discussed with the researchers and contributors who give their valuable feedback and opinion regarding the same. The feedback is then collaborated with the researches and they are edited in a comprehensive manner to aid the understanding of the subject.

Apart from the editorial board, the designing team has also invested a significant amount of their time in understanding the subject and creating the most relevant covers. They scrutinized every image to scout for the most suitable representation of the subject and create an appropriate cover for the book.

The publishing team has been involved in this book since its early stages. They were actively engaged in every process, be it collecting the data, connecting with the contributors or procuring relevant information. The team has been an ardent support to the editorial, designing and production team. Their endless efforts to recruit the best for this project, has resulted in the accomplishment of this book. They are a veteran in the field of academics and their pool of knowledge is as vast as their experience in printing. Their expertise and guidance has proved useful at every step. Their uncompromising quality standards have made this book an exceptional effort. Their encouragement from time to time has been an inspiration for everyone.

The publisher and the editorial board hope that this book will prove to be a valuable piece of knowledge for researchers, students, practitioners and scholars across the globe.

# List of Contributors

C. Aprea and A. Maiorino
Dipartimento di Ingegneria Industriale, Università di Salerno, via Ponte Don Melillo, Fisciano, Salerno, Italia

A. Greco
DETEC, Università degli Studi di Napoli Federico II, P.le Tecchio, Napoli, Italia

Juan Cagiao Villar and Breixo Gómez Meijide
Department of Mathematical Methods and Representation, Civil Engineering School, University of A Coruña, A Coruña, Spain

Sebastián Labella Hidalgo
Atos Consulting and Technology Services, Atos Spain S.A. Barcelona, Spain

Adolfo Carballo Penela
Department of Business Management and Commerce, University of Santiago de Compostela, Santiago de Compostela, Spain

Bharat Raj Singh
School of Management Sciences, Technical Campus, Lucknow, Uttar Pradesh, India

Onkar Singh
Harcourt Butler Technological Institute, Kanpur, Uttar Pradesh, India

Amjad Anvari Moghaddam
School of Electrical and Computer Engineering, College of Engineering, University of Tehran, Tehran, Iran

Gabrielle Decamous
Faculty of Language and Cultures, Kyushu University, Fukuoka, Japan

Sotoodehnia Poopak
Institute of Biological Science, Faculty of Science, University of Malaya, Kuala Lumpur, Malaysia

Amiri Roodan Reza
Department of Knowledge Management, Faculty of Creative Multimedia, Cyberjaya, Malaysia

Silvia Duhau
Departamento de Física, Facultad de Ingeniería, Universidad de Buenos Aires and Consejo Nacional de Investigaciones Científicas y Técnicas, Argentina

**Ernesto A. Martínez**
Dirección General de Cultura y Escuelas, Buenos Aires, Argentina

**Oluwatosin Olofintoye, Josiah Adeyemo and Fred Otieno**
Department of Civil Engineering and Surveying, Durban University of Technology, Durban, South Africa

**Hiroshi Ujita**
Tokyo Institute of Technology, Department of Nuclear Engineering, Japan
The Canon Institute for Global Studies, Japan

**Fengjun Duan**
The Canon Institute for Global Studies, Japan
The University of Tokyo, Japan

**Zsuzsa A. Mayer, Andreas Apfelbacher and Andreas Hornung**
European Bioenergy Research Institute (EBRI), Aston University, Birmingham, United Kingdom

**Karl Cheng**
Chairman, Innotest Inc., Science Park, Hsinchu, Taiwan

**Bharat Raj Singh**
Director (R&D), School of Management Sciences, Lucknow, India

**Alan Cheng**
Senior Scientist, Innotest Inc., Science Park, Hsinchu, Taiwan

**D. Vatansever, E. Siores and T. Shah**
University of Bolton, Institute for Materials Research and Innovation, United Kingdom

**Robson Ryu Yamamoto, Paulo Celso de Mello-Farias and Flavio Gilberto Herter**
Federal University of Pelotas, 'Eliseu Maciel' Faculty of Agronomy, Brazil

**Fabiano Simões**
Rio Grande do Sul State University, Vacaria Unidity, Brazil